Three Dimensional Space-Time Analysis Theory of Geotechnical Seismic Engineering

Changwei Yang · Guotao Yang
Jianjing Zhang · Hongsheng Ma

Three Dimensional Space-Time Analysis Theory of Geotechnical Seismic Engineering

Science Press
Beijing

Springer

Changwei Yang
Southwest Jiaotong University
Chengdu, Sichuan, China

Jianjing Zhang
Southwest Jiaotong University
Chengdu, Sichuan, China

Guotao Yang
China Railways Corporation
Beijing, China

Hongsheng Ma
Sichuan Provincial Department
of Transportation
Chengdu, Sichuan, China

ISBN 978-981-13-3355-2 ISBN 978-981-13-3356-9 (eBook)
https://doi.org/10.1007/978-981-13-3356-9

Jointly published with Science Press, Beijing, China
The print edition is not for sale in China Mainland. Customers from China Mainland please order the print book from: Science Press, Beijing.

Library of Congress Control Number: 2018963294

This Springer imprint is published by the registered company Springer Nature Singapore Pte Ltd.
The registered company address is: 152 Beach Road, #21-01/04 Gateway East, Singapore 189721, Singapore

Contents

Chapter 1
Basis of Seismology

1.1 Introduction

Seismology is a subject which studies the happening of earthquakes, the propagation of seismic waves, and the internal structure of the earth, and it is a branch of geophysics. More specifically, it mainly studies the happening of earthquake and the propagation rules of seismic waves by using the knowledge of physics, mathematics, and geology with reference to the information of natural or artificial earthquake, so as to achieve the goal of seismic prediction and control. At the same time, it studies on the lithosphere and the internal structure of the earth through the propagation rules of seismic waves. Earthquake is the perceptible shaking of the surface of the earth, which is a form of geotectonic movement. An intense earthquake is always accompanied by ground deformation and stratum dislocation, which has strong devastating potential. That is to say, earthquake is one of the most serious natural disasters. While, at the same time, it helps to reveal the internal mystery of the earth and becomes one of the most important instruments of people to know and use nature.

The main content of seismology includes the following three parts:

(1) Macro-seismology: It mainly refers to the research and study of seismic hazard and the partition of the regional basic intensity, which aims at providing reasonable information and indexes for seismic design of buildings and macro-data for seismic prediction.
(2) The propagation of seismic waves and physics of the earth's interior: It studies on the happening and propagation rules of seismic waves according to the information of earthquake observed by seismic network and uses them to study the internal structure, composition, and state of crust and the earth.
(3) Seismometry: Its content includes the development of seismic apparatus, distribution of seismic network and analysis, process and explanation of seismogram.

Sometimes, (2) and (3) are referred to as micro-seismology.

C. Yang et al., *Three Dimensional Space-Time Analysis Theory of Geotechnical Seismic Engineering,* https://doi.org/10.1007/978-981-13-3356-9_1

The description and record of earthquake start from a long time ago. There was ancient mythology like "the turtle wrestling its body" as a metaphor for earthquake, which describes the fear of ancient people to it. The earliest record about earthquake knowledge can date back to 2000 BC, and China is one of the countries which has the earliest historical documentation of earthquakes. For example, there was record of "ground movement in the eighth year under the rule of King Wen in the Zhou Dynasty" in 1200 BC by Lv's Spring and Autumn Analects. For another example, in 132 AD, Zhang Heng, scientist of ancient China, made the first seismic apparatus in the world—Houfeng Seismograph, which could detect the time and direction of earthquake, and it was over one thousand years earlier than the mercury seismoscope in Europe, 1848. There is also an example that there was concrete seismic hazard record in China for the large earthquake in Guanzhong, Shaanxi, on January 23, 1556. While in Europe, it was until the middle of the eighteenth century that there was detailed hazard record of Lisbon earthquake on November 1, 1755. These precious historical materials play an important role in the study of earthquake happening and the earth's history.

1.2 The Internal Structure of the Earth

The earth is of spheroid shape, which is slightly like a pear, and its radius is about 6400 km. From its surface to the core, the earth can be divided into three layers with different natures. In Fig. 1.1, the outer layer is the crust, which is quite thin and its thickness is from several kilometers to a few tens of kilometers; then there is the mantle, whose thickness is about 2900 km, and the interface of crust and mantle is called Moho discontinuity; the inner ball is the centrosphere, and its radius is about 3500 km. The surface of the crust is made up of uneven rocks. Several kilometers within the continent surface are various kinds of sedimentary rock, magmatic rock, metamorphic rocks, and loose sediments; while in the sea, the rock under the sea sediment is of quite single nature, that is, basalt. It is generally recognized that the continental crust is made up of both the granitic layer and the basaltic layer, while oceanic crust is only made up of the basaltic layer without the granitic layer. The thickness of crust varies greatly. Generally, it is only several kilometers under the sea; while the average thickness of crust under the continent is 30–40 km, it is much thicker under the mountains, for example, the Qinghai–Tibet Plateau in China, whose crustal thickness could amount to 70 km. Most earthquakes happen inside the crust.

It is generally taken that the mantle is made up of uniform peridotite, while the composition is still complex in its upper hundreds of kilometers. Under the Moho discontinuity, it is a layer of rock from 40 to 70 km, which forms the so-called lithosphere or rocky shell together with the crust. There exists an asthenosphere under the lithosphere with thickness of about hundreds of kilometers. The wave speed inside the asthenosphere is much slower than that inside the upper or lower rock layers, which may be because of its viscoelastic effect or rheological property due to high temperature and high pressure. The lithosphere and the asthenosphere are

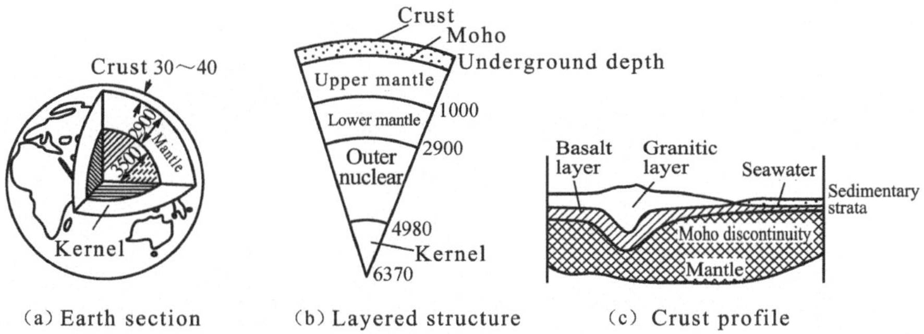

Fig. 1.1 Layered structure inside the earth

together called the upper mantle. Under the upper mantle is the lower mantle, which is quite uniform. The centrosphere is composed of the outer core and the inner core. The radius of the inner core is about 1400 km, while the outer core is in liquid state. It is found through seismic wave observation that the seismic shear wave cannot get through the outer core and the inner core is in solid state. The major core materials are nickel and iron.

The temperature of the earth increases with the depth. The temperature of 20 km underground is about 600 °C, that of 100 km underground is about 1000–1500 °C, and that of 700 km underground is about 2000 °C, while the temperature of the inner core can be up to 4000–5000 °C. The high temperature inside the earth comes from the heat emitted by radioactive substance inside it. The distribution of radioactive substance is quite different between the bottom of the sea and of the continent. Therefore, there is a theory that the temperature difference leads to creep-flow of the substance inside the mantle and causes convection.

In the nineteenth century, the Continental Fixism was in the dominant position. In the later half of the nineteenth century, people began to find out that biotic population, paleontological fossils, and even geological formation on different continents separated by the ocean have very similar genetic relationship; for example, if South America was pieced together with Africa, the rock layers with different ages since several billion years could fit; Europe, North America, and Asia all could find the fossil of the ancestor of the same animal from rock layer of one hundred million years ago, while after separation of the original continent, the animal variety changed with different natural environment. That is hard to explain with the Continental Fixism. In 1910, when Alfred Lothar Wegener, German meteorologist, read the world map, he found out that the topography of the east and west bank of the Atlantic Ocean are overlapped, and in particular, because of the gravitation of South America and the centrifugal force of the earth, cracks appeared on Pangea, which started to split and drift (Fig. 1.2).

In 1915, he published *The Origin of the Continent and Ocean* (*Die Entstehung der Kontinente und Ozeane*), which provided evidence for the continental drift, but could not explain the dynamics problems of it. Arthur Holmes, British geologist, proposed

Fig. 1.2 Splice chart of the continent

"the theory of mantle convection" in 1928. While due to the limit of scientific level then, and especially the physical mechanism of continental drift not being well solved, the sensational hypothesis was soon forgotten.

It was until the middle of the twentieth century that paleomagnetism study of rocks provided the theory with more scientific supports, and it became acceptable of more people. The magnetization direction of rocks was not susceptible to the changes of the earth's magnetic field, which could be proved by the lava study of volcanic eruption. The north magnetic pole of the earth now is at the northern part of Canada, which is quite far away from the geographical North Pole (pole of rotation). It could be known from the study of paleomagnetism that the north magnetic pole of the earth had gone through slow and consecutive changes, but its average relative position to the pole of rotation stays stable, with track of floating curve.

During 1950s–1960s, marine geological researches, especially the development of ocean drilling and exploration, made great progress, one of which was the discovery of oceanic ridges and trenches. That confirmed the existence of mantle convection and seafloor spreading, and the speed of seafloor spreading and continental drift was measured by method of radio ranging. In 1967, Xavier Le Pichon from France, W. Jason Morgan from America, and McKenzie from Britain built the model of "plate tectonics." They divided the lithosphere into six major plates: the Eurasian plate, the America plate, the Africa plate, the Pacific Ocean plate, the Australian plate, and the Antarctica plate, and many other subplates as shown in Fig. 1.3. The plates were separated by mid-oceanic ridges, subduction zones, and transform faults, and they continued to expand at mid-oceanic ridges but began to subside and subduct

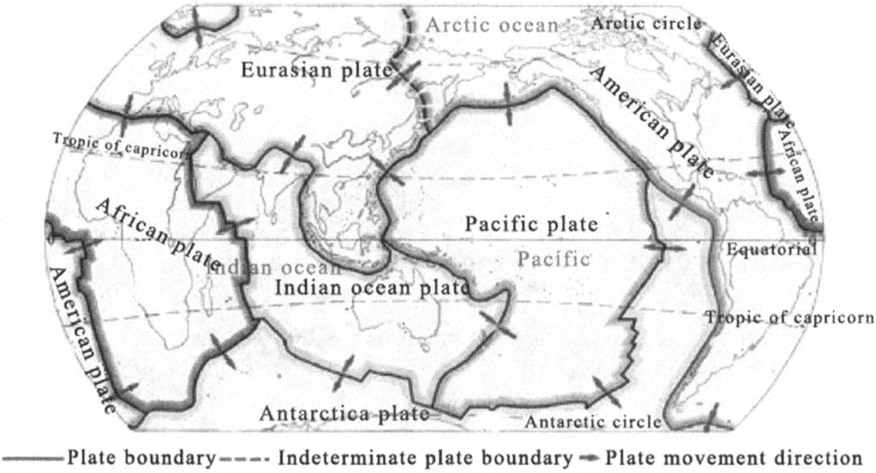

Fig. 1.3 Sketch of division of major plates and subplates in the world

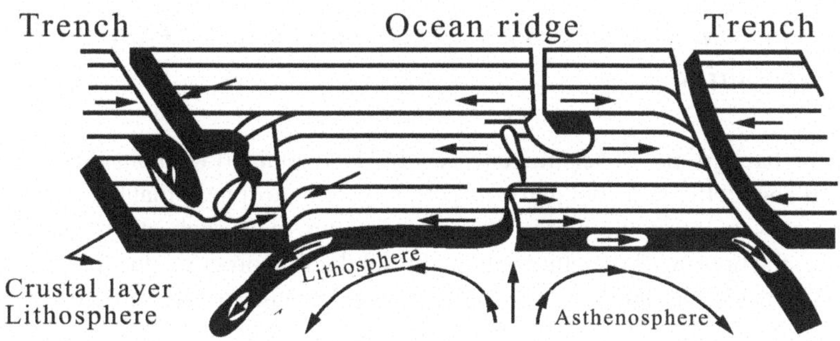

Fig. 1.4 Plate movement

at subduction zones, which were tectonic upheaval position and the major place of earthquakes and volcanic activities.

The theory of plate tectonics holds that the rock layer of crust and upper mantle constitute the global lithosphere, and substance of the asthenosphere of the upper mantle pours out of the ridges, pushes the lithosphere of 100 km on the asthenosphere to move horizontally, and forms new seabed which leads to oceanic expansion, and most substances at the lower part of the ridges forms upwelling as shown in Fig. 1.4. One part of lithosphere inserts into the bottom of another part of lithosphere in the trench, turning back to the asthenosphere and simultaneously forming downwelling.

Figure 1.4 shows the plate movement. Thus, mantle convection cell comes out of ridge belts and trench belts. The effects of mantle convection cell on plates are like setting them on conveyor belts, making them drift slowly by transportation.

The continental drift and the plate tectonics can not only explain the evolution history of the earth's continent, but also predict its future development, which is the in-depth recognition of people to the wholeness, kinematics, and dynamics of the movement pattern of the solid earth and is the great discovery in modern geology, which could be called the greatest achievement of geoscience in the twentieth century.

1.3 Seismogenesis and Earthquake Classification

Researches of seismogenesis has been nearly a hundred years. The early theories are mostly about fault rupture, while the recent arguments focus on plate tectonics. Actually, the two theories are not contradictory in nature. The author thinks that seismogenesis can be discussed from two different levels: macro-background and local mechanism.

1.3.1 The Macro-background of Seismogenesis—Plate Tectonics

The macro-background of seismogenesis could be explained with plate tectonics. As discussed above, the pour-out and convection of asthenosphere matter of the mantle push the movement of the plates. When two plates meet each other, one of them exerts into the bottom of the other, and in the subduction process, due to the complex stress state inside the plates, brittle fracture happens with the plates themselves, the nearby crust and the lithosphere, which leads to earthquakes. The historical data shows that most earthquakes happen near the border of the plates and their vicinity in the globe. While on the other side, the boundary surface between the asthenosphere and the plate is uneven, and the asthenosphere itself is of great rigidity, which leads to complex stress state and uneven deformation inside the states, and that is the basic cause of intraplate earthquakes. The rock faults inside the plates are the internal condition for the earthquake.

Figure 1.5 shows the sketch map of baseline change across San Andreas Fault before and after the San Francisco earthquake. According to the statistics, about 85% of global earthquakes happen on the plate boundary belts, and only 15% happen within continental interior or inside the plates.

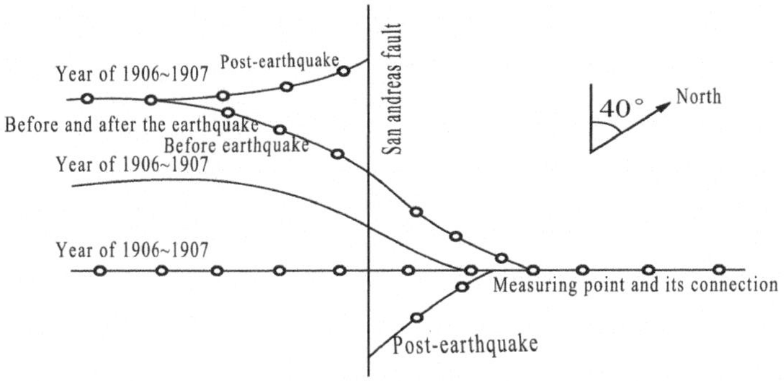

Fig. 1.5 Schematic of changes across the baselines of the San Andreas Fault before and after the San Francisco earthquake

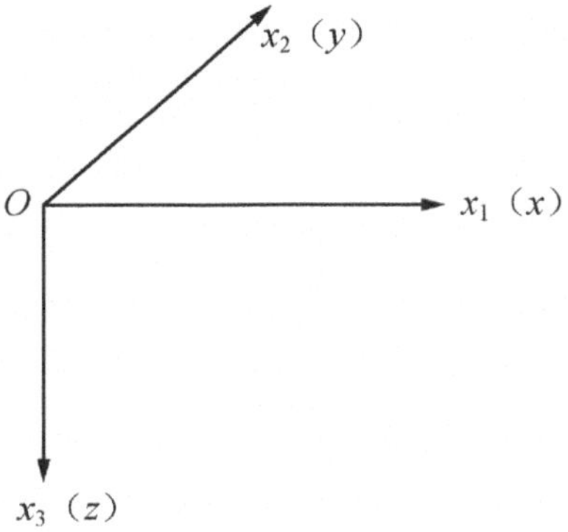

Fig. 1.6 Rectangular coordinate system of wave problem

1.3.2 Local Mechanism of Seismogenesis—Elastic Rebound Hypothesis

Elastic rebound hypothesis was proposed by Reid at the beginning of the twentieth century. The hypothesis was firstly concluded by the measured data of displacement across San Andreas Fault before and after the 8.3 magnitude earthquake in San Francisco in 1906, as shown in Fig. 1.6. The hypothesis spanning more than half a century found out that the monitoring points on the two sides of the fault slowly moved, and retest after the earthquake discovered that diastrophism with maximum fault displacement of 6.4 m appeared on the measuring line along the fault. This provides a clear and powerful evidence for the explanation that San Francisco earthquake is the result of diastrophism along San Andreas Fault with length of 960 km.

The elastic rebound hypothesis proposed by Reid argues that (1) the crust is composed of elastic rock mass with faults; (2) the energy generated from crust movement accumulates chronically at faults or nearby rock mass in the form of elastic strain energy; (3) when elastic strain energy and the deformation of the rock mass reach certain extent, relative displacement rupture occurs to the rock mass on the two sides of some point on the fault and causes displacement of proximal points along the fault, which leads to sudden slides of rock mass on the two sides of the fault to opposite directions and earthquakes happen. At this time, the elastic strain energy chronically accumulated on the fault suddenly releases; (4) after the earthquake, the rock mass deformed under the effect of elastic strain energy returns to the state before deformation. The elastic rebound hypothesis explains the macro-reasons for why the crust moves and how the elastic strain energy gets accumulated and plate tectonics offsets the deficiency of it on this point.

1.3.3 Improvement of the Elastic Rebound Hypothesis—Stick-Slip Theory

In the middle of 1960s, the elastic rebound hypothesis was mended by the test results of rock mechanics, which improved the viewpoint of earthquakes causing by faults that explained local mechanism of seismogenesis. The elastic rebound hypothesis holds that when diastrophism happens to faults, which will release all the accumulated elastic strain energy, and after the earthquake, the focus is basically in non-stress state. The stick-slip theory holds that each time diastrophism happens to faults, it only releases a small part of the total accumulated strain energy, while the rest energy is balanced by the large kinetic force of friction on the faults. After earthquakes, there is still friction force on the two sides of faults making them solidify and can accumulate stress for another time to cause bigger earthquakes. These viewpoints from the stick-slip theory are supported by type of earthquake sequence.

1.3.4 Earthquake Classification and Seismic Sequence

Earthquakes mainly belong to natural earthquakes and human-triggered earthquakes. Human-triggered earthquakes refer to earthquakes caused by artificial explosion, mining, and engineering activities (e.g. construction of a reservoir). Generally, human-triggered earthquakes are less powerful, except isolated cases (e.g. reservoir earthquake) that could cause large damage. Natural earthquakes mainly include tectonic earthquakes and volcanic earthquakes, and the latter is caused by volcanic explosion, which is generally not strong, so less attention is placed on it.

Tectonic earthquakes are the major object studied by seismic engineering. It refers to earthquakes caused by plate tectonic activities and fault tectonic activities, and it counts for over 90% of earthquakes happened in the world.

According to focal depth (h), earthquakes can be divided into (1) shallow-focus earthquakes ($h < 70$ km), which mainly concentrates in the depth range of $h < 33$ km and accounts for 72% of the total number of earthquakes; (2) intermediate-focus earthquakes (70 km $< h <$ 300 km), which accounts for 23.5% of the total number of earthquakes; and (3) deep-focus earthquakes ($h > 300$ km). Up to now, the maximum observed depth of focus is 720 km, and deep-focus earthquakes only account for 4% of the total number of earthquakes.

Seismic sequence refers to the results of arranging a series of related earthquakes in time order. According to the previous seismic record, seismic sequence mainly includes three types: (1) main shock–after shock-type earthquake, in which the main shock releases the most energy, accompanied with a number of aftershocks and incomplete foreshock. The typical examples are Tangshan earthquake in 1976 and Haicheng earthquake in 1975; (2) seismic type of earthquake swarm, in which energy is released by strong earthquakes of many times, accompanied with large amounts of small seisms, such as Xingtai earthquake in 1966 and Lancang–Gengma earthquake in 1988; (3) single-shock-type earthquake, in which the main shock is prominent, accompanied with few foreshock and aftershocks, for example, Heringer earthquake in Inner Mongolia in 1976. For the above three types of earthquakes, the main shock–after shock-type earthquake accounts for about 60%, the seismic type of earthquake swarm about 30%, and the single-shock-type earthquake only about 10%.

1.4 Propagation of Seismic Waves in Infinite Elastic Solid

Rock has certain rheological properties under high temperature and high pressure, and under long-time effect of geologic stress, its viscoelasticity and rheology become dominant, which is one of the basic theories of plate movement; however, under the effect of short-time and rapidly changing motive power, the rock shows its rheology. The influence of viscous effect could be amended by the concept of energy loss, which is the basic theoretical assumption of seismic wave propagation. Therefore, it could be assumed that earth medium is homogeneous, isotropic, and perfect elastic. Use of seismograph to observe the seismic motion of particles under seismic effects promotes development of seismic wave theories and understanding of seismic focus and the earth's structure, and provides support for the above assumption.

1.4.1 Wave Equation

Particles must satisfy stress–strain relation, continuity condition, and Newton's second law of motion of the media within uniform, isotropic, and undamped elastomer, and the basic equation of motion could be derived from theory of elastic mechanics under small deformation, which is as follows:

$$\rho \frac{\partial^2 u_i}{\partial t^2} = (\lambda + \mu) \frac{\partial \theta}{\partial x_i} + \mu \nabla^2 u_i, \quad i = 1, 2, 3 \tag{1.1}$$

In the formula, x_1, x_2, x_3 represents the three dimensions in rectangular coordinates;

μ_1, μ_2, μ_3 represents the displacement of particles along three dimensions in rectangular coordinates;

ρ represents the density of the media;

λ, μ represents lame constant of the media;

E, G represents the elasticity modulus and shear modulus of the media;

ν represents Poisson's ratio of the media;

θ represents volumetric strain of the media;

∇ represents Laplace operator, $\nabla^2 = \frac{\partial^2}{\partial x_1^2} + \frac{\partial^2}{\partial x_2^2} + \frac{\partial^2}{\partial x_3^2}$

In order to solve Formula (1.1), two potential functions are adopted, one of which is scalar potential φ and the other is vector potential $\psi(\psi_1, \psi_2, \psi_3)$. The relation between displacement and them is

$$u_1 = \frac{\partial \varphi}{\partial x_1} + \frac{\partial \psi_3}{\partial x_2} - \frac{\partial \psi_2}{\partial x_3} \tag{1.2a}$$

$$u_2 = \frac{\partial \varphi}{\partial x_2} + \frac{\partial \psi_1}{\partial x_3} - \frac{\partial \psi_3}{\partial x_1} \tag{1.2b}$$

$$u_3 = \frac{\partial \varphi}{\partial x_3} + \frac{\partial \psi_2}{\partial x_1} - \frac{\partial \psi_1}{\partial x_2} \tag{1.2c}$$

Therefore, it can get from Formula (1.1) that

$$\nabla^2 \varphi = \frac{1}{\alpha^2} \frac{\partial^2 \varphi}{\partial t^2} \tag{1.3}$$

$$\nabla^2 \psi_i = \frac{1}{\beta^2} \frac{\partial^2 \psi_i}{\partial t^2}, \quad i = 1, 2, 3 \tag{1.4}$$

In the formula, α is velocity of longitudinal wave, $\alpha = \sqrt{\frac{\lambda+2\mu}{\rho}} = \upsilon_p$

β is velocity of transverse wave, $\beta = \sqrt{\frac{\mu}{\rho}} = \upsilon_s$.

It can be known from Formula (1.2a) that volumetric strain θ is

$$\theta = \frac{\partial u_1}{\partial x_1} + \frac{\partial u_2}{\partial x_2} + \frac{\partial u_3}{\partial x_3} \tag{1.5}$$

while distortion is

$$\omega_1 = \frac{1}{2}\left(\frac{\partial u_3}{\partial x_2} - \frac{\partial u_2}{\partial x_3}\right) \tag{1.6a}$$

$$\omega_2 = \frac{1}{2}\left(\frac{\partial u_1}{\partial x_3} - \frac{\partial u_3}{\partial x_1}\right) \tag{1.6b}$$

$$\omega_3 = \frac{1}{2}\left(\frac{\partial u_2}{\partial x_1} - \frac{\partial u_1}{\partial x_2}\right) \tag{1.6c}$$

As longitudinal wave only produces tensile displacement, i.e. distortion $\omega_i = 0$, but not rotational displacement, so according to this condition, we can take

$$u_i = \frac{\partial \varphi}{\partial x_i}, \quad i = 1, 2, 3 \tag{1.7}$$

Therefore, volumetric strain θ is

$$\theta = \nabla^2 \varphi \tag{1.8}$$

so

$$\frac{\partial \theta}{\partial x_i} = \frac{\partial}{\partial x_i} \nabla^2 \varphi \tag{1.9}$$

$$\nabla^2 \frac{\partial \phi}{\partial x_i} = \nabla^2 u_i \tag{1.10}$$

Substituting it into Formula (1.11), we can get

$$\frac{\partial^2 u_i}{\partial t^2} = \alpha^2 \nabla^2 u_i, \quad i = 1, 2, 3 \tag{1.11}$$

Due to that transverse wave only goes through pure shear deformation but not volumetric changes, so volumetric strain should be zero, i.e. $\theta = 0$. Therefore, we can get from Formula (1.11) that

$$\frac{\partial^2 u_i}{\partial t^2} = \beta^2 \nabla^2 u_i, \quad i = 1, 2, 3 \tag{1.12}$$

Thus we can see that, the wave equation of longitudinal wave and transverse wave have the same form, with only different coefficients, which provides convenience and simplifications for study of wave rules.

1.4.2 Propagation of Elastic Wave

1.4.2.1 Longitudinal Wave (Wave Expansion Wave, Primary Wave, Pressure Wave, Non-rotational Wave, and *P*-Wave)

Assuming that there is a plane wave propagating along x_1-direction in an infinite space, it is easy to prove

$$\begin{cases} \varphi \text{ or } \theta = f_1(x_1 - \alpha t) + f_2\ (x_1 + \alpha t) \\ \psi_i \text{ or } \omega_i = 0 \qquad\qquad i = 1, 2, 3 \end{cases} \tag{1.13}$$

It satisfies Formulas (1.4) and (1.5); f_1 is the wave propagating along the positive direction of x_1 and f_2 is the wave propagating along the negative direction of x_1. φ value is on a plane with x_1 = constants, so it belongs to plane wave, and it also belongs to non-rotational wave because its displacement represented by φ or stress state satisfies the condition $\nabla \times \nabla\varphi = 0$ of curl. As the φ value of this longitudinal wave is only the function of $u_1 = u_2 = 0, u_3 \neq 0$, and it is irrelevant to x_2 and x_3, and $\psi_i = 0$, so there is $u_1 \neq 0, u_2 = u_3 = 0$, which shows that the vibration direction of longitudinal wave is consistent with the direction of propagation.

1.4.2.2 Transverse Wave (Distorted Wave, Shear Wave, Body Wave Including Secondary Wave, and *S*-Wave)

Assuming that there is a plane wave propagating along x_1-direction in an infinite space, it is easy to prove

$$\psi_2 = f_1(x_1 - \beta t) + f_2(x_1 + \beta t) \tag{1.14}$$

$\psi_1 = \psi_3 = \varphi = \theta = 0$ satisfies Formulas (1.4) and (1.5); as ψ_2 is on a plane with x_1 = constants, so it belongs to plane wave, and due to $\theta = 0$, it belongs to distorted wave, which means only changes in shape but not in volume. It can be known from Formula (1.3) that $u_1 = u_2 = 0, u_3 \neq 0$, i.e. the displacement of this wave is only along x_3, so its vibration direction (axis x_3) is in perpendicular with its propagation direction (axis x_1). This type of wave is called SV wave.

We can also take

$$\psi_3 = f_1(x_1 - \beta t) + f_2(x_1 + \beta t) \tag{1.15}$$

$$\psi_2 = \psi_3 = \varphi = \theta = 0$$

At this time, there is $u_1 = u_3 = 0, u_2 \neq 0$, i.e. there is only displacement along x_2, so its vibration direction is along the axis x_2, which is in perpendicular with the propagation direction x_1. This type of wave is called SH wave. While, if we take

$$\begin{aligned} \psi_1 &= f_1(x_1 - \beta t) + f_2(x_1 + \beta t) \\ \psi_2 &= \psi_3 = \varphi = \theta = 0 \end{aligned} \tag{1.16}$$

Then $u_1 = u_2 = u_3 = 0$, i.e. all the displacement components are zero, which means there exists no transverse wave whose vibration direction is the same with its propagation direction.

1.4.2.3 Rayleigh Wave

If the media is homogeneous infinite space, there can only exist P-wave and S-wave. If there exist interface whose two sides are of different properties, while because the interface must satisfy the stress balance and deformation continuity conditions, there could be waves of different types. Rayleigh wave is a surface wave which nears the earth's surface.

Supposing that the propagation direction of wave on a free surface is the axis x_1, the origin is in the middle of the surface, the axis x_3 is vertical to the surface, and the direction into the media is positive with axis x_2 on the surface. Supposing that the potential function is

$$\varphi(x_1, x_3, t) = f_1(x_3)\mathrm{e}^{ik(x_1-ct)} \tag{1.17}$$

$$\psi_2(x_1, x_3, t) = f_2(x_3)\mathrm{e}^{ik(x_1-ct)} \tag{1.18}$$

In the formula, k—the number of waves, $k = \frac{2\pi}{l} = \frac{\omega}{c}$;
l—wavelength;
ω—π;
c—velocity of Rayleigh wave;
i—symbol of imaginary numbers, $i = \sqrt{-1}$

Bringing Formulas (1.17) and (1.18) into Formulas (1.4) and (1.5), we can get the following differential equation:

$$\frac{\mathrm{d}^2 f_1}{\mathrm{d}x_3^2} - (k^2 - K^2)f_1 = 0$$

$$\frac{\mathrm{d}^2 f_2}{\mathrm{d}x_3^2} - (k^2 - K'^2)f_2 = 0$$

In the formula, $K^2 = \frac{\omega^2}{\alpha^2} = \frac{\omega^2}{\upsilon_p^2}$, $K'^2 = \frac{\omega^2}{\beta^2} = \frac{\omega^2}{\upsilon_s^2}$.

Considering that at $x_3 \to \infty$, the amplitude of waves is finite, so we get

$$\varphi = A\mathrm{e}^{-\sqrt{k^2-K^2}x_3}\mathrm{e}^{ik(x_1-ct)} = A\mathrm{e}^{-akx_3}\mathrm{e}^{ik(x_1-ct)}$$

$$\psi_2 = B\mathrm{e}^{-\sqrt{k^2-K^2}x_3}\mathrm{e}^{ik(x_1-ct)} = B\mathrm{e}^{-bkx_3}\mathrm{e}^{ik(x_1-ct)}$$

In the formula, $a = \sqrt{1 - \frac{c^2}{\upsilon_p^2}}$, $b = \sqrt{1 - \frac{c^2}{\upsilon_s^2}}$.

The boundary condition on free surface is

$$\sigma_{x_3}\big|_{x_3=0} = 0$$

$$\tau_{x_1x_3}\big|_{x_3=0} = 0$$

From this, we can get

$$(1 + b^2)A + i2bB = 0$$
$$-i2aA + (1 + b^2)B = 0$$

In the formulas, the condition of A and B has untrivial solution as its determinant of coefficient being zero, thus we can get $(1 + b^2)^2 - 4ab = 0$.

Or this can be written as $\left(2 - \frac{c^2}{\upsilon_s^2}\right)^2 = 4\sqrt{1 - \frac{c^2}{\upsilon_s^2}}\sqrt{1 - \frac{c^2}{\upsilon_p^2}}$.

Through squaring the two sides of the formula and arrangement, we can get

$$\left(\frac{c}{\upsilon_s}\right)^6 - 8\left(\frac{c}{\upsilon_s}\right)^4 + \left(24 - 16\frac{\upsilon_s^2}{\upsilon_p^2}\right)\left(\frac{c}{\upsilon_s}\right)^2 - 16\left(1 - \frac{\upsilon_s^2}{\upsilon_p^2}\right) = 0 \tag{1.19}$$

Formula (1.19) can be further written as

$$\left(\frac{c}{\upsilon_s}\right)^6 - 8\left(\frac{c}{\upsilon_s}\right)^4 + 8\frac{2 - \upsilon}{1 - \upsilon}\left(\frac{c}{\upsilon_s}\right)^2 - \frac{8}{1 - \upsilon} = 0 \tag{1.20}$$

Formula (1.20) is a cubic equation of $\left(\frac{c}{\upsilon_s}\right)^2$, so there at least exists one positive root in $0 < c < \upsilon_s < \upsilon_p$. If the value of the Poisson's ratio v is given, the velocity value c of the correspondent Rayleigh wave could be found, which is denoted as υ_R, and the solution of Formula (1.20) can be approximately expressed as

$$\upsilon_R \approx \frac{0.862 + 1.14\nu}{1 + \nu}\upsilon_s \tag{1.21}$$

If the potential function is given, we can get that the displacement is

$$u_1 = if_1(x_3)e^{ik(x_1 - ct)} \tag{1.22a}$$
$$u_2 = 0 \tag{1.22b}$$
$$u_3 = if_2(x_3)e^{ik(x_1 - ct)} \tag{1.22c}$$

in which,

$$f_1(x_3) = -Ak\left(e^{-akx_3} - \frac{1 + b^2}{2b}e^{-bkx_3}\right)$$
$$f_2(x_3) = Ak\left(-ae^{-akx_3} - \frac{1 + b^2}{2b}e^{-bkx_3}\right)$$

If consideration is only paid to the real component of displacement components, there is

$$\frac{u_1^2}{f_1^2(x_3)} + \frac{u_3^2}{f_2^2(x_3)} = 1 \tag{1.23}$$

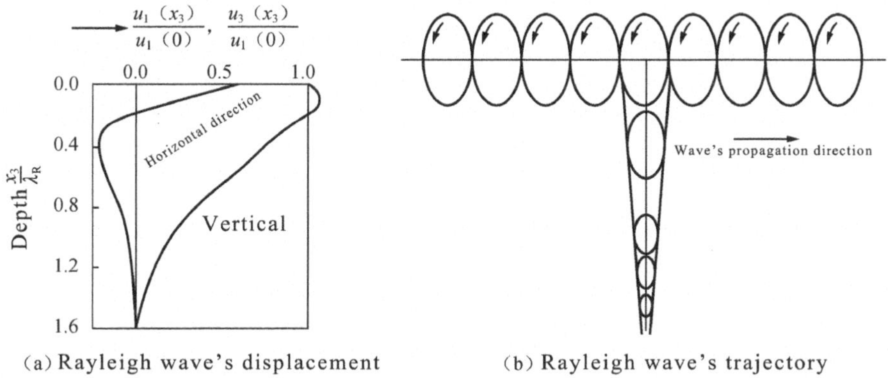

Fig. 1.7 Changes and trajectory of the displacement of Rayleigh wave in horizontal and vertical directions along vertical direction

which shows that the trajectory of a particle is an oval in a plane surface of $x_1 - x_3$, and its axial lengths in horizontal direction x_1 and vertical direction x_3 are, respectively, $f_1(x_3)$ and $f_2(x_3)$. Therefore, Rayleigh wave is a kind of elliptic polarized wave.

When the Poisson's ratio v is 0. 25, the changes and trajectory of the displacement of Rayleigh wave in horizontal and vertical directions along vertical direction are shown in Fig. 1.7. It can be seen from the figure that the changes of horizontal displacement along the vertical direction are accompanied with reversal, which means that, hereupon, the trajectory of the particle changes from retrograde ellipse to forward ellipse; the attenuation of Rayleigh wave is quite fast, with 1/5 attenuation after each wavelength. Rayleigh wave is formed by reflection stack after body wave arriving at the earth's surface, which is rare around the focus, and appears until the epicentral distance over VRh/(V2p − V2s)$^{1/2}$ (h is the focal depth).

1.4.2.4 Love Wave

Love wave is another kind of surface wave, which was discovered by real seismic observation, and its existence was then proved by A. E. H. Love in theory. The condition for the existence of Love wave is a loose horizontal cover layer existing in semi-infinite space. Love wave is a kind of SH wave.

Suppose that the origin is on the surface between the cover layer and the lower semi-infinite body, where the axis x_1 is the wave propagation direction, x_3 is the vertical direction, the positive direction is forward inside the infinite body, and the width of the cover layer is$H$. Suppose that the displacement function is

$$\begin{cases} u_2(x_1, x_3, t) = f_1(x_3)e^{ik(x-ct)}, \ -H \leq x_3 \leq 0 \\ u_2(x_1, x_3, t) = f_2(x_3)e^{ik(x-ct)}, \ x_3 \geq 0 \\ u_1 = u_3 = 0 \end{cases} \tag{1.24}$$

In the function, c is velocity of Love wave, $c = \frac{\omega}{k}$.

According to the boundary conditions of surface of $x_3 = 0$ between the free surface $x_3 = -H$ with the cover layer and the lower infinite body, and the amplitude ($f_2(\infty)$) should be bounded at infinite depth ($x_3 = \infty$), the physical conditions of the existence of Love wave can be worked out as

$$\nu_2\sqrt{1 - \frac{c^2}{\upsilon_{s2}^2}} = \nu_1\sqrt{\frac{c^2}{\upsilon_{s2}^2} - 1}\tan\left(\frac{\omega H}{c}\sqrt{\frac{c^2}{\upsilon_{s1}^2} - 1}\right) \tag{1.25}$$

In the formula, υ_{s1}, ν_1 is the shear wave speed and the Poisson's ratio of the cover layer;

υ_{s2}, ν_2 is the shear wave speed and the Poisson's ratio of the lower semi-infinite body.

Thus it can be concluded that if $\upsilon_{s1} < c < \upsilon_{s2}$, Formula (1.25) could be satisfied. Therefore, Love wave could only exist when the shear speed of cover layer is less than that of the lower semi-infinite body.

1.4.2.5 Dispersion Relation and Group Velocity

It could be seen from above that there is only one speed of S-wave or P-wave on homogeneous elastic surface, which completely depends on the property of the media. However, the propagation of surface wave could not be described solely with velocity in the layered elastic media, and there is relation $c = \frac{\omega}{k}$ among the wave velocity c, frequency ω, and wavenumber k, which is called dispersion relation. Thus, it is clear that the simple harmonic surface wave with different wavenumber or frequency propagates with different velocity. The velocity c of each simple harmonic surface wave is called phase velocity.

However, c is not the ideal wave parameter to describe the propagation of surface wave, because the key feature of wave motion is the propagation of wave motion energy, which is not the phase velocity. If the propagation of a number of simple harmonic surface wave with similar wavenumber k or frequency ω is investigated, the result of them superimposing together is to form a wave packet, and the propagation velocity of the wave packet is different from the phase velocity of single simple harmonic surface wave. Due to that the energy of wave motion depends on the amplitude when the frequency is the same, so the propagation velocity of wave packet is that of wave motion energy. Thus, the propagation of wave packet is called group velocity c_g.

1.5 Reflection and Refraction of Seismic Wave

1.5.1 Reflection of Seismic Wave on Free Surface

1.5.1.1 SH Wave

Suppose that the uplink SH plane harmonic wave u_{2i} in homogeneous elastic media with $x_3 > 0$ enters the free surface $x_3 = 0$ with incidence angle ϕ (see Fig. 1.8), u_{2i} could be written as

$$u_{2i} = E\mathrm{e}^{i(\omega t - k_1 x_1 + k_2 x_3)} \tag{1.26}$$

In the formula,

$$k_1 = k\sin\phi = \frac{\omega}{v_s}\sin\phi$$
$$k_2 = k\cos\phi = \frac{\omega}{v_s}\cos\phi$$

The general solution of plane harmonic wave in the homogeneous elastic media given by Formula (1.26) could be expressed as

$$u_2 = E\mathrm{e}^{i(\omega t - k_1 x_1 + k_2 x_3)} + F\mathrm{e}^{i(\omega t - k_1 x_1 - k_2 x_3)} \tag{1.27}$$

In Formula (1.27), the first item is the incident wave u_{2i}, and the second wave is the reflected wave u_{2r}. The amplitude coefficient of u_{2r}, F could be determined by condition $\tau_{23}\big|_{x_3=0}$ of the free surface. From this, we can get

$$E = F \tag{1.28}$$

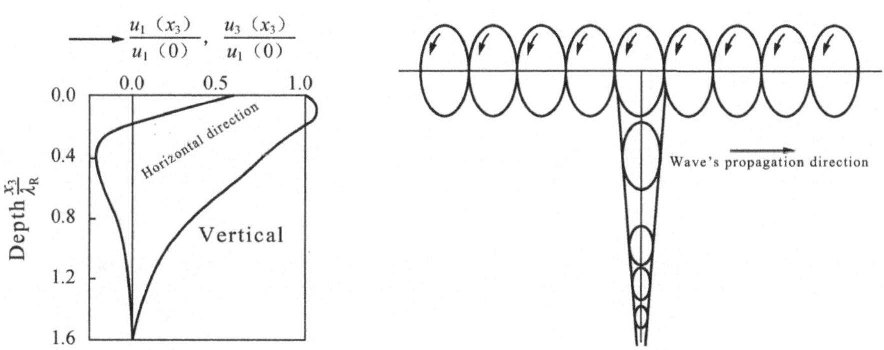

(a) Rayleigh wave's displacement (b) Rayleigh wave's trajectory

Fig. 1.8 SH waves are reflected on free surface

Thus, substituting Formulas (1.28) into (1.27), we can get

$$u_2 = 2E\mathrm{e}^{i(\omega t - k_1 x_1)} \cos(k_2 x_3) \tag{1.29}$$

When $x_3 = 0$, there is

$$u_{2i} = E\mathrm{e}^{i(\omega t - k_1 x_1)} \tag{1.30}$$

$$u_2 = 2E\mathrm{e}^{i(\omega t - k_1 x_1)} \tag{1.31}$$

It can be known from Formulas (1.30) and (1.31) that

$$\frac{\ddot{u}_2}{\ddot{u}_{2i}} = \frac{\dot{u}_2}{\dot{u}_{2i}} = \frac{u_2}{u_{2i}} = 2 \tag{1.32}$$

It can thus be seen that, no matter how much the incident angle ϕ and incident wave frequency are, when wave enters the free surface, the motion of the free surface always doubles that of the incident wave.

1.5.1.2 *P*-Wave

Suppose that the uplink P plane harmonic wave φ_i in homogeneous elastic media enters the free surface $x_3 = 0$ with incidence angle θ (see Fig. 1.9), φ_i could be written as

$$\varphi_i = E_p \mathrm{e}^{i(\omega t - k_{1p} x_1 + k_{2p} x_3)} \tag{1.33}$$

In the formula,

$$k_{1p} = k_p \sin\theta = \frac{\omega}{\upsilon_p} \sin\theta \tag{1.34}$$

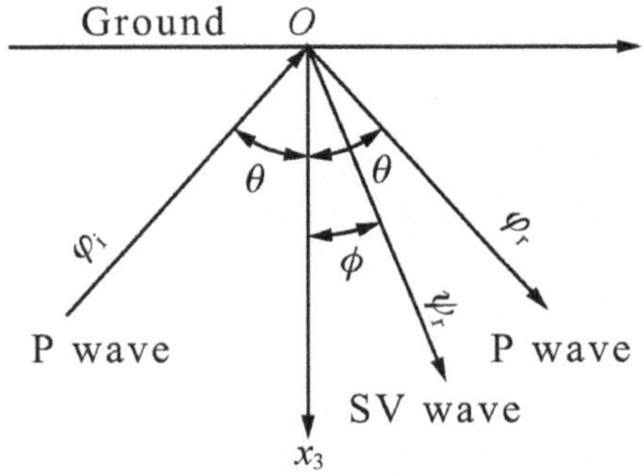

Fig. 1.9 *P*-waves are reflected on free surface

$$k_{2p} = \begin{cases} \sqrt{\left(\frac{\omega}{\upsilon_p}\right)^2 - k_{1p}^2} & k_{1p} \leq \frac{\omega}{\upsilon_p} \\ -i\sqrt{k_{1p}^2 - \left(\frac{\omega}{\upsilon_p}\right)^2} & k_{1p} > \frac{\omega}{\upsilon_p} \end{cases} \tag{1.35}$$

With respect to in-plane wave motion, the total wave field is generally composed of potential of P-wave φ and that of S-wave ψ. The general solution of their plane harmonic wave in the homogeneous elastic media could be expressed as

$$\varphi = E_p \mathrm{e}^{i(\omega t - k_{1p}x_1 + k_{2p}x_3)} + F_p \mathrm{e}^{i(\omega t - k_{1p}x_1 - k_{2p}x_3)} \tag{1.36a}$$

$$\psi = F_s \mathrm{e}^{i(\omega t - k_{1s}x_1 - k_{2s}x_3)} \tag{1.36b}$$

In the formula,

$$k_{1s} = k_s \sin\phi = \frac{\omega}{\upsilon_s}\sin\phi \tag{1.37}$$

$$k_{2s} = \begin{cases} \sqrt{\left(\frac{\omega}{\upsilon_s}\right)^2 - k_{1s}^2} & k_{1s} \leq \frac{\omega}{\upsilon_s} \\ -i\sqrt{k_{1s}^2 - \left(\frac{\omega}{\upsilon_s}\right)^2} & k_{1s} > \frac{\omega}{\upsilon_s} \end{cases} \tag{1.38}$$

In Formula (1.36a), the first item is the incident wave φ_i, and the second wave is the reflected P-wave φ_r. In the general solution Formula (1.36b) of S-wave potential ψ, the uplink item is omitted because there exists no SV wave entering the free surface. According to the conditions of free surface:

when $x_3 = 0$,

$$\sigma_3 = 0 \quad \tau_{13} = 0 \tag{1.39}$$

It is easy to see that if the conditional Formula (1.39) of free surface of $x_3 = 0$ is to be satisfied at any time, it must meet $k_{1p} = k_{1s}$; that is, the apparent velocity of all harmonious wave propagating along the axis x_1 must the same, i.e.

$$\frac{\upsilon_p}{\sin\theta} = \frac{\upsilon_s}{\sin\phi} \tag{1.40}$$

Formula (1.40) is a form of Snell law in elastic media. Therefore, it could be recorded as. Thus, it can be concluded from Formula (1.40) that

$$\left(\frac{\upsilon_s}{\upsilon_p}\right)^2 \sin 2\theta(E_p - F_p) - \cos 2\phi F_s = 0 \tag{1.41a}$$

$$\cos 2\phi(E_p + F_p) - \sin 2\theta F_s = 0 \tag{1.42b}$$

From this, we can get

$$\frac{F_\mathrm{p}}{E_p} = \frac{\upsilon_s^2 \sin 2\theta \sin 2\phi - \upsilon_p^2 \cos^2 2\phi}{\upsilon_s^2 \sin 2\theta \sin 2\phi + \upsilon_p^2 \cos^2 2\phi} \tag{1.43a}$$

$$\frac{F_s}{E_p} = \frac{2\upsilon_s^2 \sin 2\theta \sin 2\phi}{\upsilon_s^2 \sin 2\theta \sin 2\phi + \upsilon_p^2 \cos^2 2\phi} \tag{1.43b}$$

In the formula,

$$\begin{cases} \sin 2\phi = 2\sin\phi\cos\phi = 2\frac{\upsilon_s}{\upsilon_p}\sin\theta\sqrt{1-\left(\frac{\upsilon_s}{\upsilon_p}\right)^2\sin^2\theta} \\ \cos 2\phi = 1 - 2\sin^2\phi = 1 - \left(\frac{\upsilon_s}{\upsilon_p}\right)^2\sin^2\theta \end{cases} \tag{1.44}$$

It can be seen that when plane P-wave enters the free surface, it will produce both reflected P-wave and reflected SV wave. This phenomenon is called wave-type conversion. Due to that $\frac{F_p}{E_p}$ and $\frac{F_s}{E_p}$ is only related to the incident angle θ and $\frac{\upsilon_p}{\upsilon_s}$, and as $\frac{\upsilon_p}{\upsilon_s}=\sqrt{\frac{2-2\upsilon}{1-2\upsilon}}$, thus the reflected coefficient only depends on the incident angle θ and the value of the Poisson's ratio v. Figure 1.10 demonstrates the relation between the reflected coefficient and the incident angle θ. It thus can be seen that, under most conditions, $\frac{F_p}{E_p}$ is negative, i.e. the incident P-wave belongs to pressure wave, and the reflected P-wave on free surface belongs to tension wave, and vice versa. While for some Poisson's ratio values v, there may exist certain interval of θ, and when the incident angle of P-wave is within this interval and the incident P-wave is pressure wave on free surface, then the reflected P-wave is also pressure wave. Besides, there may exist certain incident angle θ_c that makes all P-wave convert into SV wave. It can be known from (1.43a) that, θ_c can be determined by Formula (1.45):

$$\upsilon_s^2 \sin 2\theta_c \sin 2\phi - \upsilon_p^2 \cos^2 2\phi = 0 \tag{1.45}$$

This condition could be satisfied for some Poisson's ratio values v, for example, when $v = 0.25$, $\theta_c = 60°$ and $77.5°$.

It can also be known from Fig. 1.10 that when P-wave vertically upward ($\theta = 0°$) enters the free surface, $\frac{F_p}{E_p} = -1$, $\frac{F_s}{E_p} = 0$, i.e. there is only reflected P-wave without reflected SV wave, and the amplitude of reflected P-wave is the same with that of the incident P-wave, but difference in phase is 180°.

Due to that $\varphi = \varphi(x_1, x_3, t)$ and $\psi = \psi(x_1, x_3, t)$ are irrelevant to the coordinate ψ_i, therefore

$$u_1 = \frac{\partial\varphi}{\partial x_1} + \frac{\partial\varphi}{\partial x_3} \tag{1.46a}$$

$$u_2 = \frac{\partial\psi}{\partial x_1} - \frac{\partial\psi}{\partial x_3} \tag{1.46b}$$

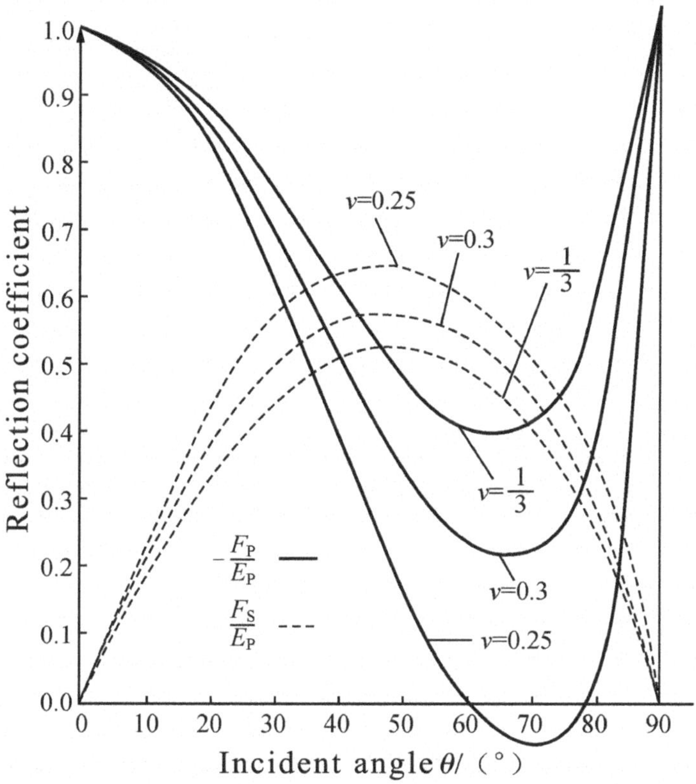

Fig. 1.10 Reflected coefficient of the incident P-wave on free surface

$$u_3 = \frac{\partial \varphi}{\partial x_3} - \frac{\partial \psi}{\partial x_1} \tag{1.46c}$$

1.5.1.3 SV Wave

Suppose that the uplink SV plane harmonic wave ψ_i enters the free surface $x_3 = 0$ with incidence angle ϕ (see Fig. 1.11):

$$\psi_i = E_s e^{i(\omega t - k_1 x_1 + k_{2s} x_3)} \tag{1.47}$$

Then the general solution of the potential of total wave field is

$$\varphi_r = F_p e^{i(\omega t - k_1 x_1)} e^{-\frac{\omega}{v_S}\sqrt{\sin^2\phi - \left(\frac{v_S}{v_P}\right)^2} x_3} \tag{1.48}$$

$$\varphi = E_s e^{i(\omega t - k_1 x_1 + k_{2s} x_3)} + F_s e^{i(\omega t - k_1 x_1 - k_{2s} x_3)} \tag{1.49}$$

Fig. 1.11 SV waves are reflected on free surface

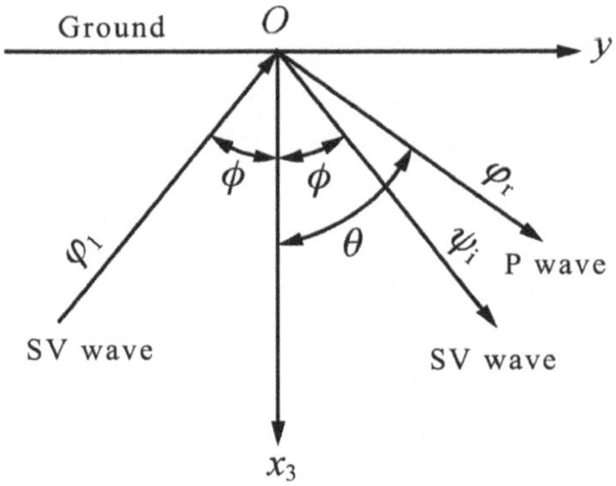

In the formulas, $k_1 = \frac{\omega \sin\phi}{\upsilon_s}$, k_{2s}, and k_{2p} are determined, respectively, by Formulas (1.40) and (1.37). According to the conditions of free surface, the reflected coefficient could be worked out:

$$\frac{F_p}{E_s} = \frac{-2\upsilon_p^2 \sin 2\phi \cos 2\phi}{\upsilon_s^2 \sin 2\theta \sin 2\phi + \upsilon_p^2 \cos^2 2\phi} \tag{1.50a}$$

$$\frac{F_s}{E_s} = \frac{\upsilon_s^2 \sin 2\theta \sin 2\phi - \upsilon_p^2 \cos 2\phi}{\upsilon_s^2 \sin 2\theta \sin 2\phi + \upsilon_p^2 \cos 2\phi} \tag{1.50b}$$

in which, $\sin\theta$ and $\sin\phi$ need to meet the Snell law of Formula (1.50), and $\sin 2\theta$ and $\cos 2\theta$ should not be taken as normal trigonometric function but a mark, whose numerical value is determined by $\sin\theta$ and $\cos\theta$; $\cos\theta$ is defined as

$$\cos\theta = \begin{cases} \sqrt{1 - \left(\frac{\upsilon_p}{\upsilon_s}\right)^2 \sin^2\phi} & \sin\phi \le \frac{\upsilon_s}{\upsilon_p} \\ -i\sqrt{\left(\frac{\upsilon_p}{\upsilon_s}\right)^2 \sin^2\phi - 1} & \sin\phi > \frac{\upsilon_s}{\upsilon_p} \end{cases} \tag{1.51}$$

It can be seen that when plane SV wave enters the free surface, it will produce both reflected SV wave and reflected SV wave. Similarly, this conversion is also function of incident angle ϕ and the Poisson's ratio v; at the same time, there may also exist certain incident angle ϕ_c which makes SV wave converting into P-wave. It can be known from (1.50) that ϕ_c can be determined by Formula (1.52):

$$\upsilon_s^2 \sin 2\theta \sin 2\phi_c - \upsilon_p^2 \cos^2 2\phi_c = 0 \tag{1.52}$$

When the incident angle is within certain critical angle $\phi_c = \sin^{-1}\frac{\upsilon_s}{\upsilon_p}$, the reflected coefficients Fp/Es and Fs/Es are all real numbers, and the reflected o wave and SV wave are all plane waves, whose wave pattern is the same with the incident SV wave.

When the incident angle $k_{2p} = k_{1s}$ is beyond the critical angle ϕ_c, it can be known from Formula (1.51) that, $\cos\theta$ is imaginary number, and the reflected coefficients Fp/Es and Fs/Es are all complex numbers; according to Snell law, it can be known from Formula (1.37) that

$$k_{2p} = -i\frac{\omega}{\upsilon_s}\sqrt{\sin^2\phi - \left(\frac{\upsilon_s}{\upsilon_p}\right)^2} \tag{1.53}$$

Therefore, the reflected P-wave could be converted by Formula (1.47) into

$$\varphi_r = F_p e^{i(\omega t - k_1 x_1)} e^{-\frac{\omega}{\upsilon_s}\sqrt{\sin^2\phi - \left(\frac{\upsilon_s}{\upsilon_p}\right)^2} x_3} \tag{1.54}$$

That is downward-attenuated right-traveling wave. Due to that the reflected coefficients Fp/Es and Fs/Es are all complex numbers, so the phase of reflected P-wave and SV wave is different to that of the incident SV wave, and because the change of phase is only relevant to the incident angle ϕ and the Poisson's ratio ϕ, so for incident SV wave with different frequency components, the phase changes of reflected P-wave and SV wave are the same. Therefore, generally speaking, the wave pattern of incident SV wave in free surface $x_3 = 0$ transient state is different from that of the reflected P-wave and SV wave (Fig. 1.12).

When SV wave vertically upward $\phi = 0°$ enters the free surface, Fp/Es equals to 0 and Fs/Es equals to 1, i.e. there is only reflected SV wave without reflected P-wave, and the amplitude of reflected SV wave is the same with that of the incident SV wave, but difference in phase is 180°.

1.5.2 *Reflection and Refraction of Seismic Wave on Interface Between Media*

1.5.2.1 SH Plane Harmonious Wave on Upper and Lower Media

Suppose that $x_3 = 0$ is the interface between the upper and the lower media, the density and lame parameter of the upper medium are, respectively, ρ', λ' and μ', and that of the lower medium are, respectively, ρ, λ, and μ. Suppose that SH plane harmonious wave enters from lower medium into the interface with incident angle of ϕ (see Fig. 1.13), the general solutions of the plane harmonious wave in upper and lower media are

$$\mu_2' = E_s' e^{i(\omega t - k_1' x_1 + k_2' x_3')} \tag{1.53}$$

$$\mu_2 = E_s e^{i(\omega t - k_1 x_1 + k_2 x_3)} + F_s e^{i(\omega t - k_1 x_1 - k_2 x_3)} \tag{1.54}$$

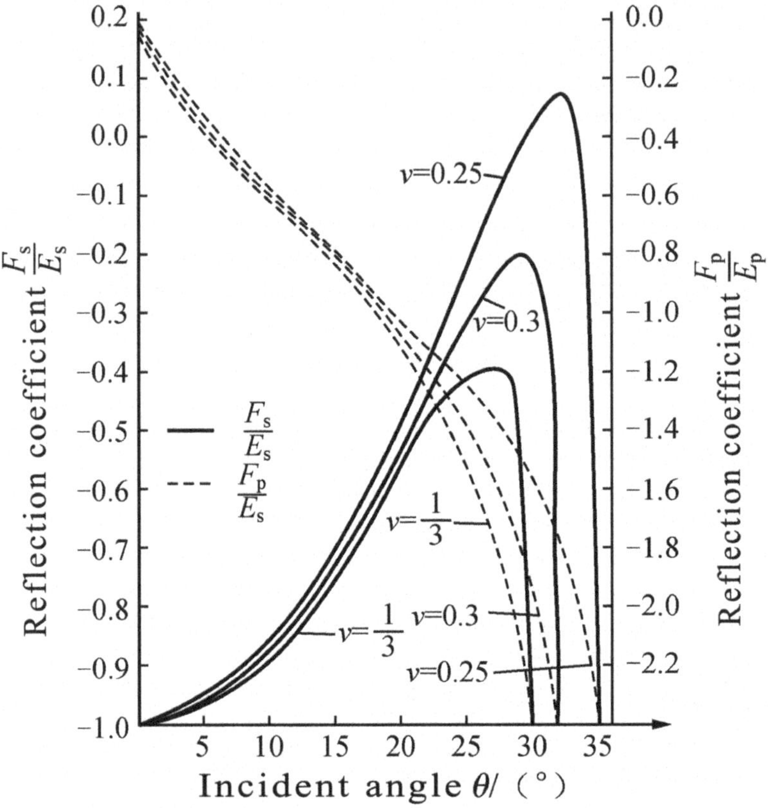

Fig. 1.12 Reflected coefficient of the incident SV wave on free surface

In Formula (1.54), the first item is the incident SH plane harmonious wave, and the second item is the reflected SH plane harmonious wave. According to the interface $x_3 = 0$ or $x_3' = h'$ of the upper and lower media and displacement and stress continuity conditions, there should be $k_1 = k_1'$.

For the upper medium:

$$\psi' = E_s' e^{i(\omega t - k_{1s}' x_1 + k_{2s}' x_3')} \tag{1.55}$$

For the lower medium:

$$\tau_{23} = i\mu k_2 \left[E_s e^{i(\omega t - k_1 x_1 + k_2 x_3)} - F_s e^{i(\omega t - k_1 x_1 - k_2 x_3)} \right] \tag{1.56}$$

The amplitude vector and stress–displacement vector are defined as follows,

$$H' = \begin{pmatrix} E'_p \\ 0 \\ E'_s \\ 0 \end{pmatrix}, \quad S' = \begin{pmatrix} \tau'_{23} \\ \mu'_2 \end{pmatrix}, \quad H = \begin{pmatrix} E_s \\ F_s \end{pmatrix}, \quad S = \begin{pmatrix} \tau_{23} \\ \mu_2 \end{pmatrix} \tag{1.57}$$

The interface $x_3 = 0$ or $x'_3 = h'$ of the upper and lower media and displacement and stress continuity conditions should be:

$$S'\big|_{x'_3=h'} = S\big|_{x_3=0} \tag{1.58}$$

Thus, we can get the amplitude conversion relation

$$H = T'H' \tag{1.59}$$

In the formula,

$$T' = \begin{bmatrix} \frac{\frac{\mu'}{\mu}k'_2+k_2}{2k_2}e^{ik'_2h'} & \frac{k_2-\frac{\mu'}{\mu}k'_2}{2k_2}e^{-ik'_2h'} \\ \frac{k_2-\frac{\mu'}{\mu}k'_2}{2k_2}e^{ik'_2h'} & \frac{\frac{\mu'}{\mu}k'_2+k_2}{2k_2}e^{-ik'_2h'} \end{bmatrix} \tag{1.60}$$

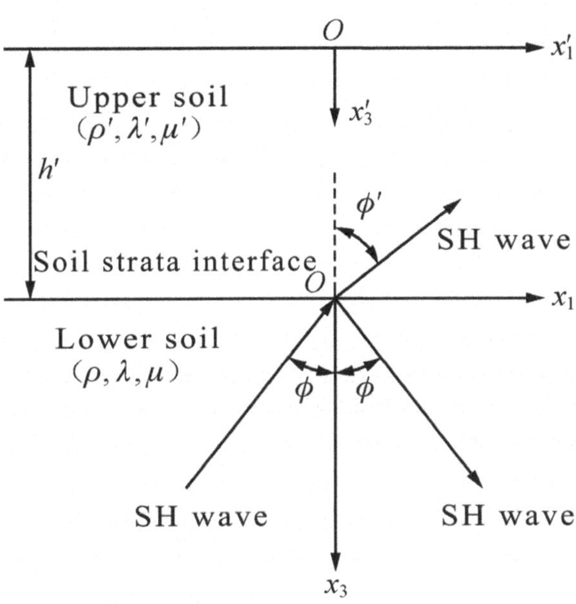

Fig. 1.13 Reflection and refraction of SH wave on interface between different media

When $k_1 = 0$, i.e. when SH wave could only propagate along axis x_3 or along axis x_3', $k_2 = \frac{\omega}{\upsilon_s}$, $k_2' = \frac{\omega}{\upsilon_s'}$, $k_{1p}' = k_{1p} = k_{1s}' = k_{1s} = k_1$, then Formula (1.69) could be written as

$$T' = \begin{bmatrix} \frac{1+\alpha'}{2} e^{ik_2'h'} & \frac{1-\alpha'}{2} e^{-ik_2'h'} \\ \frac{1-\alpha'}{2} e^{ik_2'h'} & \frac{1+\alpha'}{2} e^{-ik_2'h'} \end{bmatrix} \tag{1.61}$$

In the formula, $\alpha' = \frac{\rho'\upsilon_s'}{\rho\upsilon_s}$ is called wave impedance ratio.

Substituting Formulas (1.60) into (1.58), we can get

$$\frac{F_s}{E_s} = \frac{k_2 - \frac{\mu'}{\mu}k_2'}{k_2 + \frac{\mu'}{\mu}k_2'} \tag{1.62}$$

$$\frac{E_s'}{E_s} = \frac{2k_2}{\frac{\mu'}{\mu}k_2' + k_2} \tag{1.63}$$

In the formula,

$$k_2' = \begin{cases} \frac{\omega}{\upsilon_s}\sqrt{\left(\frac{\upsilon_s}{\upsilon_s'}\right)^2 - \sin^2\phi} & \frac{\upsilon_s}{\upsilon_s'} \geq \sin\phi \\ -i\frac{\omega}{\upsilon_s}\sqrt{\sin^2\phi - \left(\frac{\upsilon_s}{\upsilon_s'}\right)^2} & \frac{\upsilon_s}{\upsilon_s'} < \sin\phi \end{cases} \tag{1.64}$$

It can be seen from the above formula that, when the shear wave speed of the lower medium υ_s is beyond that of the lower medium υ_s', the refracted wave also belongs to plane wave. While if that of the lower medium υ_s is less than that of the upper medium υ_s', then there exists a critical angle of incidence.

$$\phi_c = \arcsin^{-1}\left(\frac{\upsilon_s}{\upsilon_s'}\right) \tag{1.65}$$

When the refracted wave $\phi < \phi_c$ is still the plane wave, while when $\phi > \phi_c$, it degenerates into upward attenuated and right-traveling inhomogeneous plane wave.

When the incident angle $\phi = 0$, i.e. SH wave enters vertically upward, it can be known from Formulas (1.62) and (1.63) that

$$\frac{F_s}{E_s} = \frac{1-\alpha'}{1+\alpha'} \tag{1.66}$$

$$\frac{E_s'}{E_s} = \frac{2}{1+\alpha'} \tag{1.67}$$

1.5.2.2 *P* and SV Plane Harmonious Wave in Upper and Lower Media

Supposing that the *P* and SV harmonious wave enters from lower medium into the interface of $x_3 = 0$, then the general solution of the plane wave in upper and lower media are:

For the upper medium:

$$\varphi' = E'_p \mathrm{e}^{i(\omega t - k'_{1p}x_1 + k'_{2p}x'_3)} \tag{1.68}$$

$$\psi' = E'_s \mathrm{e}^{i(\omega t - k'_{1s}x_1 + k'_{2s}x'_3)} \tag{1.69}$$

For the lower medium:

$$\varphi = E_p \mathrm{e}^{i(\omega t - k_{1p}x_1 + k_{2p}x_3)} + F_p \mathrm{e}^{i(\omega t - k_{1p}x_1 - k_{2p}x_3)} \tag{1.70}$$

$$\psi = E_s \mathrm{e}^{i(\omega t - k_{1s}x_1 + k_{2s}x_3)} + F_s \mathrm{e}^{i(\omega t - k_{1s}x_1 - k_{2s}x_3)} \tag{1.71}$$

According to Snell law, there is $k'_{1p} = k_{1p} = k'_{1s} = k_{1s} = k_1$. As it should satisfy the following displacement and stress continuity conditions at the interface $x_3 = 0$ or $x'_3 = h'$ of upper and lower media:

$$\begin{cases} \mu_1|_{x_3=0} = \mu'_1|_{x_3=h'} \\ \mu_3|_{x_3=0} = \mu'_3|_{x'_3=h'} \\ \tau_{13}|_{x_3=0} = \tau'_{13}|_{x'_3=h'} \\ \sigma_3|_{x_3=0} = \sigma'_3|_{x'_3=h'} \end{cases} \tag{1.72}$$

The amplitude vector and stress–displacement vector are defined as follows,

$$H = \begin{pmatrix} E_p \\ F_p \\ E_s \\ F_s \end{pmatrix}, \quad S' = \begin{pmatrix} \mu'_1 \\ \mu'_3 \\ \tau'_{13} \\ \sigma'_3 \end{pmatrix}, \quad H = \begin{pmatrix} E_p \\ F_p \\ E_s \\ F_s \end{pmatrix}, \quad S = \begin{pmatrix} \mu_1 \\ \mu_3 \\ \tau_{13} \\ \sigma_3 \end{pmatrix} \tag{1.73}$$

Therefore, the stress–displacement relation can be expressed as

$$S = TH, \quad S' = T'H' \tag{1.74}$$

The continuity condition Formula (1.74) of the interface between the upper and lower media can be written as

$$S|_{x_3=0} = S'|_{x'_3=h'} \tag{1.75}$$

Thus, we can get the amplitude conversion relation

$$H = T''H' \tag{1.76}$$

In the formula,

$$T'' = T\big|_{x_3=0}^{-1} \cdot T'\big|_{x_3'=h'} \tag{1.77}$$

$$T\big|_{x_3=0} = \begin{bmatrix} ik_1 & -ik_{2p} & ik_{2s} & -ik_{2s} \\ ik_{2p} & -ik_{2p} & ik_1 & ik_1 \\ 2\mu k_1 k_{2p} & -2\mu k_1 k_{2p} & \mu k_{33} & \mu k_{33} \\ \mu k_{33} & \mu k_{33} & -2\mu k_1 k_{2s} & 2\mu k_1 k_{2p} \end{bmatrix} \tag{1.78}$$

$$T'\big|_{x_3'=h'} = \begin{bmatrix} ik_1 e^{ik_{2p}'h'} & -ik_{2p}' e^{-ik_{2p}'h'} & ik_{2s}' e^{ik_{2s}'h'} & -ik_{2s}' e^{-ik_{2s}'h'} \\ ik_{2p}' e^{ik_{2p}'h'} & -ik_{2p}' e^{-ik_{2p}'h'} & ik_1 e^{ik_{2s}'h'} & ik_1 e^{-ik_{2s}'h'} \\ 2\mu k_1 k_{2p}' e^{ik_{2p}'h'} & -2\mu k_1 k_{2p}' e^{-ik_{2p}'h'} & \mu k_{33}' e^{ik_{2s}'h'} & \mu k_{33}' e^{-ik_{2s}'h'} \\ \mu k_{33}' e^{ik_{2p}'h'} & \mu k_{33}' e^{-ik_{2p}'h'} & -2\mu k_1 k_{2s}' e^{ik_{2s}'h'} & 2\mu k_1 k_{2p}' e^{-ik_{2s}'h'} \end{bmatrix} \tag{1.79}$$

In the formula, $k_{33} = k_1 - k_{2s}^2, \quad k_{33}' = k_1' - k_{2s}'^2$.
The above relation can be shortened into the form

$$E_p' = a_{pp}' E_p + a_{ps}' E_s \tag{1.80a}$$

$$E_s' = a_{sp}' E_p + a_{ss}' E_s \tag{1.80b}$$

$$F_p = a_{pp} E_p + a_{ps} E_s \tag{1.80c}$$

$$F_s = a_{sp} E_p + a_{ss} E_s \tag{1.80d}$$

It can thus be seen that the propagation of P and SV plane wave is much more complex than that of SH wave in the upper and lower media. When the incident wave only includes the P plane wave, then there is $E_s = 0$, and the horizontal wavenumber k_1 can be determined by the incident angle $\theta = 0$ of P-wave, $k_1 = \frac{\omega \sin\theta}{v_p}$; when the incident wave only includes SV plane wave, then there is $E_p = 0$, and the wavenumber k_1 can be determined by the incident angle ϕ of SV wave: $k_1 = \frac{\omega \sin\phi}{v_s}$.

If the incident angle of P plane wave $\theta = 0$, i.e. it enters vertically upward, then it could prove:

Refraction coefficient

$$a_{pp}' = \frac{2}{1+\alpha} \tag{1.80}$$

Reflection coefficient

$$a_{pp} = \frac{1-\alpha}{1+\alpha} \tag{1.81}$$

In the formula, $\alpha = \frac{\rho' v_p'}{\rho v_p}$. Besides, $a_{sp}' = a_{sp} = 0$; i.e. there is no conversion of wave pattern.

If the incident angle of SV plane wave $\phi = 0$; i.e. it enters vertically upward, then it could prove:

Refraction coefficient

$$a_{ss}' = \frac{2}{1+\alpha'} \tag{1.82}$$

Reflection coefficient

$$a_{ss} = \frac{1-\alpha'}{1+\alpha'} \tag{1.83}$$

In the formula, $\alpha' = \frac{\rho' v_s'}{\rho v_s}$. Besides, $a_{ps}' = a_{ps} = 0$; i.e. there is no conversion of wave pattern. It can thus be seen that the results of vertically upward incidence of SV and SH plane wave are exactly the same.

Bibliography

Chen Guoxing. Principle of seismic design of engineering structures [M]. Beijing: China Water & Power Press, 2002.

Chen Guoxing. Geotechnical earthquake engineering [M]. Beijing: Science Press, 2007.

Hu Yuxian. Earthquake engineering (second edition) [M]. Beijing: Seismological Press, 2006.

Liao Zhenpeng. An introduction to engineering vibration theory [M]. Beijing: Science Press, 1996.

Liu Huixian. Intensity standard and intensity revolution [A]. the Institute of Engineering Mechanics, State Seismological Bureau. Liu Huixian Earthquake Engineering Symposium [C]. Beijing: Seismological Press, 1992:81–86.

Liu Huixian. Seismic damage of Tangshan earthquake (I) [M]. Beijing: Seismological Press, 1985.

Li Jie, Li Guoqiang. An introduction to earthquake engineering [M]. Beijing: Seismological Press, 1992.

Min Ziqun. Strong earthquake records in China (23rd Century BC-1977 A.D.) [M] Beijing: Seismological Press, 1995.

Wu Shiming, et al. Soil dynamics [M]. Beijing: China Architecture & Building Press, 2000.

Xu Jian. Construction vibration engineering manual [M]. Beijing: China Architecture & Building Press, 2002.

Chapter 2
Analysis Technics of Seismic Signals

The main contents of seismic signals are its expression, conversion, and operation as well as the contained information. The typical seismic signal process generally contains preprocess of seismic signals, time-domain analysis, frequency-domain analysis, and modal analysis.

During the observation of seismic motion, the signals get by seismic motion tests on detected objects with sensors, amplifiers, intermediate converters, or data acquisition instruments must mix with many unwanted components in the output process because of the influence of all factors inside and outside of the test system. Under most circumstances, the directly obtained data are not the real signals, or in other words, there is difference between them and the real seismic signals. Therefore, taking the seismic signals directly obtained from tests without analysis, process, and amendments as results is easy to cause errors and sometimes even leads to wrong conclusions. Therefore, the originally get signal needs preliminary process, the deformation of wave pattern needs amendments, and the noise and disturbance mixed in the signal needs to be eliminated. Besides, the unwanted information needs to be weakened and the valued part needs to be strengthened so as to restore the real seismic signal as truly as possible, i.e. to restore the nature of earthquakes as authentically as possible. The preprocess of seismic signals is, with some mathematical methods, to process and eliminate the false and retain the truth of the observed signals in the seismic tests and further work out the wanted contents so as to move on to further analysis and research. Generally, the preprocessing methods of seismic signals include calibration transformation, i.e. converting the quantitative number of the collected seismic signal data into the tested physical quantities; eliminating the signal trend, i.e. amending the distortion of wave pattern caused by baseline deviation and eliminating the disturbance and interference mixed in the signals with filtering; and digital filtering could eliminate the unwanted spectral components from the seismic signals so as to get rid of the influence of noise on seismic signals.

C. Yang et al., *Three Dimensional Space-Time Analysis Theory of Geotechnical Seismic Engineering*, https://doi.org/10.1007/978-981-13-3356-9_2

2.1 Calibration Transformation

For data of the digital value of voltage, the calibration transformation could be completed by multiplying the calibration value of the sensors, i.e. the ratio of physical quantity with the output voltage of the sensors. For data of the digital value of integer, it needs to firstly multiply the resolution ratio of the collector, i.e. quantization unit, to convert the data into voltage data, and then go through calibration transformation of physical units. For example, for a 16-bit data collector with output voltage range of −10 to +10 V, its full-scale voltage is 20 V, so the resolution ratio could be worked out with 20 divided by 2^{16}, and the result is 0.000305175 V. Use the value of the resolution ratio to multiply each integer value of the collected seismic signal to get the seismic signal data using voltage as a unit, and then use the calibration value of the sensors to multiply the voltage signal to get the seismic signal data with real physical unit.

2.2 Eliminating the Signal Trend

During seismic motion test, due to factors such as zero drifts of amplifiers caused by temperature changes, instability of low-frequency performance beyond the frequency range of sensors, and interference of environment around sensors, the collected signal data often deviate from the baseline, and the deviation degree even changes with time. The whole process of baseline deviation changing with time is called the trend of the signal. The trend is the frequency components whose vibration cycle is longer than the signal sampling length, and often due to reasons of the signal systems, a linear or slowly varying trend error comes out in time sequence.

There are mainly four reasons that cause trend: (1) The original signals do not get through proper process during sampling like high-pass filtering to make low-frequency components with cycles longer than the sampling length involved in the collected signals; (2) external causes include zero drifts of sensors or instruments, or offset of signal wave pattern caused by base motion; (3) the trend caused by signals goes through integrating amplifiers due to mis-operation, for example, the constants caused by setting wrong zero point become a straight line after integral, and the low-frequency components become slowly varying trend after going through integrating amplifiers; (4) the sampling length is improperly selected when the record inception is done.

The existence of trend will lead to large errors in relevant analysis of time-domain and power spectrum analysis of frequency, and even make low-frequency spectrum completely lose its authenticity. Therefore, the trend is often eliminated in test signal analysis. The existing methods to eliminate trend are:

(1) Fitting method: The most commonly used are method of least squares, stepwise regression method, G (1,1) and GM (2,1). These methods are commonly used for eliminating linear trend, higher-order polynomial trend, and periodic

oscillation trend with changing amplitudes. The major defect is that it needs to predict the trend type in advance for signal fitting, such as linear function, exponential function, and power function. With the increase of trend complexity, the difficulty of fitting also increases, so they are not fit for analyzing complex real signals.

(2) Wavelet method: It is a quite effective method that develops in recent years to extract and eliminate trend from non-stationary signals. It decomposes and reconstitutes the random non-stationary signals to get the trend sequence and eliminate the zero-mean random stationary signals. The method is simple and general to use which needs no prior knowledge of the trend. While the selection of wavelet basis function and determination of wavelet order need pre-prediction, different features of signals require different wavelet basis and wavelet order.

(3) Empirical Mode Decomposition (EMD) is the newly developed time–frequency analysis method to deal with nonlinear non-stationary signals. This method takes the multi-resolution advantage of wavelet transform and overcomes the difficulty of selecting wavelet basis in wavelet transform. It decomposes the signals into finite IMF with different feature scales according to the inner features of the data. These IMF components not only show the real physical information of the inner data, but also satisfy the conditions of Hilbert transform. The Hilbert spectrum (HS) get from this could exactly reflect the distribution rules of energy in time and other various scales in this physical process.

(4) Filtering methods such as sliding average method and Fourier transform are easy and effective, while their apparent defect is frequency leakage, i.e. the trend after filtering has obvious random signal components which reduce the analysis accuracy of dominant frequency signals and even cause loss of some important information.

(5) Model methods such as auto-regressive integrated moving average (ARIMA) model method, seasonal model method, and X-12-ARIMA method eliminate the trend by some process, and the processed signals can be treated as stationary and random signals so it can be studied as random stationary process. The modeling of these methods is easy, but the concrete form of trend cannot be obtained, so it fits for prediction purpose but not systematic analysis.

In the real engineering, the technicians always use linear trend to substitute the trend because of the complexity of the fitting calculation of the latter. Such process is acceptable for circumstances with little requirement on accuracy, while it must adopt nonlinear trend to conduct high-accuracy analysis to precision analysis signals for better trend elimination effects.

2.2.1 Least-Squares-Fit Method

The methods to eliminate trend are diverse, which mainly depend on the features of the signals and the physical models of the tested objects. For random signals and stationary signals, least-squares-fit method is generally adopted, which could eliminate the baseline offset with linear state and higher-order polynomial trend. It is also the commonly used method in practical engineering. Its modeling steps are: First, suppose a polynomial trend and list the solving equation with principle of least squares; second, work out the trend coefficient matrix with matrix method and obtain the trend fitting curve; last, work out the wanted signals by using the original signals to subtract the trend.

Suppose that data using sequence $\{u_n\}(n = 1, 2, \ldots, N)$, sampling time interval is Δt, and adopt K-order polynomial U_n to fit the trend. Suppose the expression of U_n is as follows:

$$U_n = \sum_{k=0}^{K} b_k (n\Delta t)^k \quad (n = 1, 2, \ldots, N) \tag{2.1}$$

in which b_k is the coefficient of the polynomial.

According to the principle of least squares, suppose the estimation error of $\{U_n\}$ to $\{u_n\}$ is $E(\Delta t)$, and $E(\Delta t)$ is the error between the estimation value and the true value:

$$E(\Delta t) = \sum_{n=1}^{N} (u_n - U_n)^2 \tag{2.2}$$

Substitute Formula (2.1) and we can get:

$$E(\Delta t) = \sum_{n=1}^{N} \left(u_n - \sum_{k=0}^{K} b_k (n\Delta t)^k \right)^2 \tag{2.3}$$

Work out the minimal value of $E(\Delta t)$ and make the partial derivative of Formula (2.3) to b_j zero;

$$\frac{\partial F}{\partial b_j} = \sum_{n=1}^{N} 2\left[u_n - \sum_{k=0}^{K} b_k (n\Delta t)^k \right]\left[-(n\Delta t)^j \right] = 0 \tag{2.4}$$

After arrangement, we can get:

$$\sum_{n=1}^{N} u_n (n\Delta t)^j = \sum_{k=0}^{K} b_k \sum_{n=1}^{N} (n\Delta t)^{k+j}, \quad j = 0, 1, 2, \ldots, K \tag{2.5}$$

Working out the fitting trend coefficient by Formula (2.5), we can get the estimated polynomial of the trend. In practical engineering, if the order of the fitting polynomial is high, i.e. the value of K is large, it is difficult to solve Formula (2.5) with common algebraic methods and easy to make mistakes. Therefore, it commonly uses the method of matrix combined with MATLAB program, which can obtain the trend coefficient in an easy and fast way and finally get the trend.

Make $\sum = \sum_{n-1}^{N}$, so:

When K is 0, the matrix of the trend coefficient could be obtained:

$$b_0 = \frac{u_n}{N} \tag{2.6}$$

When K is 1, the matrix of the trend coefficient could be obtained:

$$\begin{bmatrix} b_0 \\ b_1 \end{bmatrix} = \begin{bmatrix} N & \sum n\Delta t \\ \sum n & \sum n^2\Delta t \end{bmatrix}^{-1} \begin{bmatrix} \sum u_n \\ \sum nu_n \end{bmatrix} \tag{2.7}$$

When K is 2, the matrix of the trend coefficient could be obtained:

$$\begin{bmatrix} b_0 \\ b_1 \\ b_2 \end{bmatrix} = \begin{bmatrix} N & \sum n\Delta t & \sum n^2\Delta t^2 \\ \sum n & \sum n\Delta t^2 & \sum n^2\Delta t^3 \\ \sum n^2 & \sum n\Delta t^3 & \sum n\Delta t^4 \end{bmatrix}^{-1} \begin{bmatrix} \sum u_n \\ \sum nu_n \\ \sum n^2 u_n \end{bmatrix} \tag{2.8}$$

Deduce the rest from this, and the matrix of the trend coefficient when $K=s$ could be obtained:

$$\begin{bmatrix} b_0 \\ b_1 \\ b_2 \\ \ldots \\ b_n \end{bmatrix} = \begin{bmatrix} N & \sum n\Delta t & \sum n^2\Delta t^2 & \ldots & \sum n^s\Delta t^s \\ \sum n & \sum n^2\Delta t & \sum n^3\Delta t^2 & \ldots & \sum n^{s+1}\Delta t^s \\ \sum n^2 & \sum n^3\Delta t & \sum n^4\Delta t^2 & \ldots & \sum n^{s+2}\Delta t^s \\ \ldots & \ldots & \ldots & \ldots & \ldots \\ \sum n^s & \sum n^{s+1}\Delta t & \sum n^{s+2}\Delta t^2 & \ldots & \sum n^{s+s}\Delta t^s \end{bmatrix}^{-1} \begin{bmatrix} \sum u_n \\ \sum nu_n \\ \sum n^2 u_n \\ \ldots \\ \sum n^s u_n \end{bmatrix} \tag{2.9}$$

Formula (2.9) is the matrix of the trend coefficient, and substitute it into Formula (2.1) to get the general expression of the polynomial of the trend.

2.2.2 Wavelet Method

Express the signal $f(t)$ in wavelet series, and their forms are as follows:

$$f(t) = \sum_{j=-\infty}^{\infty} \sum_{k=-\infty}^{\infty} (f, \psi_{j,k})\psi_{j,k}(t) = \sum_{j=-\infty}^{\infty} d_{j,k}\psi_{j,k}(t) \tag{2.10}$$

in which $\psi_{j,k}(t)$ is the wavelet function, $d_{j,k}$ is the wavelet coefficient, and the expression is as follows:

$$d_{j,k} = (f, \psi_{j,k}) \tag{2.11}$$

Through the "time–frequency window" similarity analysis of windowed Fourier transform, we can get the "time–frequency window" Cartesian product of wavelet transform:

$$\left[b + at^* - a\Delta_\psi, b + at^* + a\Delta_\psi\right] \times \left[\frac{\omega^*}{a} - \frac{1}{a}\Delta^\cap, \frac{\omega^*}{a} + \frac{1}{a}\Delta^\cap\right] \tag{2.12}$$

In the above formula, $a = 2^j$, the width of time window is $2a\Delta_\psi$, which narrows with the increase of frequency and widens with the decrease of the frequency. According to this, it can satisfy the time localization problem of high frequency and low frequency if information of high and low frequencies are not checked simultaneously. At the maximum frequency level (i.e. according to the time interval Δt, the maximum detected frequency is Nyquist frequency, which is half of the sampling frequency), select the minimum width of "time–frequency window," detect the signals with maximum frequency in the original signals, and strip these signals from the original signals, putting them in space W_{N-1}, while putting the total of rest low-frequency signals in another space V_{N-1} after stripping the signals with maximum frequency. Then, increase the width of "time–frequency window," detect once again the high-frequency information in the space V_{N-1}, and strip them away from space V_{N-1}, leaving them in the space W_{N-2}, while putting the total of rest low-frequency signals in another space V_{N-2} after stripping. Repeat this process, and the high-frequency components could be stripped away from the low-frequency ones in the signals. From the stripping process, we can see that it requires two interconnected space:

$$\begin{aligned} V_N &= V_{N-1} + W_{N-1} = V_{N-2} + W_{N-2} + W_{N-1} = \cdots \\ &= \cdots + W_0 + W_1 + W_2 + W_3 + \cdots + W_{N-2} + W_{N-1} \\ &N \in ZZ \end{aligned}$$

The nature of subspace shows that only if proper wavelet base is selected, the original signals could be decomposed into signals with different disjoint frequency bands, and the total energy of signals stays unchanged. The so-called trend term is the linear terms or slowly varying nonlinear terms with cycle length longer than the record length. Even if in deterministic signals, there are always trend terms, i.e. the low-frequency components of signals. For the signals with trend terms, through decomposition and reconstruction of signals, the high-frequency components could be stripped away from the low-frequency ones in the signals so as to reach the purpose of eliminating or extracting the trend.

2.2.3 Sliding Average Method

Sliding average method is a classical data process method, which needs to assume the form of trend function in advance, but need not to solve the expression of the trend. It is proved to be an effective method in practical engineering, which thus obtains widespread application.

The fundamental calculation formula of sliding average method is:

$$y_i = \sum_{n=-N}^{N} h_n x_{i-n}, \quad i = 1, 2, 3, \ldots, m \tag{2.12}$$

in which

- X sampling data;
- Y processed data;
- m the high number of data points;
- $2N + 1$ the average points number;
- H weighted average factors, whose sum must be 1.

For simple sliding average method, its weighted average factors are:

$$h_n = \frac{1}{2N + 1}, \quad n = 0, 1, 2, \ldots, N \tag{2.13}$$

The correspondent fundamental expression is:

$$y_i = \frac{1}{2N + 1} \sum_{n=-N}^{N} x_{i-n} \tag{2.14}$$

For the weighted average method, if weighted average of five points ($N = 2$) is adopted, we can adopt

$$\{h\} = \{h_{-2}, h_{-1}, h_0, h_1, h_2\} = \frac{1}{9}\{1, 2, 3, 2, 1\} \tag{2.15}$$

The method using the principle of least squares to do sliding average method to discrete data is called linear sliding average method. The calculation formula of weighted average of five points ($N = 2$) is:

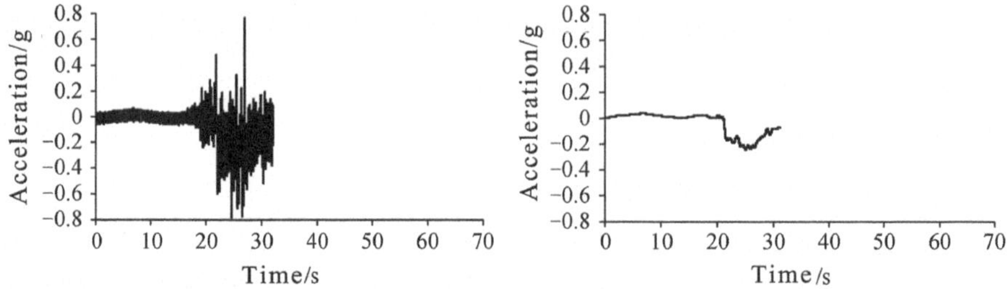

Fig. 2.1 Oscillograph and trend before process

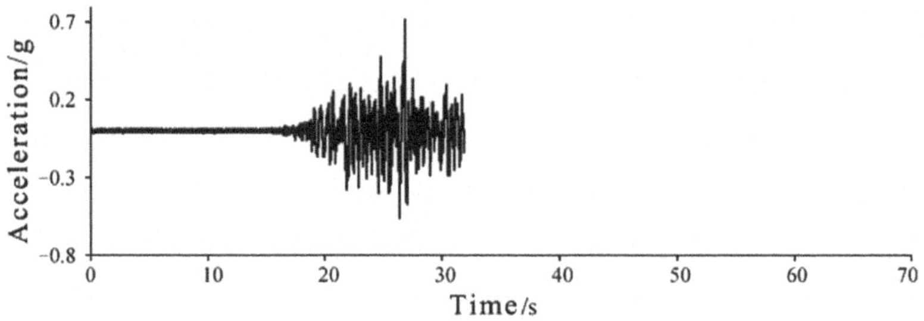

Fig. 2.2 Oscillograph after process

$$\left.\begin{array}{l} y_1 = \frac{1}{5}(3x_1 + 2x_2 + x_3 - x_4) \\ y_2 = \frac{1}{10}(4x_1 + 3x_2 + 2x_3 + x_4) \\ \ldots \\ y_i = \frac{1}{5}(x_{i-2} + x_{i-1} + x_i + x_{i+1} + x_{i+2}) \\ \ldots \\ y_{m-1} = \frac{1}{10}(x_{m-3} + 2x_{m-2} + 3x_{m-1} + 4x_m) \\ y_m = \frac{1}{5}(-x_{m-3} + x_{m-2} + 2x_{m-1} + 3x_4) \end{array}\right\} \quad (i = 3, 4, \ldots, m-2) \tag{2.16}$$

The above calculation process could based on MATLAB software, and according to the basic theories of the sliding average method, we could do the relevant calculation program and extract the trend of seismic signals so as to eliminate them. For the calculation results, see Figs. 2.1 and 2.2.

2.3 Digital Filtering

The seismic signal data collected by data collector are always overlapped with noise signals, which include not only periodic interference signals with power frequency of 50 Hz or multiple times of the frequency, but also random irregular interference signals. Therefore, the original signals need getting through filtering process to eliminate the interference of noise signals. In the analysis of seismic signals, digital filtering is the processing method to select the most wanted signals from the collected discrete signals with mathematical methods. The main effects of this method include filtering the noise or false components from tested signals, improving signal-to-noise ratio, going through smooth analysis of data, and restraining the interference signals as well as stripping the frequency components.

After the signals entering the filter, some frequency components could pass through, while others could be blocked. The frequency range that could pass through the filters is called passband. The frequency range that is blocked or reduced to very small is called stop band. The junction point between the passband and the stop band is called cutoff frequency. In the design of the filters, there often keeps a frequency range that gradually changes from passband to stop band, which is called transition band.

If the filters are classified by function, there are low-pass filter (LPF), high-pass filter (HPF), band-pass filter (BPF), band-stop filter (BSF), and comb filter. If considered by mathematical methods, digital filtering includes frequency-domain filtering method and time-domain filtering method.

According to the definition of filtering, the expression of the relation between input time signal $x(t)$ and output time signal $y(t)$ in Domain Z is:

$$Y(z) = H(z)X(z) \tag{2.17}$$

The calculation link of achieving filtering function is called filter, and the digital filter could be expressed by transfer function as:

$$H(z) = \frac{Y(z)}{X(z)} \tag{2.18}$$

2.3.1 Frequency-Domain Method of Digital Filtering

The frequency-domain method of digital filtering is to conduct discrete Fourier transform on input sampling signal by FFT fast algorithm, analyze its frequency spectrum, and set zero the frequency part that needs filtering or set zero after adding gradual transition frequency band according to the filtering demands. For example, add a transition frequency band of cosine window function between the passband and the stop band and conduct discrete Fourier transform on filtered data by FFT fast algorithm to restore the time-domain signals. The frequency signals have good frequency selectivity and flexibility. Besides, due to that the Fourier spectrum of signals and the frequency features of filters are in simple multiplicative relationship, its calculation speed is much more faster than calculation of equivalent time-domain convolution and will not cause time shift as method of time-domain filtering. In addition, the frequency-domain method of digital filtering could be used to conduct integral or differential transform processing only with slight extension. The expression of frequency-domain method of digital filtering is:

$$y(r) = \sum_{k=0}^{N-1} H(k)X(k)e^{j2k\pi r/N} \tag{2.19}$$

In the expression, X is the discrete Fourier transform of input signal x; H is the frequency response function of filters, which determines the form and features of filtering.

Suppose the upper cutoff frequency is f_u, the lower cutoff frequency is f_d, the frequency resolution ratio is Δf, and in ideal situation, the frequency response function of LPF is:

$$H(k) = \begin{cases} 1 \ k\Delta f \leq f_u \\ 0 \ \text{others} \end{cases} \tag{2.21}$$

The frequency response function of HPF is:

$$H(k) = \begin{cases} 1 \ k\Delta f \leq f_d \\ 0 \ \text{others} \end{cases} \tag{2.22}$$

The frequency response function of BPF is:

$$H(k) = \begin{cases} 1 \ f_d \leq k\Delta f \leq f_u \\ 0 \ \text{others} \end{cases} \tag{2.23}$$

The frequency response function of BSF is:

$$H(k) = \begin{cases} 1 \ k\Delta f \leq f_d, \quad k\Delta f \geq f_u \\ 0 \ \text{others} \end{cases} \tag{2.24}$$

The features of frequency-domain method of digital filtering are easy to use, fast to calculate and high in control accuracy of filtering band, which could be used to design any response filters include multi-band comb filters. In particular, with the rapid development of computer technology, the calculation speed and internal memory are no longer problems, and the FFT length also increases from dozen times of power of 2 to dozens of times, which means the frequency-domain method of digital filtering could be widely applied. However, because spectrum leakage caused by sudden inception of frequency-domain data will lead to distortion of time-domain signals after filtering, so if under the circumstance of paying no attention to adding smooth attenuation transition band, this method fits more the signals with larger data length or signals with gradual decreasing vibration amplitude, for example, the process of seismic signals like impulse response and seismic response.

2.3.2 Time-Domain Method of Digital Filtering

The time-domain method of digital filtering is to complete filtering through mathematical calculation of difference equation. There are mainly two implementation methods of typical digital filters: One is IIR digital filter, which is called infinite impulse response filter, and the other is FIR filter, which is called finite impulse response filter.

The design of digital filters is always finished in three steps. Before considering certain filters, it must firstly confirm some technical indexes, which need to refer to practical engineering. Once the technical indexes are confirmed, we could adopt the fundamental principle of mathematics and digital signal process to propose a filtering model to approximate the given indexes, which is the key problem of the filter design and the top priority in the whole design process. The result of the above two steps is the filter described by difference equation (or system function, or impulse response). The filter could be accomplished on the computer according to this description.

For a lot of practical applications, such as the process of speech signals, audio signals, or seismic signals, the digital filters are always adopted to achieve frequency-selecting operation. Therefore, the index forms of digital filters are generally determined as amplitude in frequency-domain and phase response. For passband, it often hopes that the system has linear phase response, and exact linear phase could be obtained through FIR filter. However, the linear phase in passband could not be obtained through IIR filter, thus its design only considers the indexes of amplitude, which could be given in two ways. The first one is absolute indexes, which provide requirements on amplitude response function $\left|H(\mathrm{e}^{j\omega})\right|$, and they are usually applied to design of FIR filter. The indexes are given in a different approach in the design of IIR filter, and they are relative indexes, which raise requirements in the form of decibel (dB). Decibel is defined as:

$$\mathrm{dB} = -20\lg\frac{\left|H(\mathrm{e}^{j\omega})\right|}{\left|H(\mathrm{e}^{j\omega})\right|_{\max}} \geq 0 \tag{2.24}$$

These indexes are most accepted in practical engineering.

There are other important technical indexes in the design of filters, and they are passband cutoff frequency ω_p, stop band cutoff frequency ω_s, passband fluctuation coefficient R_p, stop band fluctuation coefficient R_s, and the order M of filters. If a LPF is designed, its passband is 0- ω_p, the permitted fluctuation is R_p, the stop band is $R_s f_s/2$, f_s is the sampling frequency, the required stop band attenuation is R_s, and $\omega_p - \omega_s$ is called transition band. Generally speaking, the bigger the order M of filters is, the narrower of the band width of transition band becomes.

2.3.2.1 IIR Digital Filter

The feature of IIR filter is that it has impulse response in infinite duration of time. As this filter is generally accomplished by recursive model, it is also called recursive filter. This is because IIR filter not only calculates with input finite terms, but also reinputs and recalculates the finite terms output by the filter before, which is called feedback in engineering. The filtering expression of IIR filter could be defined as a difference equation:

$$y(n) = \sum_{k=0}^{M} a_k x(n-k) - \sum_{k=1}^{N} b_k y(n-k) \tag{2.25}$$

In the formula, $x(n)$ and $y(n)$ are input and output time-domain signal sequences; a_k and b_k are filtering coefficients. Its systematic transfer function could be represented with the following expression:

$$H(z) = \frac{\sum_{k=0}^{M} a_k z^{-k}}{1 + \sum_{k=1}^{N} b_k z^{-k}} \tag{2.26}$$

In the formula, N is the order of the filter, or it could be called pole number of systematic transfer function of the filter; M is the zero number of the filter; a_k and b_k are both weight function coefficients.

IIR digital filter is always designed with the analog filter prototype, for converting the analog filters into digital filters. The design of analog filters is more mature with both complete design formula and charts that are available for query, so taking advantage of these existing resources will definitely bring convenience to the design of digital filters. The prototype generating functions of commonly used analog LPF are prototype of Butterworth filter, Modes I and II of prototype of Chebyshev filter, prototype of elliptic filters, and prototype of Bessel filter, etc.

The design of IIR filter includes the production of analog LPF prototype, the conversion from analog LPF prototype into low-pass, high-pass, band-pass, and band-stop analog filters, and the maximum order selection of analog and digital filters as well as realization of conversion from analog filters to digital filters through Blinear and Impinvar. The design steps of IIR digital filter are as follows:

(1) Converting the technical parameters of digital filters into technical indexes of analog filters according to certain rules.
(2) Designing the analog LPF $H(s)$ according to the conversed technical indexes.
(3) Converting the analog filters $H(s)$ into digital filters $H(z)$ according to certain rules.
(4) If high-pass, band-pass, and band-stop filters are to be designed, it needs firstly converting the high-pass, band-pass, and band-stop technical parameters into technical indexes of low-pass analog filters, then designing the LPF $H(s)$ according to certain rules, and finally, converting $H(s)$ into $H(z)$.

MATLAB signal processing toolbox provides several functions that could directly design IIR digital filter (butter, chehyl, che6y2, and ellip are, respectively, low-pass, high-pass, band-pass, and band-stop filter design functions of Butterworth filter, Mode I of Chebyshev filter, Model II of Chebyshev filter, and elliptic filters), and they integrate the above steps into a entirety, which provides great convenience for designing of universal filter.

2.3.2.2 FIR Digital Filter

The feature of finite impulse response digital filter (FIR digital filter) is that the impulse response could only last for a period of time, which, in practical engineering, mainly adopts non-recursive algorithm to realize. Generally speaking, the design of FIR filter focuses on design of filters with linear phase. The main advantage is that because it has unit impulse response with finite length, it is always stable and is easy to make the filters have accurate linear phase. In addition, the design only includes real algorithm but not complex operation, so there will be no delayed distortion but only delay with fixed number. For filters with length of M, its computation amount is the magnitude of $M/2$. The major defect is that when the filtering performance is given, the order of FIR filter is much more higher than the IIR filter, and at the same time, its time delay is much longer than filters with same performance. The filtering expression of IIR filter could be defined in the form of difference equation as follows:

$$y(n) = \sum_{k=0}^{N-1} b_k x(n-k) \tag{2.27}$$

In the formula, $x(n)$ and $y(n)$ are input and output time-domain signal sequences; b_k is the filtering coefficient.

The z transform of the impulse response function $h(n)$ of FIR filter is systematic transfer function, which could be expressed as:

$$H(z) = b_0 + b_1 z^{-1} + \cdots + b_{N-1} z^{1-N} = \sum_{n=0}^{N-1} b_n z^{-n} \tag{2.28}$$

Then its impulse response should be:

$$h(n) = \begin{cases} b_n & 0 \le n \le N \\ 0 & \text{others} \end{cases} \tag{2.29}$$

The design methods of FIR digital filter mainly include method of window function and frequency sampling method. Method of window function is one of the main methods to design FIR digital filter, and due to its simple computation and intuitive physical meaning, it has become the mostly used method in practical engineering. The frequency response function of an ideal digital filter could be expressed as:

$$H_{\mathrm{d}}(\mathrm{e}^{j\omega}) = \sum_{n=-\infty}^{\infty} h_d(n)\mathrm{e}^{-j\omega n} \tag{2.30}$$

In the formula, $h_d(n)$ is the impulse response sequence.

It could be obtained from the inverse Fourier transform that:

$$h_d(n) = \frac{1}{2\pi}\int_{-\pi}^{\pi} H(\mathrm{e}^{j\omega})\mathrm{e}^{j\omega n}\mathrm{d}\omega \tag{2.31}$$

As $h_d(n)$ is not causal, and the duration of $h_d(n)$ is $(-\infty, +\infty)$, it could not be realized in physics. Therefore, the direct way is to cut down this ideal impulse response sequence and to approximate by sequence with finite length. Therefore, a new impulse response sequence with finite length could be expressed as:

$$h(n) = \begin{cases} h_d(n) \; 0 \le n \le M \\ 0 \qquad \text{others} \end{cases} \tag{2.32}$$

$H(n)$ could be seen as the product of the ideal impulse response and a window function with finite length, that is:

$$h(n) = h_d(n)\omega(n) \tag{2.33}$$

In the formula, $\omega(n)$ is the rectangular window function constituted by simple inception, and its definition is:

$$\omega(n) = \begin{cases} 1 \; 0 \le n \le M \\ 0 \text{ others} \end{cases} \tag{2.34}$$

By complex convolution theorem, we can get:

$$H(\mathrm{e}^{j\omega}) = \frac{1}{2\pi}\int_{-\pi}^{\pi} H_d(\mathrm{e}^{j\theta})W(\mathrm{e}^{j(\omega-\theta)})\mathrm{d}\theta \tag{2.35}$$

It could be known from the feature of discrete Fourier transform with finite length that the sudden inception of sequence by rectangular window function will lead to spectrum leakage and cause Gibbs phenomenon. In order to reduce the influence of Gibbs phenomenon, a proper window function could be selected to make the inception happen slowly, i.e. making the sequence gradually reduces to zero. The commonly used window functions in practical engineering are rectangular window function, Bartlett window, Hanning window, Hamming window, Blackman window, Kaiser window, etc.

2.4 Singular Point Rejecting

The data-collecting system may import some false data in operation, and the reasons are of various kinds, such as serious instant distortion of signals in the transmission line, the instant breakdown or accident of digital installation or different kinds of interference. The resultant singular points of data will cause a great deal of problems in the following data analysis; thus, it will be better to carry out detection and rejection of singular points before the whole process of data analysis. But it will be hard to reach a perfect extent. Through long-term efforts of scientific and technical people, there have already been many approaches to reject the singular points, but none of which could do it to a perfect extent. The process of singular points rejection is based on the thought that the normal data sequence is "smooth," while the singular points are "mutable." In this sense, it will firstly use data process method of digital LPF, mean value of the "median," weighted sliding average to get smoothing estimation of input sampling data sequence. Then, according to the statistical features of input sampling data sequence (which is usually taken as a random stationary process), determine the normal value range of data of each sampling point. If the data exceeds that range, then it will be rejected and will be substituted with intrapolation or extrapolation. Obviously, the times will be limited.

2.5 Time-Domain Analysis of Seismic Signals

The time-domain analysis of seismic signals mainly deals with wave pattern process of time-domain signals. Wave filtering is the key part of time-domain signal process, and according to the needs, whether filtering or retaining some frequency components of measured signal wave pattern could be realized by filtering process. The maximal value, mean value and effective value of wave pattern, relevant functions for analyzing the similarity among wave patterns, and the integral and difference transform to convert displacement, velocity, and acceleration all belong to the range of time-domain process of seismic signals. For the time-domain process of random seismic signals, there are some more commonly used measures besides the above-mentioned ones, such as probability distribution function, probability density function, mean value, mean square value, variance, and correlation analysis.

Seismic signals are typical random vibration. In the process and analysis of random vibration, it is generally call one piece of signal record a sampling function, and infinite sampling functions constitute the set function of random vibrating signals. If, for one random vibration, the set average of one moment of all sampling functions is same with that of any other moment, then it will be called random stationary vibration. Generally speaking, the statistical feature of random stationary vibration is that it does not change with the time, that is to say, the statistical feature of random stationary vibration is not time function. If the set average of a random stationary vibration equals to the time average of any sampling function, then it will be called ergodic random vibration.

Most of the random vibrating signals in practical engineering are assumed as ergodic random vibration to process and analyze. And according to large amount of statistics, most random vibrations satisfy the ergodic assumption. However, even if for ergodic random vibrations, as the point numbers of single sampling function need to be of infinite length, thus it will be impossible to conduct in practical engineering. Generally, it could only adopt finite point numbers to calculate, and the resulted statistical features are not the real value of random signals, but an estimated value is approximate to the real one. The following listed methods for random signal processing are all estimation made by taking finite length of time coordinate for random stationary vibrating signals.

2.5.1 Probability Distribution Function and Probability Density Function

2.5.1.1 Probability Distribution Function

The probability distribution function of random vibrating signals refers to that a random vibration is the set of N sampling functions: $X = \{x(n)\}$, which at the moment of t_1. The function value N_1 sampling function cannot exceed the specified value x; then, the estimation of its probability distribution function is:

$$P(X \leq x, t_1) = \lim_{N \to \infty} \frac{N_1}{N} \tag{2.36}$$

The probability distribution function of instant values is real numbers between 0 and 1, and it is the non-decreasing function of variable x. It must specify that only when the number of sampling functions is big enough that $\frac{N_1}{N}$ will approach a stable value, i.e. probability.

2.5.1.2 Probability Density Function

The probability density function is the first-order derivative of the probability distribution function to variable x, which represents the probability of the amplitude of some random vibrating signals which falls into some range, and it changes with the different ranges. The estimation of the probability density function of random vibrating signals is:

$$p(x) = \frac{N_x}{N\Delta x} \tag{2.37}$$

In the formula, Δx is the interval taking x as the center, and N_x is the number of data group $\{x(n)\}$ whose value falls into the range $x \pm \frac{\Delta x}{2}$, and N is the total data number.

2.5.2 *Mean Value, Mean Square Value and Variance*

2.5.2.1 Mean Value

The mean value of random vibrating signals represents the integral average of the sampling function $x(k)\,(k = 1, 2, \ldots, N)$ on the whole time coordinate, and its physical meaning is the central tendency of this random vibrating signal change, or it can be called zero drift. The mean value estimation of random vibrating signals is:

$$u_x = \frac{1}{N}\sum_{k=1}^{N} x(k) \tag{2.38}$$

2.5.2.2 Mean Square Value

The mean square value estimation of random vibrating signals is the integral average of the square of sampling function record $x(k)\,(k = 1, 2, \ldots, N)$ for finite length on time coordinate. The mean square value expression of discrete random vibrating signals is:

$$\psi_x^2 = \frac{1}{N}\sum_{k=1}^{N} x^2(k) \tag{2.39}$$

2.5.2.3 Variance

Obviously, the definition of variance is the mean square value after removal of mean value. As it removes zero-frequency components, variance is an expression of pure dynamic component strength of vibrating signals. The expression of variance of discrete random signals is:

$$\sigma_x^2 = \frac{1}{N}\sum_{k=1}^{N} [x(k) - u_x]^2 \tag{2.40}$$

2.5.3 Relevant Functions

Relevance is the reflection of connection intimacy among characteristic quantities of objective things or of some processes. It describes the correlation degree of random sampling functions with instant values at different moments, which could be put simply as the similarity degree of random vibration wave pattern and other wave patterns with the change of time coordinate. That could conduct similarity degree calculation for the same random vibrating sampling function changing with time coordinate, and the result is called auto-correlative function. The process could also be carried out with two sampling functions, and the result is called cross-correlative function. The relevant functions could deeply reveal the wave pattern structure of random vibrating signals.

2.5.3.1 Auto-Correlative Function

The auto-correlative function describes the dependency relationship among different instant amplitudes of the same sampling function of random vibration. That is, the function reflects the correlative intimacy of the same piece of random vibrating signal moving along the time coordinate. The auto-correlative function expression of discrete random vibrating signals is:

$$R_{xx}(k) = \frac{1}{N}\sum_{i=1}^{N-k} x(i)x(i+k) \tag{2.41}$$

In the formula, $x(i)$ is the sampling function of random vibrating signals.

The auto-correlative function is an important parameter for the analysis of random vibrating signals. The rapidity of the convergence of the auto-correlative functions, to some extent, reflects the quantity of each frequency component contained and the gentle or steep degree of the wave patterns. In practical engineering, the auto-correlative functions are usually used to detect whether periodic vibration components are contained in random vibrating signals. That is because the auto-correlative functions of random components are always approaching infinity or zero or some constant value with the change of time coordinate, while the auto-correlative functions of periodic components keep the original periodicity without attenuation, and it could qualitatively know the quantity of frequency components contained in the vibrating signals.

2.5.3.2 Cross-correlative Function

The cross-correlative function describes the dependency relationship among different instant amplitudes of two sampling functions of random vibration, which reflect the correlative intimacy of two pieces of random vibrating signals moving along the time coordinate. The size of cross-correlative functions directly reflects the correlation of two signals, and it is the measure of wave pattern similarity. The cross-correlative function expression of discrete random vibrating signals is:

$$R_{xy}(k) = \frac{1}{N-k}\sum_{i=1}^{N-k} x(i)\,y(i+k) \quad (k = 0, 1, 2, \ldots, m) \tag{2.42}$$

2.6 Frequency-Domain Analysis of Vibrating Signals

Frequency-domain analysis is also called frequency spectrum analysis, which is the time–frequency transform based on Fourier transform, and its result is the function with frequency as a variable, which is called spectral function. The major method of frequency-domain process is Fourier transform. Through the real part and imaginary part of the Fourier transform results, it could be converted to amplitude spectrum and phase spectrum. The frequency-domain analysis of random vibrating signals takes the power spectral density function (PSDF) as the fundamental function, and frequency response function (FRF) and coherence function could be derived through auto-power spectrum and cross-power spectrum. FRF is the basic data of frequency-domain identification of experimental modal parameters. In addition, the other methods of frequency-domain analysis include refined Fourier transform, real cepstrum, complex cepstrum, 1/3 octave spectrum, and response spectrum.

2.6.1 Power Spectral Density Function (PSDF)

2.6.1.1 Auto-power Spectral Density Function

The PSDF of single random vibrating signal is called auto-PSDF, which is the Fourier transform of auto-correlative function of this random vibrating signal, and its expression is:

$$S_{xx}(k) = \frac{1}{N}\sum_{r=0}^{N-1} R_{xx}(r)\mathrm{e}^{-j2\pi kr/N} \tag{2.43}$$

2.6.1.2 Cross-power Spectral Density Function

The PSDF of two random vibrating signals is called cross-PSDF, which is the Fourier transform of auto-correlative function of this random vibrating signal, and its expression is:

$$S_{xy}(k) = \frac{1}{N}\sum_{r=0}^{N-1} R_{xy}(r)\mathrm{e}^{-j2\pi kr/N} \tag{2.44}$$

2.6.2 Frequency Response Function (FRF)

FRF is the quotient of cross-PSDF divided by auto-PSDF:

$$H(k) = \frac{S_{xy}(k)}{S_{xx}(k)} \tag{2.45}$$

It belongs to complex function, and it is the expression form of the dynamic characteristics of the tested system in frequency-domain, which is to say that it is the description of transmission characteristics of the input signals in frequency-domain given by the tested system itself. When all frequency components of the input signals pass this system, FRF amplifies some of the frequency components and weakens others, and distribution of new frequency components of output signals will be obtained through process. Therefore, FRF plays a particularly important role in the test of dynamic characteristics of the structures.

2.6.3 Coherence Function

The coherence function is the quotient of modular square of cross-PSDF divided by product of excitation and response auto spectrum.

$$C_{xy}(k) = \frac{\left|S_{xy}(k)\right|^2}{S_{xx}(k)\,S_{yy}(k)} \tag{2.46}$$

It is the index representing the correlation of two random vibrating signals in frequency-domain. For a random vibrating system, the evaluation of causality of input signals and output signals, i.e. how much of the frequency response of output signals is caused by excitation of input signals, could be expressed by coherence function. Usually, in the random vibrating tests, the values of coherence functions resulted are positive real numbers in the range of 0–1. In practical engineering, the coherence function is often adopted to evaluate whether the FRF is good or bad. The closer its result approaches 1, the less influence of noise is, which means the better the evaluation results is. It is taken generally that when the value exceeds 0.8, the results are accurate and reliable.

2.7 Time–Frequency Analysis of Seismic Signals

Seismic signals are complex non-stationary signals whose amplitudes and frequency change with time. Simple adoption of time-domain or frequency-domain information is hard to express complex seismic signals. Therefore, it is quite necessary to build the three-dimensional analysis system of time–frequency–amplitude. Up to now, the most commonly used and maturest time–frequency analysis techniques of seismic signals are short-time wavelet transform and Hilbert–Huang transform (HHT). The former has gone through long development history, while the latter is a brand-new signal analysis technique. In this manner, this book will only give detailed elaboration on the latter.

2.7.1 The Basic Principle of Hilbert–Huang Transform (HHT)

HHT is a brand-new signal analysis technique, which is composed of EMD and Hilbert transform, with its core of EMD. It is more suitable than Fourier transform and wavelet transform to analyze local properties of time and frequency. This section will focus on elaboration and discussion of the principle, algorithm, and completeness of HHT and the relevant existing problems, and further introduce HHT theorem into the field of seismic signal analysis, so as to lay a reliable theoretical foundation for following creative study work.

2.7.1.1 Principle and Algorithm of EMD

In order to study transient state and non-stationary phenomenon, frequency must be the function of time. Instantaneous frequency (IF for short) is an intuitive and basic physical concept in HHT, while there have been disputes on the definition of IF. At present, after the researchers get rid of the deep-rooted interference of Fourier analysis, most arguments hold that IF only exists under particular conditions like monocomponent signal. However, the overall definition is meaningless to the non-stationary signals whose frequency changes every moment. In order to get meaningful IF, it must modify overall constraints into local constraints.

N.E. Huang et al. propose that to define a meaningful IF in physics, it must satisfy: The function is in symmetrical with local zero mean, and they have same extremum and zero crossing point; they further define functions satisfying the following two conditions as intrinsic mode function (IMF), as shown in Fig. 2.3.

(1) In the whole data sequence, the quantity of extremum points equals to that of zero crossing points or differs at least by 1.

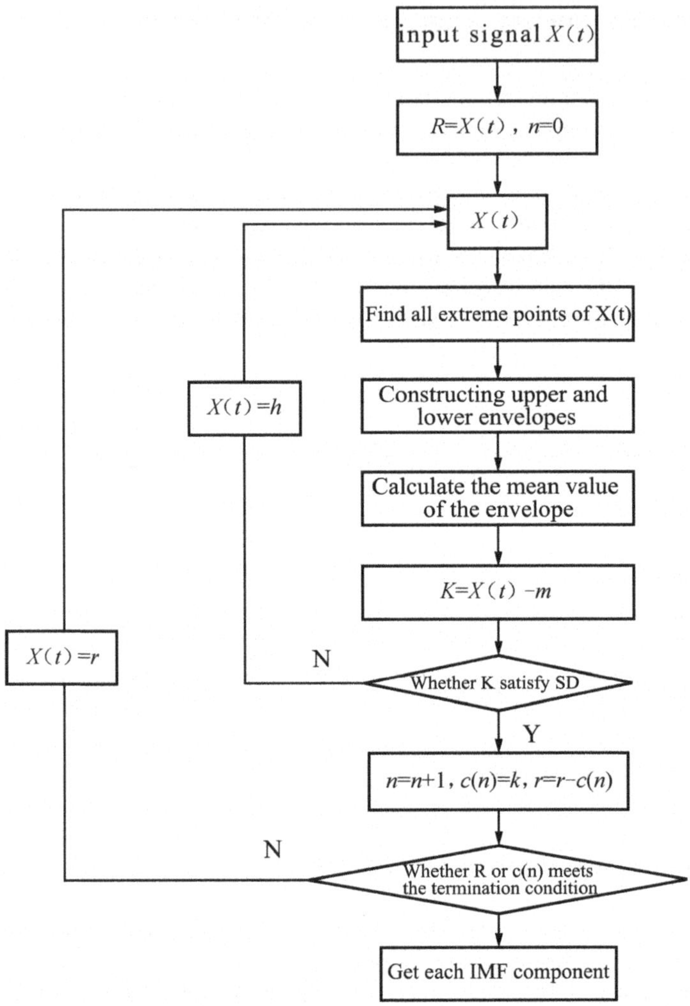

Fig. 2.3 Flowchart of EMD decomposing signals

(2) For any point of the signals, the mean value of envelop determined by local maximum point and envelop determined by local minimum point is 0, i.e. the signals are in local symmetry of the time axis.

There exists IF for IMF, which is obtained through Hilbert transform. The ordinary signals are always complex signals, which do not satisfy the IMF condition, and whose IF could not be solved. Therefore, Huang Norden E. et al. creatively propose the following hypothesis: Any complex signals are composed of IMF components of mutually different, simple, and nonlinear sine functions; each IMF component could be linear or nonlinear; at any time, one signal could contain a great deal of IMF; if the modes are overlapped, they could form compound signals. Based on this, Huang Norden E. et al. also propose EMD, i.e. Huang transform. EMD is the key of HHT, and it also becomes the sifting processing, which plays two major roles: One is to eliminate the superposition waves, and the other is to make the wave patterns more symmetrical.

The detailed steps of EMD algorithm are as follows:

For a original signal $X(t)$, it needs firstly finding out the extremum points on it. Then adopting three spline function curve to conduct interpolation for all maximum values so as to fit the upper envelop $X_{\max}(t)$ of the original signal $X(t)$. In like manner, the lower envelop $X_{\min}(t)$ could be obtained. The upper and lower envelops include all the signal data, and connect them by order. Then connect the mean value of upper and lower envelops by order, a mean value line $m_1(t)$ could be got, and use $X(t)$ to subtract $m_1(t)$ to get $h_1(t)$:

$$h_1(t) = X(t) - m_1(t) \tag{2.47}$$

For different signals, $h_1(t)$ may be an IMF component and may be not. Generally speaking, it does not satisfy the conditions needed by IMF, so take $h_1(t)$ as the original signal and repeat the above step, and we will get:

$$h_{11}(t) = h_1(t) - m_{11}(t) \tag{2.48}$$

In the formula, $m_{11}(t)$ is the mean value of upper and lower envelops of $h_1(t)$; if $h_1(t)$ is not the IMF component, then continue the sifting. Repeat the above method for k times, and get the sifting data $h_k(t)$:

$$h_{1k}(t) = h_{1(k-1)}(t) - m_{1(k-1)}(t) \tag{2.49}$$

$H_k(t)$ is not an IMF component, and there must be a sifting process and a stop criterion, which could take the value of standard deviation (SD) between two consecutive process results as criterion:

$$\mathrm{SD} = \sum_{i=0}^{T} \left| \frac{\left| h_{1(k-1)}(t) - h_{1k}(t) \right|^2}{h_{1(k-1)}^2(t)} \right| \tag{2.50}$$

The select of SD value must be careful to determine whether the sifting process should stop. The criterion should not be too strict, so as to avoid the IMF components becoming pure frequency-modulated signal and causing constant amplitude; at the same time, it should also avoid over loose criterion, which may result in components too much different from the requirements of IMF components. It could repetitively sift signals to finally decide the SD value, and experience shows that it is most appropriate when the SD value is in the range of 0.2–0.3, which could not only ensure the linearity and stability of IMF, but also endow correspondent physical meanings to it. Taking the FFT spectrum of 1024 points and FFT spectrum resulted from 1024 points removing 5 points as comparison, the resulted SD value by point-to-point calculation is also in the range of 0.2–0.3. Thus, selecting the SD value in the range of 0.2–0.3 is a quite strict requirement for two consecutive sifting processes.

When $h_k(t)$ satisfies the requirement of SD value, then $h_k(t)$ is the first-order IMF, and it can be denoted as $c_1(t)$, i.e.

$$c_1(t) = h_{1k}(t) \tag{2.51}$$

Subtracting $c_1(t)$ from $X(t)$, we can get the remaining signals, i.e. residual $r_1(t)$:

$$r_1(t) = X(t) - c_1(t) \tag{2.52}$$

Taking $r_1(t)$ as a group of new signals, repeat the above EMD process, and we can get the total residual $r_i(t)$ through multiple times:

$$r_{i-1}(t) - c_1(t) = r_i(t), \quad i = 2, 3, \ldots, n \tag{2.53}$$

When $r_1(t)$ satisfy the condition, $c_n(t)$ or $r_n(t)$ become monotonic function as they are smaller than the predesigned error or residual $r_n(t)$, i.e. when we cannot extract IMF component from that, stop the EMD process. The determination of this condition should also be appropriate, as if the conditions are too strict, the last couple of obtained IMF components is of little meaning and that consumes time; while if the conditions are too loose, it will lose useful signals. The selection of concrete termination conditions could be finally determined by repetitive decomposition of signals with reference to the prior knowledge of the original signals. Thus, the original signal $X(t)$ could be composed of n-order IMF components and residual $r_n(t)$:

$$X(t) = \sum_{i=1}^{n} c_i(t) + r_n(t) \tag{2.54}$$

Figure 2.3 is the flowchart of getting IMF components through decomposing signals with EMD.

2.7.1.2 Hilbert Transform and Hilbert Spectrum

① Hilbert transform

By EMD, we can get combination of many IMF, and by conducting Hilbert transform to these IMF, we can get the instant frequency spectrum of each IMF. Integrating the instant frequency spectrum of all IMF components, we can get a new time–frequency description method, i.e. Hilbert spectrum. Hilbert transform is a kind of linear transform, which emphasizes on local properties. The instant frequency obtained by this method is the best definition, which avoids many actually non-existing high- and low-frequency components caused by Fourier transform, and it has intuitive physical meanings.

Conducting Hilbert transform to IMFc(t):

$$H[c(t)] = \frac{1}{\pi}\text{PV}\int_{-\alpha}^{+\alpha} \frac{c(t')}{t - t'}\text{d}t' \tag{2.55}$$

In the formula, PV represents Cauchy principal value, and conduct structural analysis to $z(t)$ according to this:

$$z(t) = c(t) + jH[c(t)] = a(t)\text{e}^{j\phi(T)} \tag{2.56}$$

In the formula, $a(t)$ represents amplitude function.

$$a(t) = \sqrt{c^2(t) + H^2[c(t)]} \tag{2.57}$$

$\Phi(t)$ represents phase function.

$$\phi(t) = \tan^{-1}\frac{H[c(t)]}{c(t)} \tag{2.58}$$

The above formulas, taking the form of polar coordinates, clearly express the instant amplitudes and instant phases, and well reflect the instant features of data. Therefore, Hilbert transform provides a unique function to define amplitude and phase. In addition, Hilbert spectrum emphasizes on the local characteristics of functions.

2.7.1.3 Instantaneous Frequency (IF)

Based on Formula (2.58), the IF is defined as:

$$f(t) = \frac{\text{d}\phi(t)}{\text{d}t} \tag{2.59}$$

The above formula shows that IF is the function of time, and it reveals the measure of concentration extent of signal energy in short frequency at some moment, i.e. the IF of the signal, which is identical with the definition given by typical wave pattern theorem to frequency. IF and Fourier frequency are relevant and different in concept. Some scholars like Mandle firmly holds that IF and Fourier frequency are completely different concepts. Fourier frequency is defined by the following formula:

$$F(w) = \int_{-\alpha}^{\alpha} f(t)\text{e}^{-jwt}\text{d}t \tag{2.60}$$

The above two formulas show up three major differences of Fourier frequency and IF in concept:

(1) Fourier frequency is an independent quantity, while IF is the function of time.
(2) Fourier frequency is relevant to Fourier transform, while IF is relevant to Hilbert transform.
(3) Fourier frequency is a global reference defined with the total signal length, while IF is the local frequency description mode of some moment.

However, these two concepts are relevant in statistics when their weighted mean value and variance are calculated. If the IF of mean value and variance of Fourier frequency are identical, then they are similar in statistics. After Hilbert transform of each IMF component, the original signal could be expressed as the following formula:

$$X(t) = \mathrm{Re}\sum_{i=1}^{n} n_i(t)\mathrm{e}^{i\phi(t)} = \mathrm{Re}\sum_{i=1}^{n} a_i(t)\mathrm{e}^{\int wj(t)\mathrm{d}z} \tag{2.61}$$

Here residual function r is omitted, and Re represents selection of real part. Formula (2.61) expresses that amplitude and frequency are all the total of n components of time function, while for the same signal, it could be expressed by the order of Fourier frequency as:

$$X(t) = \sum_{i=1}^{n} a_i(t)\mathrm{e}^{\int wj(t)\mathrm{d}z} \tag{2.62}$$

In the formula, a_i and w_i are constants.

Comparing the above two formulas, Formula (2.62) represents generalized Fourier expression, in which each IMF component could modulate amplitude or frequency. The variable amplitude and frequency greatly improve the efficiency of signal decomposition or expansion. This decomposition method could also be applied to non-stationary data. Thus, through expansion of IMF component signal, its amplitude and frequency modulation could be clearly split, and the signal description method of variable amplitude and variable frequency could be get, so as to break through the limits of Fourier transform with fixed amplitude and frequency, which makes HHT transform successfully be applied to process of nonlinear and non-stationary signals.

IF defined based on phase stationary value principle expresses as the frequency value best approximating some standard sine signal, which could define the frequency value of any moment through the point-to-point change without referring to the whole signal cycle. IMF could be understood as a new signal type obtained through modulation of a sine signal with single frequency. The frequency modulation includes interwave modulation and intra-wave modulation. Interwave modulation could get the modulated signal through gradual frequency change of a sine signal, while the intra-modulation is relatively strange, but it is actually a ubiquitous physical phenomenon. In practice, its frequency could change within the period of one wave, for example, in one period of sine wave, its frequency changes with time, and then distortion occurs to the wave pattern. In the past, People often ignore it just because that the lack of exact expression of this change form. While actually, this kind of distortion will be of more practical meaning if taken as intra-modulation. Each IMF component could modulate amplitude or frequency.

2.7.1.4 Hilbert Spectrum

Formula (2.62) could express signal amplitude as the function of time and IF in three-dimensional space, and signal amplitude could also be expressed as contour line on time–frequency plane. This processed amplitude distribution on time–frequency plane becomes the Hilbert time–frequency spectrum, i.e. Hilbert spectrum. It has various expression forms: the grayscale pattern with or without smooth process, contour line pattern, or three-dimensional pattern. The expression of Hilbert spectrum is:

$$H(\omega, t) = \sum_{i=1}^{n} a_i(t) e^{\int wj(t)dz} \quad (2.63)$$

If $H(w, t)$ is integral to time, then we can get Hilbert marginal spectrum:

$$h(\omega, t) = \int_0^T H(w, t)dt \quad (2.64)$$

The marginal spectrum expresses the amplitude (or energy) of each IMF in overall sense, and it represents the total accumulated amplitude in statistical sense. In Fourier expression, the energy existence of some frequency at w represents the existence of sine wave or cosine wave in the whole length of time, while in the marginal spectrum, the existence of some frequency only represents the possibility of the existence of signals with such frequency. In addition, as the attached results of Hilbert marginal spectrum, the Hilbert instant energy could be defined as follows:

$$IE(t) = \int_0^T H^2(w, t)dt \quad (2.65)$$

The instant energy provides situation of signal energy changing with time. Actually, if the square of amplitude is integral to time, we can get the Hilbert energy spectrum:

$$ES(w) = \int_0^T H^2(w, t)dt \quad (2.66)$$

Hilbert energy spectrum provides the energy calculation expression of each frequency and expresses the energy of each frequency which accumulates in the whole time length. N. E. Huang points out that whether for Hilbert marginal spectrum or Hilbert energy spectrum, the physical meaning of the obtained frequency is completely different to that of frequency obtained through Fourier analysis. And then on the base of Hilbert energy spectrum, define $E(w)$ as:

$$E(w) = \int_0^T H^2(w, t)\mathrm{d}t \tag{2.67}$$

Here, $E(w)$ is called Hilbert marginal energy spectrum, and it describes the distribution of signal energy changing with frequency. We can also, according to the requirements, adopt some IMF in Formula (2.54) to get local Hilbert spectrum through analysis:

$$H'(w, t) = \mathrm{Re}\sum_{i=1}^{n} a_i(t)\mathrm{e}^{i\phi(t)} = \mathrm{Re}\sum_{i=j}^{n} a_i(t)\mathrm{e}^{\int wj(t)\mathrm{d}z} \tag{2.68}$$

$H'(w, t)$ describes the changing rules of signal amplitude with time and frequency in the needed frequency sections and defines $H'^2(w, t)$ the local Hilbert energy spectrum as:

$$E'(w) = \int_0^T H'^2(w, t)\mathrm{d}t \tag{2.69}$$

It is the local instant energy, which describes the energy changes of some section of signal frequency component with time.

2.7.1.5 The Superiority of HHT Transform

It could be concluded from the above that HHT transform has many advantages; it depends on analysis methods of Fourier and wavelet based on prior functions, so it is more suitable to deal with non-stationary signals, which has none fixed prior basis, so it is self-adapting; it gives the definition of IMF for the first time, points out the amplitude is adaptable, and breaks through the limits of taking the simple harmonious signal with fixed amplitude as basis in tradition, which makes signal analysis more flexible and changeable; each IMF could be seen as a fixed vibration mode in signals, and the IF obtained through Hilbert transform has clear physical meanings which could express the local features of signals. Besides, it could accurately form the time–frequency graph, which is hard for wavelet transform and other signal analysis. IF is defined as the derivative of phase functions, which is not in need of the whole wave to define local frequency, so it can identify singular signals from low-frequency signals, which is of clear advance to wavelet transform.

2.7.2 *Application Examples of Hilbert–Huang Transform (HHT)*

In order to verify the efficiency of Hilbert transform, now we superpose three sine signals and one exponent trend term as simulation signal to conduct simulation analysis, and the simulation signal is represented by the following equation:

$$S(t) = \sin(200\pi t) + \sin(100\pi t) + \sin(50\pi t) + e \tag{2.70}$$

Conduct EEMD decomposition to this signal (whose sampling frequency is 1000 Hz and the data length is 1024 points), and get three components of $c1$, $c2$, and $c3$ as well as trend term $c4$, as shown in Fig. 2.4. The time–frequency graph of each IMF component is as shown in Fig. 2.5. It could be seen from Fig. 2.4 that EEMD decomposes signals according to different time measures, which firstly get the high frequency, then low frequency, infra-low frequency by order, and the last is trend term. The obtained three IMF components and one trend term are just the four original signals of simulation signals, which show the accuracy and high efficiency of decomposition. The EMD decomposition process is determined by the signals themselves. The time–frequency graph in Fig. 2.5 quantitatively describes the relation between time and frequency, from which we can see that its frequencies concentrate around 100, 50, 20, and 0 Hz, respectively, with only swing at the end due to pollution, but its results are highly identical with the expectation, which shows that HHT is effective.

2.7.3 *Relevant Issues Studied by Hilbert–Huang Transform (HHT)*

Because this method has only been proposed in a short time, there are many problems that need solution, for example, adoption of better methods to solve incomplete envelope of envelopes, and overshoot and undershoot caused by cubic spline interpolation, and how to reduce pollution in decomposition caused by swing of end points and how to ensure uniqueness of EMD decomposition results as well as the value selection of decomposition convergence standards.

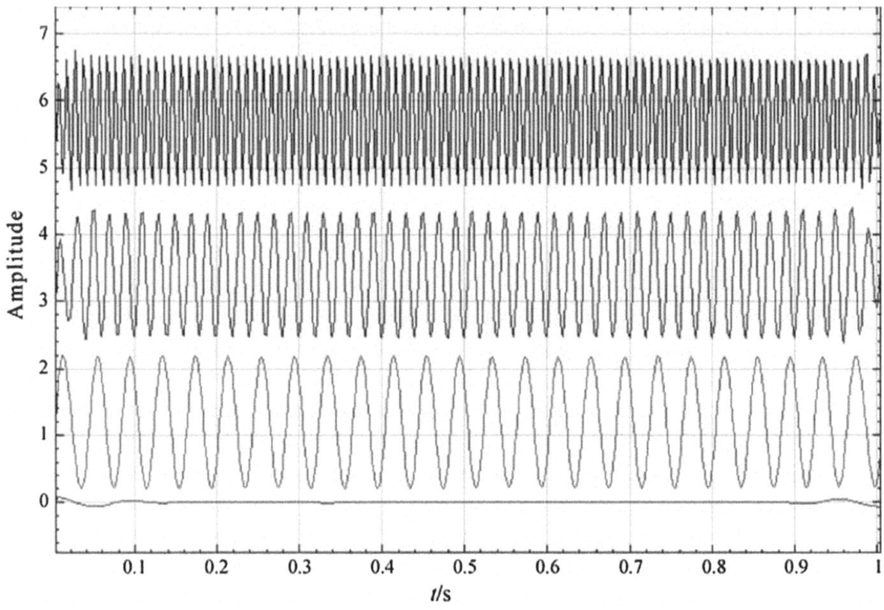

Fig. 2.4 EDM component of signals

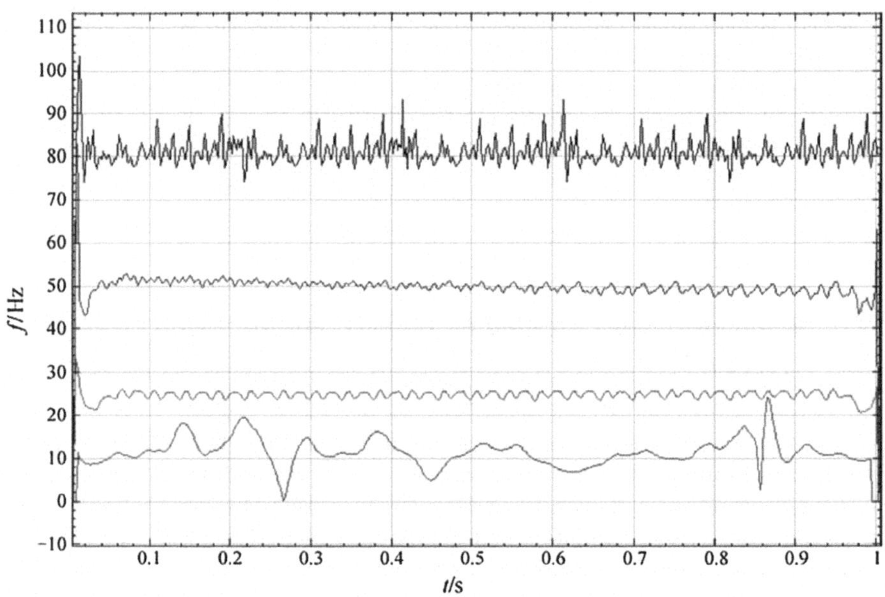

Fig. 2.5 Time–frequency graph of EDM component of signals

2.7.3.1 Decomposition Methods

In EMD decomposition, signal decomposition algorithm is the key, and its quality directly influences the accuracy of signal decomposition. The applications show that EMD decomposition technique is quite effective to nonlinear and non-stationary phenomena, while at the same time, there exist some problems. One is when used to calculate the envelope mean value defined by local extremum, it adopts twice cubic spline interpolation, which could directly cause overshoot and undershoot problems. The second is because the signal the end point of the signal may not be the extreme point, so spline interpolation may cause large swing at the two ends. Too many sifting times or too less signal data points will cause swing which might transmit to the middle section of signals and "pollute" the entire data. For EMD decomposition algorithm, scholars, domestic and abroad, have proposed many solutions. N. E. Huang himself proposed an effective data extension method at the time he proposed EMD method and has adjusted it has got patent in the USA.

2.7.3.2 Physical Explanation of Signals

The complex signals, through EMD decomposition, will get many IMF components. Although some components cannot well ensure good physical meaning, for most circumstances, it could get some useful information from each IMF. The concrete physical meaning represented by complex signal decomposition is waiting for further discussion.

2.7.3.3 End Effects

In spline interpolation of signal decomposition and in Hilbert transform of calculating IF, there all exist end effect problems. Hilbert transform is based on Fourier transform, and because the end points and last points are of different values, Fourier transform may lead to Gibbs phenomenon. In order to reduce Gibbs phenomenon, Fourier method often adopts windowed form. While EMD decomposition does not need the window technique, and it adds feature waves to the two ends of data. These waves start from the beginning of signals and stop at the signal ends to restrain the Gibbs phenomenon. However, up to now, there has no method that can completely eliminate the influence of end effects.

2.7.3.4 Selection of Signal Length

In spline interpolation of signal decomposition and in Hilbert transform of calculating IF, there all exist end effect problems. There is swing at the two ends of signal decomposition. Therefore, the signal length should not be too short in decomposition.

2.7.4 *Completeness and Orthogonality of Hilbert–Huang Transform*

2.7.4.1 Verification of Completeness

The EMD method itself determines its completeness. Firstly, decompose the original signals into IMFs and conduct original data restructuring to these IMFs with numerical methods to prove its completeness. The restructuring process is as follows: Start from the last IMF, and add in the IMFs decomposed from it. The first IMF component is the rest trend term by the original data decomposition, which represents instrument drifts or the entire trend of signals, and its value is very small. Then adds in several components of secondary level, and the change is also not obvious, which is mainly because the energy of components of these levels is small. With continuing add-in of components, the modification of original signal restructuring becomes more evident, and when the last IMF component, i.e. $c1$, is added in, the major energy of signals is totally restored. The signals restructured through each IMF components are quite the same with the original signals, which are of little error.

2.7.4.2 Verification of Orthogonality

At present, orthogonality cannot be verified in theory, but it could be satisfied in actual meaning. If the result is orthogonal, then all the IMF components are orthogonal in local parts. Rewrite Formula (2.71) into the following form:

$$X(t) = \sum_{j=1}^{t=1} c_j(t) \tag{2.71}$$

Here, the decomposed residual rn is taken as an IMF component, and firstly, square the signal and we will get:

$$X^2(t) = \sum_{j=1}^{t=1} c_j^2(t) + 2\sum_{j=1}^{t=1}\sum_{k=1}^{x=1} c_j(t)c_k(x) \tag{2.72}$$

If the IMF components are orthogonal, then the second item on right of the above formula is 0, and define the total orthogonal coefficient as:

$$\mathrm{IO} = \frac{2\sum_{j=1}^{t=1}\sum_{k=1}^{x=1} c_j(t)c_k(x)}{X^2(t)} \tag{2.73}$$

Then the orthogonality of any two components *cf* and *ck* could be defined as:

$$\mathrm{IO} = \frac{c_f c_k}{c_f^2 + c_k^2} \tag{2.74}$$

According to Formula (2.74), the orthogonal coefficient of blasting vibration signals through calculation is 0.0783. Due to that, the time sequences of same frequency component of different IMF component at different moments are various, with the influence of finite length and decomposition results of data, IO is nor completely zero but approaches 0. The above verifies the IMF components obtained through Huang transform are basically in orthogonality. Actually, orthogonality is the requirement of linear systems, while for nonlinear systems, we should pay attention to that it can reveal the nature of signals, and the requirement on orthogonality is not strict.

Bibliography

Gao Pinxian. The influence and elimination of the trend term to the parameter identification in time domain [J]. Vibration, test and analysis, 1994, 14(2):20–26.

Gong Maosheng, Xie Lili. Preliminary study on the application of HHT method in Earthquake Engineering [J], World Earthquake Engineering, 2003, 19(3):39–43.

Huang N.E Shen Z. Long S R.et al. The empirical mode decomposition and Hilbert spectrum for nonlinear and non-stationary time series analysis [J]. Proc R. Soc Lond. 1998 A 454:903–995.

Li Shaoyuan, Cao Baoding, Meng Zhaozhong, Wang Xianqiong. Study on forecasting model based on time series analysis [J], Journal of Hebei Institute of Technology, 1995, (3):7–11.

N E Huang, Zheng Shen, Steven R L. A new view of nonlinear water waves: the Hilbert spectrum [J]. Annu.Rev.Fluid Mech. 1999, (31):417–457.

Ren Daqian, Wu Zhaotong, Yan Gongbiao. Evaluation index of EMD endpoint effect and window function method to restrain the end effect [J]. Manufacturing automation, 2007, 29(1):21–24.

Song Pingjian, Zhang Jie. Application of two dimensional empirical decomposition in ocean remote sensing image separation [J]. High technical bulletin, 2001, (9):39–43.

Wang Ji. The application of MATLAB in vibration signal processing [M]. Beijing: China Water &Power Press, 2006:73–78.

Xiong Xuejun, Guo Binghuo, et al. Application and discussion of EMD method and Hilbert spectrum analysis method [J], Huang Bo Ocean, 2002, 20(2):12–21.

Yang Shixi, Hu Jinsong, Wu Shaotong et al. An analytical comparison between Hilbert Transform and Wavelet transform frequency of rotating machinery vibration signals based on EMD [J]. Journal of Chinese Electrical Engineering Science, 2003, 23(6):102–107.

Zhi-Yuang Su, Chuan-Chen Wang, Tzuyin Wu, et al. Instantaneous frequency-time analysis of physiology signals: The application of pregnant women's radial artery pulse signals. Physica A 2008, 387:485–494.

Zhu Xuefeng, Han Ning. Non stationary signal trend term elimination method based on Wavelet Transform [J]. Journal of aircraft measurement and control, 2006, 10(5):82–83.

Z.K. Peng, Peter W. Tse, F.L. Chu. An improved Hilbert-Huang transform and its application in vibration signal analysis. Journal of Sound and Vibration, 2005, 286:187–205.

Chapter 3
Analysis of Seismic Dynamic Characteristics of High and Steep Rock Slopes

Knowing the dynamic characteristics of rock slopes under seismic effects is the premise of revealing instability mechanism of bedrock and overburden slopes under earthquakes. This chapter will, with the help of observed data of field seismic motion monitoring array, large-scale shaking table test, and numerical calculation, study the influence of factors like bevel angles, seismic intensity, type of seismic input wave, type of slope body, and local slope shape on the dynamic characteristics of rock slopes from different aspects.

3.1 Seismic Array Monitoring Results Analysis of "5.12 Wenchuan Earthquake" at Xishan Park, Zigong

3.1.1 General Engineering Situation

Zigong City is located in southern Sichuan Basin, and it is about 227 km to the epicenter of Wenchuan earthquake in Yingxiu, Wenchuan. Its topography is composed of hills, valleys, and ravines. Elevation decreases from northwest to southeast, with altitude mainly in the range of 250–450 m above sea level, as illustrated in Fig. 3.1. The altitude of downtown area is in the range of 280–400 m, with small ground relative, which for shallow hilly area is in the range of 20–50 m, and for some middle hilly area is in the range of 50–100 m. The seismic array to study the topography of Zigong City is located in Xishan Park, Zigong, with elevation in the range of 330–430 m, and it generally belongs to middle hilly topography with complex topography and geological conditions as shown in Fig. 3.2.

The seismic array is built on the mountain in Xishan Park, which is composed of eight stations (0#–7#), and each of them is equipped with digital seismographs of ETNA recorder and ES-T-type accelerometer. The maximum height difference is about 72 m between station (6#) on the top of the hill and station (0#, 1#) at the

C. Yang et al., *Three Dimensional Space-Time Analysis Theory of Geotechnical Seismic Engineering*, https://doi.org/10.1007/978-981-13-3356-9_3

Fig. 3.1 Geographical position of Zigong City

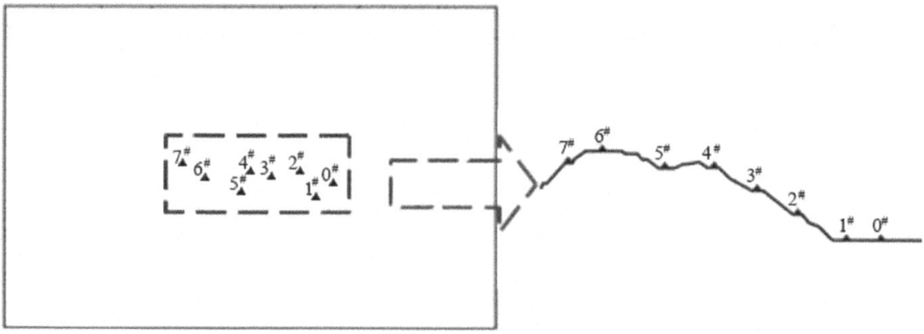

Fig. 3.2 Plane distribution map of monitoring points of Zigong topography

bottom of the hill, in which station 0# is located on soil site, station 1# is located on rock site, and other stations are located on Jurassic bedrocks with different elevations (Fig. 3.3 and Table 3.1).

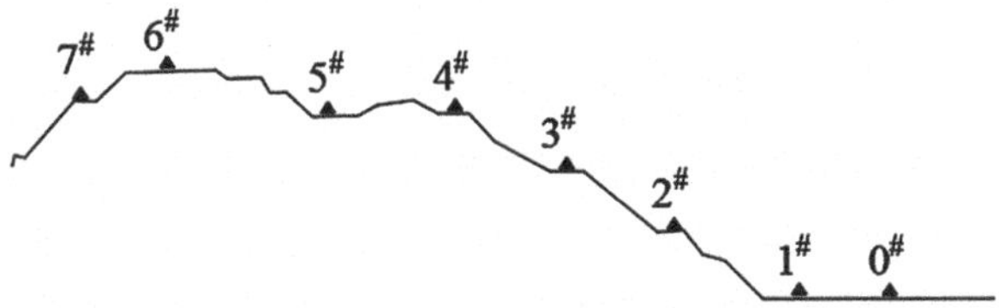

Fig. 3.3 Distribution map of seismic array monitoring points

Table 3.1 Table of array parameters of Zigong topography

Station name	Elevation	Horizontal distance to reference station	Rock property
0#	345.0	51.55	Soil layer
1#	345.0	0.00	Jurassic bedrock
2#	367.0	41.64	Jurassic bedrock
3#	385.0	116.82	Jurassic bedrock
4#	392.0	171.41	Jurassic bedrock
5#	390.0	258.98	Jurassic bedrock
6#	417.0	304.55	Jurassic bedrock
7#	397.0	345.72	Jurassic bedrock

3.1.2 Seismic Array Monitoring Results Analysis of "5.12 Wenchuan Earthquake" at Xishan Park, Zigong

Wenchuan earthquake (Ms 8.0) occurred in China on May 12, 2008, and its seismogenic fault is central Longmen Shan rupture, with 230 km of surface rupture. The front mountain fault also caused surface rupture of 70 km. The terrain is extremely complex with staggered high and steep mountains and canyons, the elevation of upward plate of central rupture is about 2000–3000 m, that of upward plate of the post-mountain fault is in the range of 4000–5000 m, and that of downward plate of the Chengdu Plain the is just about 500 m. The complex terrain of Longmen Shan fault zone definitely exerted great effects on seismic motions generated by the Wenchuan earthquake. According to the relevant data provided by Chinese seismic array network, in the "5.12 Wenchuan earthquake," the seismic station in Zigong City has monitored relatively complete main shock records, which provides the basis for the seismic response analysis of slope. The monitoring results are displayed in Figs. 3.4, 3.5, and 3.6.

Table 3.2 shows the response characteristics of PGA of all monitoring points of seismic array in Xishan Park. To study the variation of PGA along elevation, so PGA in EW, NS, and UD directions is extracted, respectively, as shown in Table 3.2, and compared with PGA of Station 1# in these three directions to get PGA amplification coefficients in three directions, as shown in Table 3.2.

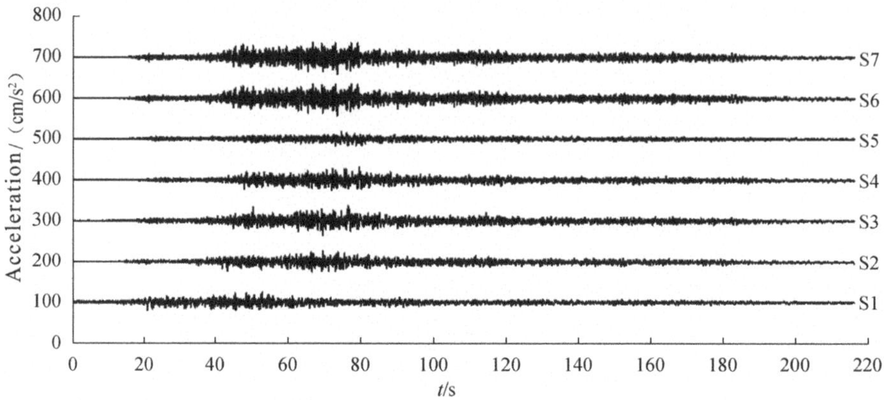

Fig. 3.4 Measured seismic ground motion time history curve in EW direction of S1–S7. *Note* All the time history curves are drawn by uniform scale, and the stations are numbered S1 (1#) to S7 (7#)

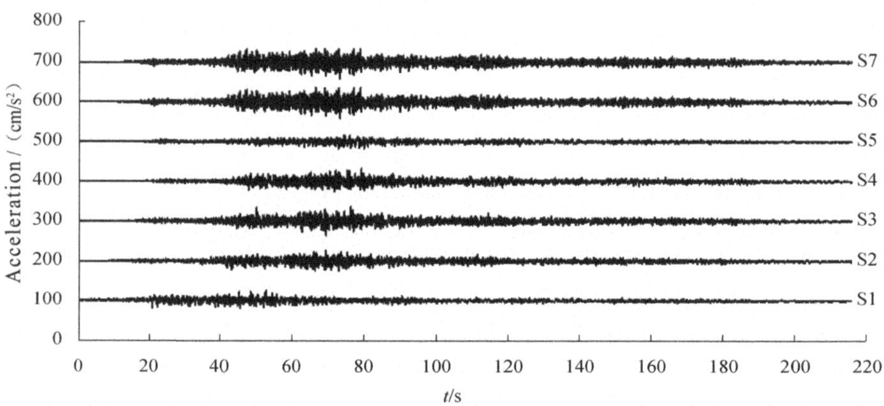

Fig. 3.5 Measured seismic ground motion time history curve in NS direction of S1–S7 *Note* All the time history curves are drawn by uniform scale, and the stations are numbered S1 (1#) to S7 (7#)

3.1.3 Amplitude Response Characteristics Analysis of Zigong Topography Seismic Array

Based on the data arrangement results of Zigong seismic array, PGA elevation amplification effect will be discussed in this section. Station 1#, the minimum elevation point of the slope, is chosen as the reference point. PGA amplification coefficient is defined as the ratio of PGA in three directions of any point to that of Station 1#, as shown in Figs. 3.7, 3.8, and 3.9. It should be pointed out that the relative elevation of each site is the absolute elevation difference to Station 1#, Station 1# is the ref-

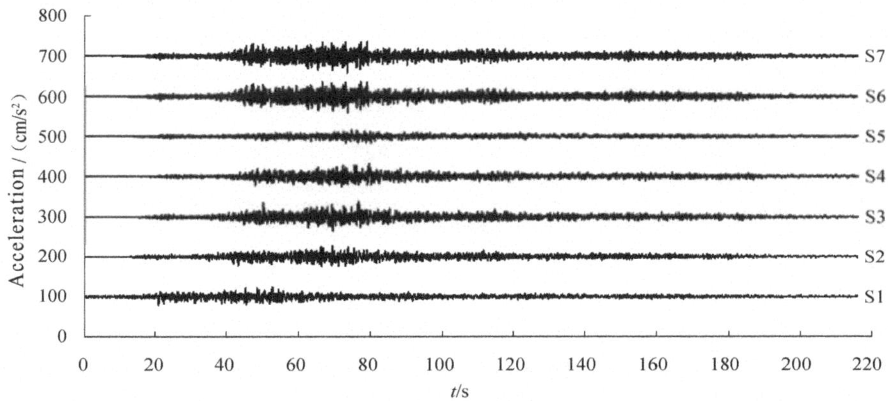

Fig. 3.6 Measured seismic ground motion time history curve in UD direction of S1–S7 *Note* All the time history curves are drawn by uniform scale, and the stations are numbered S1 (1#) to S7 (7#)

Table 3.2 PGA response characteristics of Zigong topography monitoring array (cm/s/s)

No.	Relative elevation (m)	PGA (cm/s^2)			Acceleration amplification coefficient		
		EW component	NS component	UD component	EW component	NS component	UD component
1#	0	23.036	26.489	14.726	1	1	1
2#	22	28.078	29.952	15.953	1.219	1.131	1.083
3#	40	34.863	32.533	18.336	1.513	1.228	1.245
4#	47	32.531	32.226	19.719	1.412	1.217	1.339
5#	45	33.030	42.169	17.240	1.434	1.592	1.171
6#	72	40.735	42.244	19.742	1.768	1.595	1.341
7#	52	39.299	45.489	16.049	1.706	1.717	1.090

erence station, with PGA amplification coefficient = 1, and the PGA amplification coefficients of any other stations are the ratio of their PGA with PGA of Station 1#.

From Figs. 3.7, 3.8, and 3.9, it could be known that the PGA amplification coefficients show clear increase with the increase of relative elevation in three directions. The maximum PGA amplification coefficient in EW direction is 1.768 at Station 6#, that in ES direction is 1.717 at Station 7#, and that in UD direction is at Station 6#. Therefore, the PGA elevation amplification effects in two horizontal directions of EW direction and NS direction are close, which are all over that in the vertical direction.

This may be caused by the fact that EW is close to the free surface direction and NS is close to the direction of the slope strike. The free surface amplification effect results in a more intensive amplification in EW direction than NS direction. It is noteworthy that Station 4# is higher than Station 5#, but the PGA amplification coefficient of the

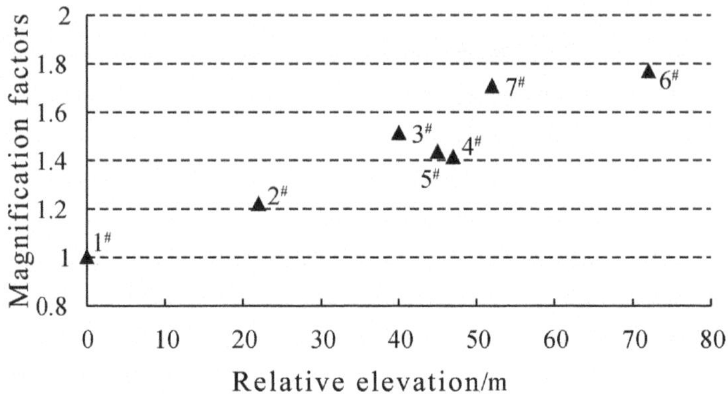

Fig. 3.7 Acceleration elevation amplification effects in EW direction

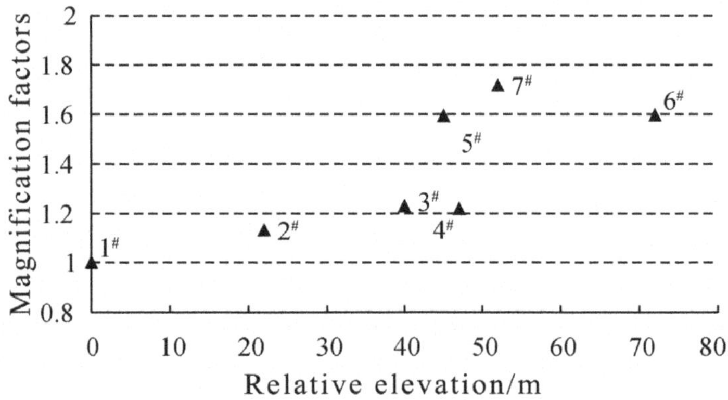

Fig. 3.8 Acceleration elevation amplification effects in NS direction

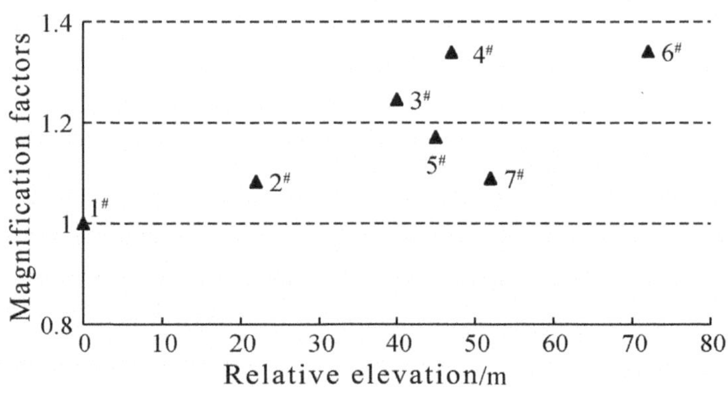

Fig. 3.9 Acceleration elevation amplification effects in UD direction

former is smaller than the latter. This can be contributed to the fact that Station 5# is located in a local canyon. The reflection, refraction, and transmission effect will be triggered, while the upward propagating seismic waves encounter the interface between the soil layer and underlying bedrock, which lead to a much more intense ground motion at Station 5#. This indicates that the acceleration amplification effects are dependent on direction; namely, the amplification effects in free surface direction are intense than in slope strike, with the weakest in the vertical direction.

3.1.4 Analysis of Frequency Spectrum Response Characteristics of Zigong Topography Seismic Array

Based on the monitoring results of seismic dynamic response of seismic array of Zigong topography, this section analyzes the frequency spectrum response characteristics of rock slopes from both Fourier spectrum and response spectrum.

① The Changing Rules of Fourier Spectrum along Elevation of Zigong Topography Seismic Array

Fourier spectrum is the method to change complex seismic acceleration time history $\alpha(t)$ into N frequency combinations by discrete Fourier transform technique:

$$\alpha(t) = \sum_{i=1}^{N} A_i(\omega)\sin[\omega_i t + \varphi_i(\omega)] \tag{3.1}$$

In the formula, $A_i(\omega)$ and $\varphi_i(\omega)$ are the amplitude and phase angle of vibration component of angular frequency ω_i.

Formula (2.1) can be rewritten into

$$\alpha(t) = \sum_{i=1}^{N} A_i(i\omega)\mathrm{e}^{i\omega_i t} \tag{3.2}$$

In the formula, $i = \sqrt{-1}$, complex function $A(i\omega)$ is Fourier Spectrum, and its module $|A(i\omega)|$ is the amplitude spectrum.

Calculate the Fourier spectrum in EW, NS, and UD directions from S1 to S7 based on the monitoring results of Zigong topography seismic array, and carry out smooth process to the Fourier spectrum, and the results are shown in Figs. 3.10, 3.11, 3.12, 3.13, 3.14, 3.15, and 3.16.

Figures 3.10, 3.11, 3.12, 3.13, 3.14, 3.15, and 3.16 show that as for Fourier spectrum of acceleration in EW direction, the dominant frequency of S1 to S4 is in the range of 1–2 Hz and that of S5 and S6 present a double peak pattern, with dominant frequency being stable in the range of 1–2 Hz and 4–5 Hz. The amplitudes of the two dominant frequency ranges increase with the elevation. As for Fourier spectrum of acceleration in NS direction, the dominant frequency ranges of S1 and

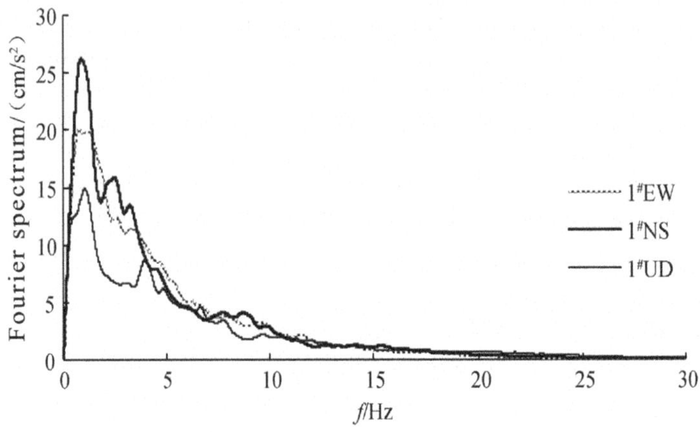

Fig. 3.10 Fourier spectrum of three-direction acceleration of S1

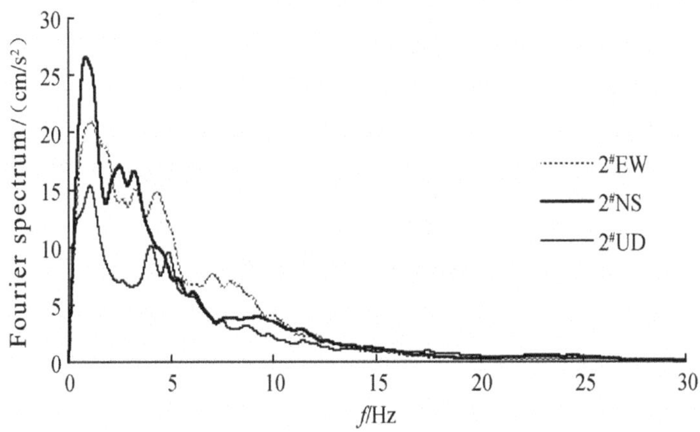

Fig. 3.11 Fourier spectrum of three-direction acceleration of S2

S2 concentrate in the range of 1.0–2.0 Hz, and apparent double peak phenomenon appears in S3 and S6, with dominant frequency being stable in the range of 1–2 Hz and 3–5 Hz. With the increase of elevation, the amplitude of first-order dominant frequency is of slight increase and that of second-order dominant frequency is of gradual increase. As for Fourier spectrum of acceleration in UD direction, with the increase of elevation, the dominant frequency range stables in the range of 1–2 Hz and the peak value of Fourier spectrum is barely changed.

This might lie in the following reasons: According to the evaluation method for seismic period of rock slopes proposed by J. Y. Xu et al. [132], conduct natural frequency calculation of slopes in this part, and the results show that the natural vibration period of this slope is $T = 0.248$ s, and the natural frequency is $f = 4.02$ Hz, which

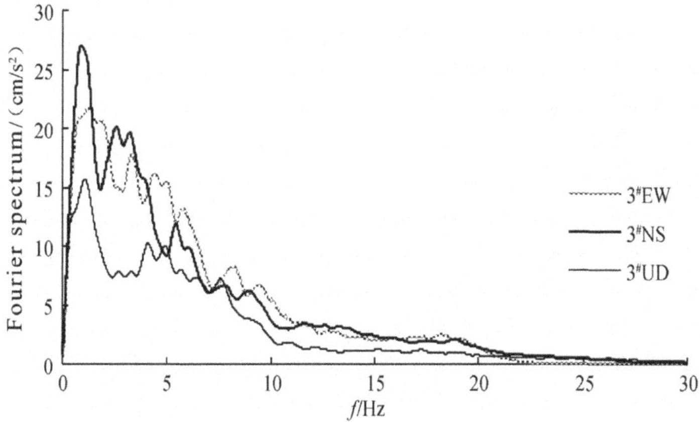

Fig. 3.12 Fourier spectrum of three-direction acceleration of S3

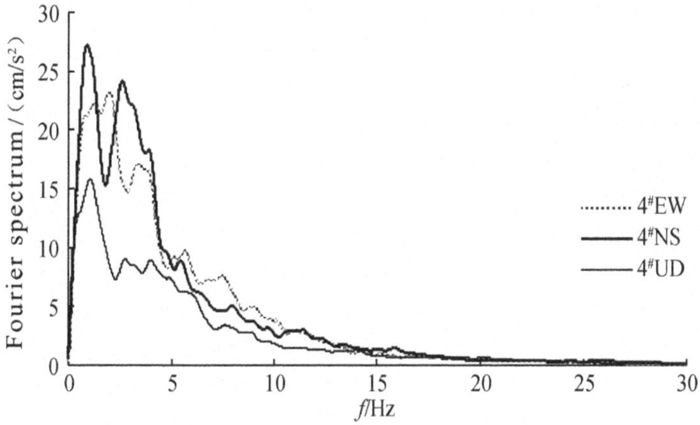

Fig. 3.13 Fourier spectrum of three-direction acceleration of S4

causes the Fourier spectrum amplification of its surrounding frequency components within the range of 3.0–5.0 Hz.

Generally, the dominant frequency range of Fourier spectrum in horizontal direction concentrates in the range of 1.0–2.0 Hz and 3.0–5.0 Hz. With the elevation increase, there are clear amplification effects of Fourier spectrum value for components close to the natural frequency of the slope (3.0–5.0 Hz), with the transformation from single mode to double peak pattern. At the same time, the dominant frequency range of acceleration in vertical direction is in the range of 1.0–2.0 Hz, which is independent of elevation.

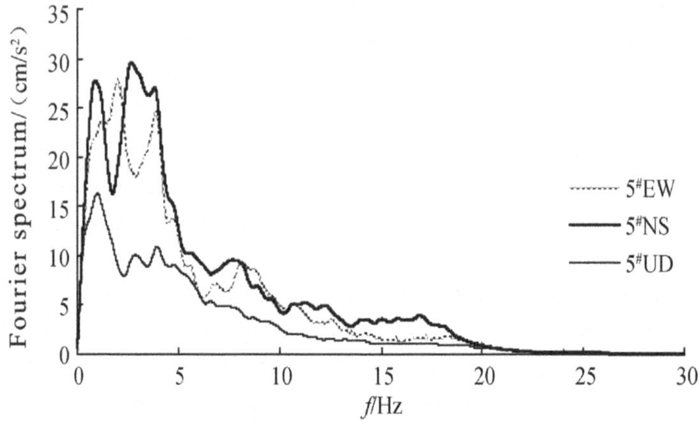

Fig. 3.14 Fourier spectrum of three-direction acceleration of S5

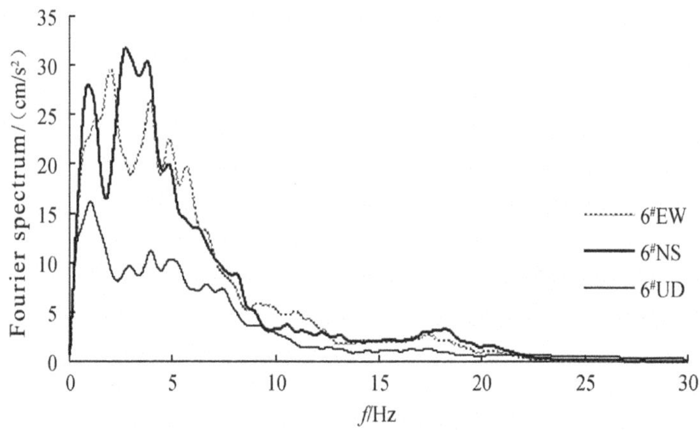

Fig. 3.15 Fourier spectrum of three-direction acceleration of S6

① The Changing Rules of Seismic Response Spectrum along Elevation of Zigong Topography Seismic Array

In essence, seismic response spectrum is the characteristic of seismic intensity and spectrum. This is because any response spectrum curve is arranging result of single degree of freedom structure with various dynamic characteristics to the maximum dynamic response in one seismic time history according to period T. Therefore, this section takes the damping ratio of 5%, and the characteristics of acceleration response spectrum of S1–S6; see Figs. 3.17, 3.18, 3.19, 3.20, 3.21, and 3.22.

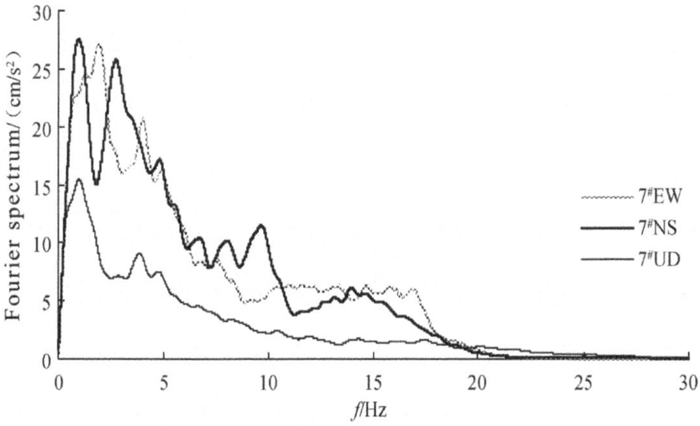

Fig. 3.16 Fourier spectrum of three-direction acceleration of S7

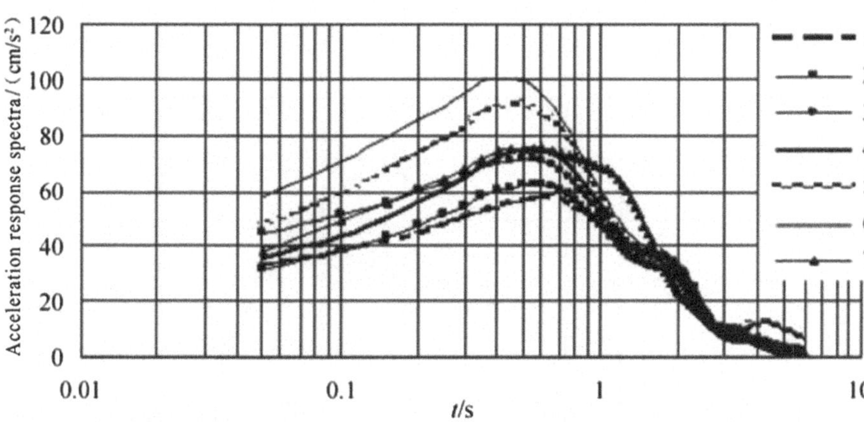

Fig. 3.17 Acceleration response spectrum of all stations in EW direction

Figures 3.17, 3.18, 3.19, 3.20, 3.21, and 3.22 show that the dominant periods of response spectrum in both EW and NS directions are in the range of 0.3–0.7 s, and the amplitudes increase with the elevation for $T \leq 1$ s, while for $T > 1$ s, the response spectrum of each point is almost the same. In terms of the vertical direction, the dominant periods of response spectrum stabilize in the range of $T = 0.6$–1.0 s, and the amplitudes increase with the elevation of $T \leq 1$ s, while for $T > 1$ s, the response spectrum of each point is almost the same. Besides, the amplitudes are about one-third to half of that in the horizontal direction, and the amplitudes of 2#, 4#, and 7# are of little different. It is noteworthy that the S5 is located in a local canyon and is influenced by local basin effects, so even though the elevation of 5# is smaller than that of 7#, the amplitude of 5# is greater than that of 7#, which indicates that the local site condition has a significant effect on seismic amplification.

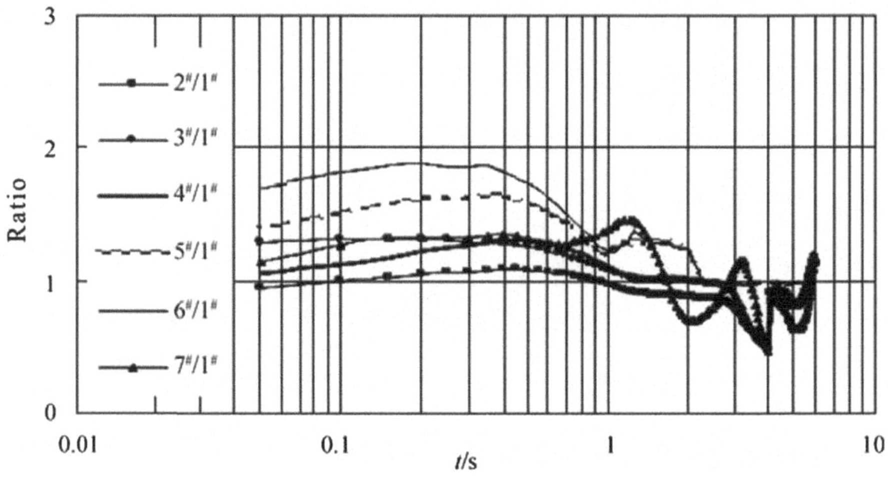

Fig. 3.18 Amplification coefficients of acceleration response spectrum of all stations in EW direction

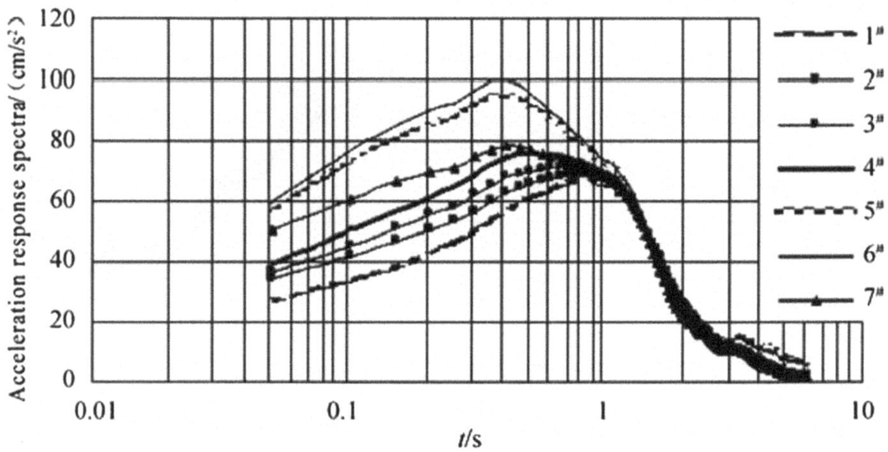

Fig. 3.19 Acceleration response spectrum of all stations in NS direction

In general, the amplification coefficient of acceleration response spectrum increases with the elevation, and the amplification value in the horizontal direction is greater than that in the vertical direction.

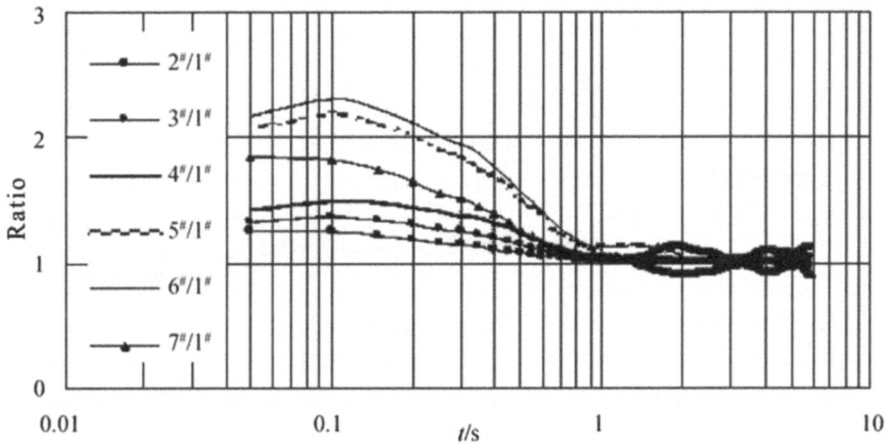

Fig. 3.20 Amplification coefficients of acceleration response spectrum of all stations in NS direction

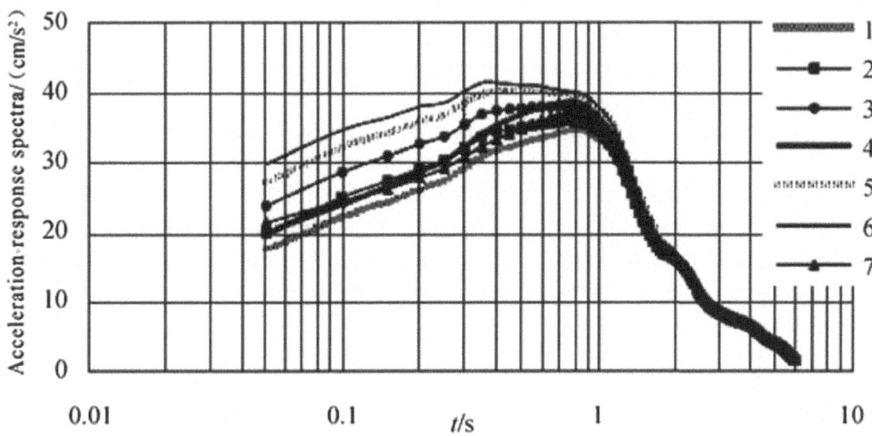

Fig. 3.21 Acceleration response spectrum of all stations in UD direction

3.1.5 *Brief Summary*

Through the arrangement and analysis of the seismic array results of Zigong topography in "5.12 Wenchuan earthquake," studies of seismic dynamic characteristics have been carried out in various respects like amplitude response, Fourier spectrum, and response spectrum, and the results could be concluded as follows:

(1) The PGA in three directions increases nonlinearly with the increase of elevation, and the amplification coefficients concentrate in the range of 1–1.768, 1–1.717, and 1–1.341 for three directions, respectively. The EW direction has the largest

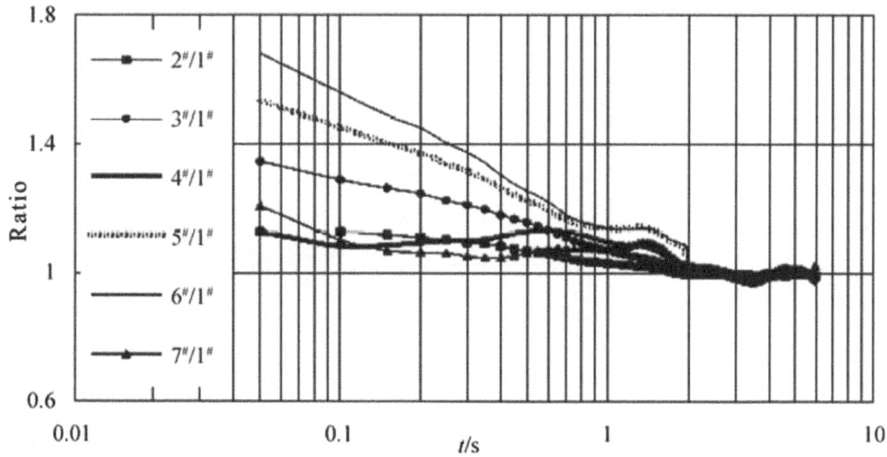

Fig. 3.22 Amplification coefficients of acceleration response spectrum of all stations in UD direction

amplification value, the vertical direction has the smallest amplification, and the NS direction is moderate.

(2) Dominant frequency range in horizontal direction concentrates in the range of 1.0–2.0 Hz and 3.0–5.0 Hz, and with the elevation increase, there is clear amplification effects of Fourier spectrum value for components close to the natural frequency of the slope (3.0–5.0 Hz), with transformation from single mode to double peak pattern. At the same time, the dominant frequency range of acceleration in the vertical direction is in the range of 1.0–2.0 Hz, which is independent of elevation.
(3) The amplification coefficient of acceleration response spectrum increases with the elevation, with the amplification effects strengthened, and the amplification coefficient in the horizontal direction is greater than that in the vertical direction.

3.2 Shaking Table Test Study on Dynamic Characteristics of Rock Slope

It is known to all that the to study on response characteristics of rock slope seismic action is to conduct field test through seismic array network, which could intuitively reflect the motion features and seismic dynamic characteristics of rock slopes so as to provide precious original data for the future work. But because of the destructiveness, occasionality of earthquake, and the uncertainty of many other factors, the field test consumes a large amount of time, labor, and cost, and it is not repetitive as it can only study on single rock slope. By contrast, the similar model test in the laboratory could effectively compensate the shortcomings of field tests, which is the model test technique, based on the similarity theory, using the likeness or similarity between or among things or phenomenon to study natural laws. The structure models are small in size, easy to make, assemble and disassemble, which save materials, time, and labor, and have high pertinency and accurate and reliable data collection. In addition, one model could be used for tests under different working conditions. It is especially applicable for research areas in which data results are difficult to obtain through theoretical analysis and field test measures, and is quite effective for analysis and comparison of theoretical research results. Therefore, with macro-analysis of rock slopes using seismic array network monitoring data, conducting the similar model test in the laboratory could enrich research results and its contents.

Currently, the shaking table test is the major means to simulate earthquake in the laboratory, which has similar ground movements and impacts on structures as real earthquake occurs. It is the important way and measure to study failure mechanism and failure model of earthquakes and to evaluate the global seismic capacity, which has been widely applied to the theoretical research and practices of earthquake engineering. Thus, this chapter will begin by deducing the similarity law of rock slope shaking table test and then, using the test results, study the response rules of rock slope acceleration under earthquake from the domain of both time and frequency, and the influence on rock slope dynamic characteristics of bevel angle, seismic intensity, and ground motion input type, etc., so as to lay a foundation for the study on instability mechanism of bedrock and overburden layer slopes under earthquakes.

3.2.1 Large-Scale Shaking Table Test Design

3.2.1.1 Purpose of Shaking Table Test

In order to study the dynamic response rules of homogeneous rock slopes, this paper, taking the 5.12 Wenchuan earthquake as background, carries out large-scale shaking table physical simulation tests. During the experimental process, tracking the dynamic response of seismic acceleration and displacement, etc. of different position and elevation of rock slopes, revealing the response rules of rock slope acceleration under earthquake, and the influence of bevel angle, seismic intensity, seismic wave input type and input wave frequency on the rock slope dynamic characteristics, which provides theoretical supports for further study on deformation and failure mechanism of bedrock and overburden slopes, and for reasonable explanation for Wenchuan earthquake inducing secondary geological disasters.

3.2.1.2 Shaking Table Test Installation

The shaking table test installation adopts the large-scale shaking table of three axles and six degrees of freedom designed by the Seismic Laboratory of the second institute of Nuclear Power Institute of China. The installation is currently the earthquake

Fig. 3.23 Model containers of shaking table test

shaking table with the largest capacity and the most advanced performance in China. The surface size is 6 m × 6 m, the maximum load is 60*t*, and the frequency range is 0.1 Hz. The shaking table could load from three axles and six degrees of freedom simultaneously, and when in full load, the maximum displacement in *X* and *Y* axle is ±150 mm with the maximum acceleration of ±1.0 g, and the maximum displacement in *Z* axle is ±100 mm with the maximum acceleration of ±0.8 g.

In the shaking table test of geotechnical engineering, the typical model containers include laminar shear model box, dish-type container, ordinary rigid container plus lining, and flexible container. However, given the requirements of model scale and rigidity, this test adopts the rigid model container made of steel plate and steel-reinforced concrete, and its inner space size is 5.0 m × 5.0 m × 2.1 m (length × width × height). The baseboard and framework of the model container are made of steel plate, angle steel, and channel steel, and the total weight of the model container is 7473.55 kg. See Fig. 3.23.

3.2.1.3 Similarity System of Shaking Table Test

For the shaking table test of rock slope model, it is crucial to keep similarity between model and the prototype. Due to difficulties in test materials and technologies, most shaking table tests could only adopt scale model for experiments, and because of the intricate background, complex terrain, and large scale of Wenchuan earthquake-induced landslides and collapses, it is hard to make the model completely consistent with the prototype. Therefore, for the main purpose of this shaking model test, we could only make one or some indicators similar to or proximately similar to that of the prototype for the experiment.

There are mainly three kinds of similarity relations in the physical simulation test: geometric similarity, dynamic similarity, and kinetic similarity. If two physical phenomena meet the similarity conditions in geometric similarity, dynamic similarity, and kinetic similarity, then they are considered similar phenomenon. Among these three kinds of similarity relations, geometric similarity is easy to achieve, while dynamic similarity is controlled by and shown with geometric similarity and kinetic similarity. Thus, dynamic similarity is the key among the three.

① Three Theorems of Similarity

Physical phenomenon is only similar to certain conditions; the results of similarity simulation test could be applied to the prototype only under certain conditions. The three theorems of similarity are the summary of such conditions.

(1) The First Theorem of Similarity

The first theorem of similarity is defined as the similarity rules of the similar phenomena are equal, the similarity index equals to 1, and the single-valued conditions are similar.

The single-valued condition refers to features in which individual phenomenon differs from the same phenomena, which include geometric condition, physical condition, boundary condition, and initial condition. Geometric condition refers to the shape and size of objects in the process, physical condition refers to the physical property of objects, boundary condition refers to the external constraints on the objects' surface, and initial condition refers to certain features of the objects at the initial time. Taking research on heat conduction process for example, the shape and geometric size of the object are the geometric condition, the specific heat capacity and the heat conductivity coefficient are the physical conditions, the heat conductivity coefficient of the surface medium belongs to boundary condition, and the temperature of the object at the initial time is the initial condition.

(2) The Second Theorem of Similarity

The second theorem of similarity is also called "π theorem," which could be defined as follows: If the phenomena are similar, the parameter relations used to describe them could be converted into functional relations of similarity rules, and the functional equation of the similarity rules is the same.

As the similarity rules are dimensionless, therefore the physical equation used to describe similar phenomenon could be converted into dimensionless equation of similarity rules.

$$f(a_1, a_2, \ldots, a_k, a_{k+1}, a_{k+2}, \ldots a_n) = 0; \tag{3.3}$$

$$F(\pi_1, \pi_2, \ldots, \pi_{n-k}) = 0 \tag{3.4}$$

In Eq. (3.3), $a_1, a_2, \ldots a_k$ are basic quantities and $a_{k+1}, a_{k+2}, \ldots$ and a_n are derived quantities, which shows there are $(n-k)$ similarity rules.

The second theorem of similarity provides theoretical support for the spread of model test results. This is because if the two phenomena are similar, according to the second theorem of similarity, the model test results could be spread into the prototype and bring perfect explanation for it. The first and second theorems of similarity state the common features of similar phenomena and provide support for the spread of model test results.

(3) The Third Theorem of Similarity

The third theorem of similarity could be defined as if two phenomena can be represented by the same function equation and are similar in single-valued condition and, at the same time, the similarity rules composed of single-valued condition are equal, then the two phenomena are similar. In the engineering practice, it is quite difficult to make model and prototype completely meet the requirements of the third theorem of similarity, and it is even impossible. In the situation, we can, according to the features of the objects, reasonably select the high-impact factors, seize the main contradiction of the phenomena, and omit the secondary factors to achieve the model test, which is the so-called approximate simulation. The success of approximate simulation lies in reasonable selection of influence factors. Although it cannot make sure all the similar conditions are satisfied, it ensures the similarity of the main factors, so the accuracy of the research results generally can meet the requirement of engineering practice.

② Selection and Adjustment of Similarity Rules

When carrying out model test similarity design, we could adopt similarity transformation method, dimensional analysis method, and matrix method to work out the similarity rule, i.e. the original rule. Ordinarily, the original rule need process through transformation and adjustment on form, which is mainly because when different methods are adopted to derive similarity rules, the form of these rules might be different. Even for some single derivation method, position arrangement changes of certain parameters might lead to form changes of these rules. It is difficult or even impossible to make model test meet all the similarity conditions when carrying out similarity tests. At this time, we should seize the main factors that influence the inherent law of phenomenon, i.e. to satisfy the major rules and omit some secondary rules, which is a common measure used by approximate simulation. It must be cautious to omit similarity rules to ensure they indeed have little influence to the change laws of phenomena, which needs in-depth analysis to the researching phenomena.

③ Determination of Related Physical Parameters

This large-scale shaking table test of rock slope relates to many physical parameters. There are 17 separate physical parameters after analysis and classification as follows: geometric dimensions L, gravitational acceleration $g(C_g = 1)$, cohesive force c, dynamic modulus E, internal friction angle φ, dynamic Poisson's ratio μ, unit weight γ, shear wave velocity V_s, input acceleration A, duration Td, frequency ω, angular displacement θ, linear displacement s (should guarantee $C_s = C_L$), response speed V, response acceleration a, stress σ, and strain ε.

Table 3.3 Dimension of main physical parameters

Physical parameters	Mass system	Physical parameters	Mass system
$[L]$	L	$[A]$	LT^{-2}
$[c]$	$ML^{-1}\ T^{-2}$	$[T_d]$	T
$[\varphi]$	1	$[\omega]$	T^{-1}
$[\Upsilon]$	$ML^{-2}\ T^{-2}$	$[s]$	L
$[E]$	$ML^{-1}\ T^{-2}$	$[\theta]$	1
$[\mu]$	1	$[\varepsilon]$	1
$[V_s]$	LT^{-1}	$[\sigma]$	$ML^{-1}\ T^{-2}$
[g]	LT^{-2}	$[V]$	LT^{-1}
$[\alpha]$	LT^{-2}		

④ Derivation of Similarity Criteria through Matrix Method

The above 17 physical parameters need to meet the physical equation (see Eq. 3.5). Then, we adopt $[M]$, $[L]$, $[T]$ as fundamental dimension and adapt it to the similarity rule equation of dimensionless parameters (see Eq. 3.6). Finally, the general expression of similarity rule is worked out (see Eq. 3.7).

$$f(L, c, \varphi, \gamma, E, \mu, v_s, g, A, T_d, \omega, s, \theta, \varepsilon, \sigma, V, a) = 0; \tag{3.5}$$

$$F(\pi_1, \pi_2, \ldots \pi_{14}) = 0 \tag{3.6}$$

In the similarity rule, the dimension of the above 17 physical parameters is as Table 3.3.

Bringing the above dimension of main physical parameters into the general expression of similarity rule (3.7), we get:

Merging the same dimensions, we can get:

$$\begin{aligned} M^0L^0T^0 = {} & L^{a_1} \cdot (ML^{-1}T^{-2})^{a_2} \cdot (1)^{a_3} \cdot (ML^{-2}T^{-2})^{a_4} \cdot (ML^{-1}T^{-2})^{a_5}(1)^{a_6} \\ & \cdot (LT^{-1})^{a_7} \cdot (LT^{-2})^{a_8} \cdot (LT^{-2})^{a_9} \cdot (T)^{a_{10}} \cdot (\omega)^{a_{11}} \cdot (L)^{a_{12}} \cdot (1)^{a_{13}} \\ & \cdot (1)^{a_{14}} \cdot (ML^{-1}T^{-2})^{a_{15}} \cdot (LT^{-1})^{a_{16}} \cdot (LT^{-2})^{a_{17}} \end{aligned} \tag{3.7}$$

According to the uniformity of dimensions, by Eq. (3.8):

$$\begin{aligned} M^0LT^0 = {} & M^{a_2+a_4+a_5+a_{15}} \cdot L^{a_1-a_2-2\cdot a_4-a_5+a_7+a_8+a_9+a_{12}-a_{15}+a_{16}+a_{17}} \cdot \\ & T^{-2a_2-2\cdot a_4-2\cdot a_5-a_7-2\cdot a_8-2a_9+a_{10}-a_{11}-2\cdot a_{15}-a_{16}-2\cdot a_{17}} \end{aligned} \tag{3.8}$$

Use matrix method to solve the 14 derived similar constants. Table 3.4 is using matrix method to derive similarity criteria.

We can get similarity criteria from Table 3.2, and further, we can get similarity rule (respectively see Tables 3.5 and 3.6).

Table 3.4 Similarity criteria derived from matrix method

Objective variables	Physical parameters																	
	c	φ	E	μ	v_s	g	A	T_d	s	θ	ε	σ	V	a	L	γ	ω	The obtained relation expression
	a_2	a_3	a_5	a_6	a_7	a_8	a_9	a_{10}	a_{12}	a_{13}	a_{14}	a_{15}	a_{16}	a_{17}	a_1	a_4	a_{11}	
π_1	1	0	0	0	0	0	0	0	0	0	0	0	0	0	−1	−1	0	$\pi_1 = \frac{C}{L\gamma}$
π_2		1	0	0	0	0	0	0	0	0	0	0	0	0	0	0	0	$\pi_2 = \varphi$
π_3			1	0	0	0	0	0	0	0	0	0	0	0	−1	−1	0	$\pi_2 = \frac{E}{L\gamma}$
π_4				1	0	0	0	0	0	0	0	0	0	0	0	0	0	$\pi_4 = \mu$
π_5					1	0	0	0	0	0	0	0	0	0	−1	0	−1	$\pi_5 = \frac{v_s}{L\omega}$
π_6						1	0	0	0	0	0	0	0	0	−1	0	−2	$\pi_6 = \frac{g}{L\omega^2}$
π_7							1	0	0	0	0	0	0	0	−1	0	−2	$\pi_7 = \frac{A}{L\omega^2}$
π_8								1	0	0	0	0	0	0	0	0	1	$\pi_8 = T_d\omega$
π_9									1	0	0	0	0	0	−1	0	0	$\pi_9 = \frac{S}{L}$
π_{10}										1	0	0	0	0	0	0	0	$\pi_{10} = \theta$
π_{11}											1	0	0	0	0	0	0	$\pi_{11} = \varepsilon$
π_{12}												1	0	0	−1	−1	0	$\pi_{12} = \frac{\sigma}{L\gamma}$
π_{13}													1	0	−1	0	−1	$\pi_{13} = \frac{V}{L\omega}$
π_{14}														1	−1	0	−2	$\pi_{14} = \frac{a}{L\omega^2}$

Table 3.5 Similarity criteria

$\pi_8 = T_d\omega$	$\pi_1 = c/(L\gamma)$	$\pi_9 = s/L$	$\pi_2 = \varphi$	$\pi_{10} = \theta$	$\pi_3 = E/(L\gamma)$	$\pi_{11} = \varepsilon$
$\pi_4 = \mu$	$\pi_{12} = \sigma/(L\gamma)$	$\pi_5 = V_s/(L\omega)$	$\pi_{13} = v/(L\omega)$	$\pi_6 = g/\left(L\omega^2\right)$	$\pi_{14} = a/\left(L\omega^2\right)$

Table 3.6 Similarity rule

$C_c/\left(C_l C_\gamma\right) = 1$	$C_\varphi = 1$	$C_E/\left(C_l C_\gamma\right) = 1$	$C_\mu = 1$	$C_{V_s}/(C_l C_\omega) = 1$	
$C_g/\left(C_l C_\omega^2\right) = 1$	$C_A/\left(C_l C_\omega^2\right) = 1$	$C_\omega/C_{T_d} = 1$	$C_s = C_l$	$C_\theta = 1$	$C_\varepsilon = 1$
$C_\sigma/(C_l C_\omega) = 1$	$C_v/(C_l C_\omega) = 1$	$C_a/\left(C_l C_\omega^2\right) = 1$			

Then, we bring the association relation into Eq. (3.7) and we can get Eq. (3.9):

$$\pi_i = \left[\frac{c}{L\gamma}\right]^{a_2} \cdot [\varphi]^{a_3} \cdot \left[\frac{E}{L\gamma}\right]^{a_5} \cdot [\mu]^{a_6} \cdot \left[\frac{v_s}{L\omega}\right]^{a_7} \cdot \left[\frac{g}{L\omega^2}\right]^{a_8} \cdot \left[\frac{A}{L\omega^2}\right]^{a_9}$$
$$\cdot [T_d\omega]^{a_{10}} \cdot \left[\frac{s}{L}\right]^{a_{12}} \cdot [\theta]^{a_{13}} \cdot [\varepsilon]^{a_{14}} \cdot \left[\frac{\sigma}{L\gamma}\right]^{a_{15}} \cdot \left[\frac{V}{L\omega}\right]^{a_{16}} \cdot \left[\frac{a}{L\omega^2}\right]^{a_{17}} \quad (3.9)$$

Comprehensively analyzing Table 3.4, we can know that some similar constants can be directly derived as follows:

$$C_\mu = 1 \quad C_\varphi = 1 \quad C_\varepsilon = 1 \quad C_\theta = 1$$

Because geometric constant must equal to displacement constant, we can get $C_s = C_L$.

Gravitational acceleration g remains unchanged in the experiment, so we take $C_g = 1$, and we can know from the similarity rule in Table 3.4:

$$C_g/(C_l C_\omega^2) = 1 \quad (3.10)$$
$$C_A/(C_l C_\omega^2) = 1 \quad (3.11)$$
$$C_a/(C_l C_\omega^2) = 1 \quad (3.12)$$

We can get the similar constants $C_g = C_A = C_a = 1$.

According to the similarity rule, we can get:

$$C_\omega = C_l^{-0.5} \quad (3.13)$$

Bringing Eq. (3.13) into the similarity rule, we can get:

$$C_\omega C_{Td} = 1 \quad (3.14)$$
$$C_{Vs}/(C_l C_\omega) = 1 \quad (3.15)$$
$$C_v/(C_l C_\omega) = 1 \quad (3.16)$$

We can get the similarity rule:

$$C_{vs} = C_V = C_l C_\omega = C_l^{0.5}, \quad C_{Td} = C_\omega^{-1} = C_l^{0.5}$$

The unit weight stays unchanged in the experiment; we take and can get the similar constants:

$$C_E = C_c = C_l C_\gamma = C_l C_\sigma = C_l C_\gamma = C_l$$

Summarize the deduced results (see Table 3.7) in which the most important factor is geometric size. According to the practical situation of prototype and model, we choose geometric similarity ratio of 10:1 in the design.

Table 3.7 Similar constants

No.	Physical parameters	Similarity relation	Similar constants	Notes
1	Geometric dimensions (L)	C_L	10	Controlled quantity
2	Unit weight (γ)	C_γ	1	Controlled quantity
3	Duration (Td)	$C_{Td} = C_L^{0.5}$	3.16	Controlled quantity
4	Cohesive force (c)	$C_C = C_L$	10	
5	Internal friction angle (φ)	$C_\varphi = 0$	1	
6	Dynamic modulus (E)	$C_E = C_L$	10	
7	Dynamic Poisson's ratio (μ)	$C_\mu = 1$	1	
8	Shear wave velocity (V_s)	$C_{V_S} = C_L^{0.5}$	3.16	
9	Gravitational acceleration (g)	$C_g = 1$	1	
10	Input acceleration (A)	$C_A = 1$	1	
11	Frequency (ω)	$C_\omega = C_L^{-0.5}$	0.316	
12	Linear displacement (s)	$C_S = C_L$	10	
13	Angular displacement (θ)	$C_\theta = 1$	1	
14	Strain (ε)	$C_\varepsilon = 1$	1	
15	Response speed (V)	$C_V = C_L^{-0.5}$	10	
16	Stress (σ)	$C_\sigma = C_L$	3.16	
17	Response acceleration (a)	$C_a = 1$	1	

3.2.1.4 Selection of Similarity Materials of Shaking Table Test

When selecting the similar materials, we not only need to make all their physical parameters meet the requirements of similarity, but also need to make them meet certain selecting principles. Based on years of research results of the predecessors, we summarize several principles that the similar rock materials shall meet:

(1) The raw materials shall be safe and harmless to humans.
(2) Try to buy the ordinary, accessible, and low-price raw materials.
(3) The rock materials shall have stable chemical, mechanic, and physical character, and try to make them not affected by external conditions like time, humidity, and temperature.

Table 3.8 Name and function of raw materials

Type	Name	Function
Aggregate	River sand	As coarse aggregate to change the internal friction angle of materials
Cementing materials	Barite powder	It is of high unit weight and can change the unit weight of materials. As filling materials, it is cohesive
Admixtures	Clay	It can preserve moisture and decrease mechanical properties
Binder	Water and gypsum	Different ratio can simulate different parameters. It plays the role of cementation

(4) The fabrication process of the similar materials needs to be as easy and convenient as possible.
(5) The finished similar materials need to have high insulation.
(6) The materials need to be plastic, easy for grinding and shaping, and have short solidification time.
(7) Adjusting the ration of materials could largely change their mechanic characteristics.

There are mainly two kinds of stratum lithology of rock slope in Wenchuan earthquake region: One is hard rock with the representative of thick massive limestone, and the other is soft rock with the representative of mudstone. The shaking model test mainly selects these two representative materials: hard rock and soft rock. The raw similarity materials used to configure the shaking table need to be of different type and performance. The model simulation materials are commonly composed of aggregate, cementing materials, and admixtures. According to the features of shaking table test and several commonly used similarity materials in model test as well as the mechanic performance indexes of various proportion ratios, this experiments adopt barite powder, river sand, gypsum, clay, and water as raw materials, and the detail performance and function of them are shown in Table 3.8.

According to the similarity rule, the density of the model material should be the same with that of the prototype. The proportional ratio of the similarity materials is unknown before we do the configuration. Referring to the similarity material proportional ratio and the commonly used similarity raw materials summarized by *The Similarity Theory and the Statics Model Test* and *Similarity Materials and Similar Model*, we regulate the main parameters of the similarity materials through using different raw materials and adjusting their ratio to finally work out the proportional ratio of the similarity materials needed by the shaking table test: The proportional ratio of materials constituting soft rock is river sand:plaster:water:clay = 1.0:0.27:0.12:0.27, and the proportional ratio of materials constituting hard rock is river sand: plaster: water: clay: barite powder = 1.0:0.5:0.28:0.60:0.80. The concrete physical and

Table 3.9 Physical and mechanical parameters of hard rock and soft rock used in the experiment

Lithology	Density (g/cm^3)	Moisture content (%)	Cohesive force (kPa)	Internal friction angle (°)
Soft rock	2.2	9.655	104	41
Hard rock	2.3	7.826	200	45

Fig. 3.24 Strain-controlled direct shear apparatus

Fig. 3.25 Cutting ring method for unit weight of soil test

mechanical indexes of similarity materials are listed in Table 3.9, and the photographs of the laboratory configuration test are shown in Figs. 3.24 and 3.25.

3.2.1.5 Selection of Test Instruments and Monitoring Points Disposition Principle

This experiment focuses on the responses of rock slope to earthquakes, and we mainly collect the acceleration and displacement response values of up and down parts of rock slopes under vibration loads, especially the acceleration responses. The acceleration sensor and laser displacement sensor are used to record the acceleration and displacement values of different parts of slopes, and high-frequency digital photography system is used to record the slope displacement changes during the earthquake.

① **Measured information and selection of sensor types**

The measurement indexes and monitoring contents are the acceleration and displacement values of different parts of slopes and the slope displacement changes during the earthquake.

(1) The acceleration sensor

The acceleration sensor mainly adopts the three-direction acceleration sensors produced by Donghua Test Technology Company Ltd., Jiangsu Province. Its measuring range is 0–20 m/s/s, weight 8 g, resonant frequency 12 kHz, sensitivity 60 mV/m/s/s (X, Y, Z), frequency range 0–1500 Hz (X), 0–1500 Hz (Y), 0–800 Hz (Z), transverse sensitivity ratio less than or equal to 5%, and external size 12 mm × 14 mm × 8 mm.

(2) Laser Displacement Sensor

The laser displacement sensor used the dynamic displacement sensor for earthquakes produced by Xi'an Bohua Measurement and Control electronic technology co., LTD. Its detailed parameters are as follows: form: LVDT, measuring range: ±50 mm, frequency response: 0–500 Hz, accuracy: ≥0.2%, output voltage: ±5 V, output interface: BNC interface, and motion parts: non-spring rebound type M4 thread.

(3) High-frequency digital photography system

The high-frequency digital photography system was independently researched and developed by our research group, which could be used to monitor the displacement field of model in real time. At present, it has already formed complete hardware and software corollary equipment.

② **Disposition Principles of Monitoring Points in Shaking Table Tests**

In the shaking table test, the disposition of the sensors should meet the following two principles:

The sensors should be disposed with full reference to the dynamic response value calculation results of rock slopes under earthquake and the existing research results and with close relation to the experiment purpose.

With basic measurement information collection requirements of the experiment purpose met, the number of sensors should be as fewer as possible, so that to keep the completeness of models and avoid to make artificial geological structures due to instrument disposition so as not to affect the accuracy of research results.

③ **Monitoring points disposition principle of shaking table test**

In order to study on the dynamic characteristics and dynamic response rules of rock slopes under strong earthquakes, in the different position of rock slope models should arrange single-direction acceleration sensors, three-direction acceleration sensors, differential displacement sensors, and laser displacement sensors so as to get synchronous acceleration and displacement response data in the loading process. The single-direction acceleration sensors are charged with AC, and they are mainly used to test the acceleration field distribution of bedrock. Three-direction acceleration sensor is charged with DC, and they are mainly used to monitor the acceleration field inside the rock. Differential displacement sensors and laser displacement sensors are charged with AC, and they are mainly used to monitor displacement changes in the earthquake. At the same time, in order to accurately describe the displacement changes of the whole model surface in the earthquake, this experiment adopts the independently researched and developed high-frequency digital photography system to monitor.

(1) Distribution of single-direction acceleration sensors

The single-direction acceleration sensors are mainly used to monitor the free-field acceleration response in the earthquake, and two of the free-field points are arranged in this experiment. Three-direction acceleration sensor is charged with DC, and they are mainly used to monitor the acceleration field inside the rock. Differential displacement sensors and laser displacement sensors are charged with AC, and they are mainly used to monitor displacement changes in the earthquake (Figs. 3.26, 3.27, and 3.28).

(2) Distribution of three-direction acceleration sensors

Three-direction acceleration sensors are mainly used to monitor the acceleration amplification effect along elevation in the earthquake. In this experiment, four acceleration monitoring points are laid out in the central line on the 30°, 45°, 50°, and 60° surfaces from top down. See Figs. 3.29, 3.30, 3.31, and 3.32 for detailed information.

(3) The arrangements of laser displacement sensors and differential displacement sensors

Laser displacement sensors and differential displacement sensors are mainly used to monitor the slope deformation in the earthquakes. In this experiment, one acceleration monitoring point was, respectively, laid out at the central line on the 30° and 60° surfaces, and a displacement sensor was laid out at the central line on the 50° surface.

(4) The monitoring points arrangements of high-frequency digital photography system

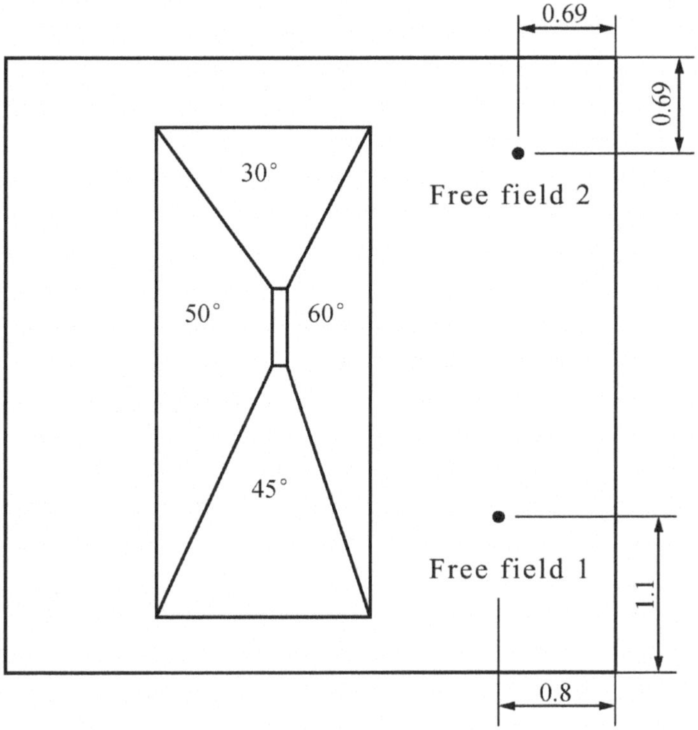

Fig. 3.26 Floor plan

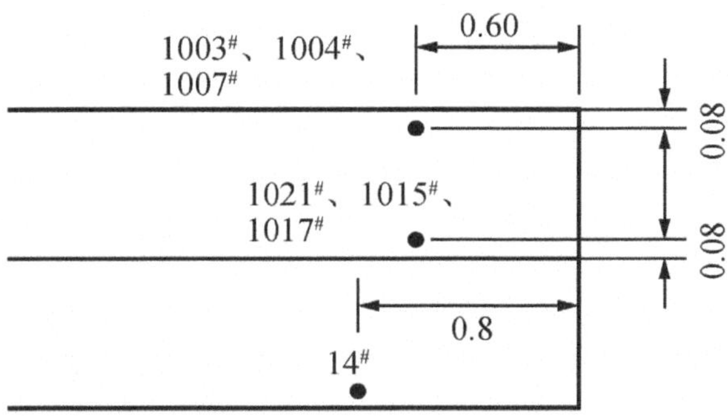

Fig. 3.27 Sectional drawing of free-field one

High-frequency digital photography system was mainly used for research of the slope deformation in the earthquakes and the displacement field distribution of the whole model surface. In this experiment, the monitoring points are set up on the whole model surface with 0.2 m × 0.2 m network distribution.

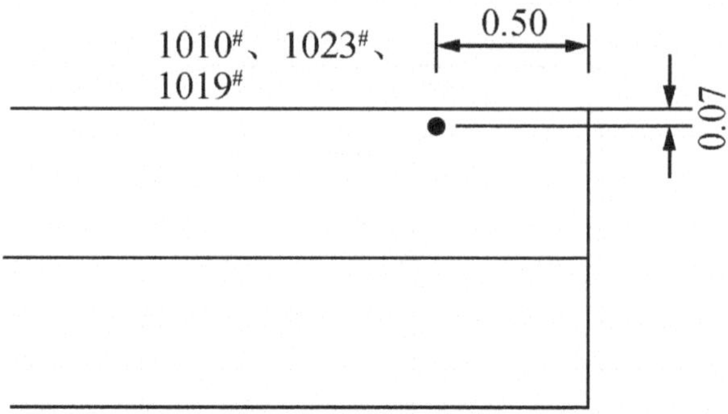

Fig. 3.28 Sectional drawing of free-field two

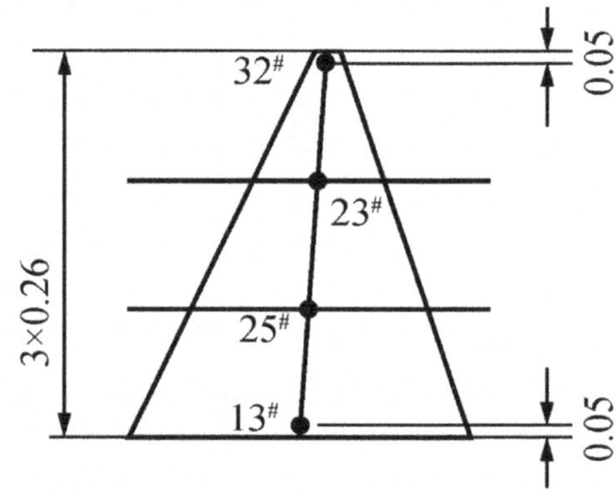

Fig. 3.29 Acceleration monitoring point arrangement plan on 30° surface

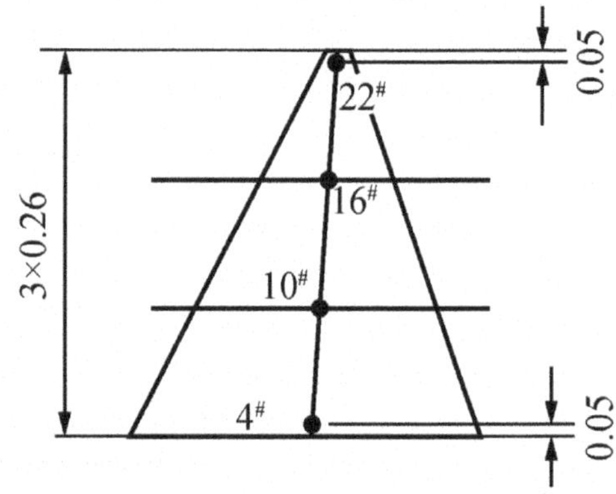

Fig. 3.30 Acceleration monitoring point arrangement plan on 45° surface

3.2.1.6 Design and Manufacture of Shaking Table Test

The factors influencing the slope dynamic response characteristics are various and complex, which include slope structure, topography, and geomorphology, lithologic character as well as loading conditions, while a model with various influencing factors mixing together would be difficult for not only modeling but also analysis. Therefore,

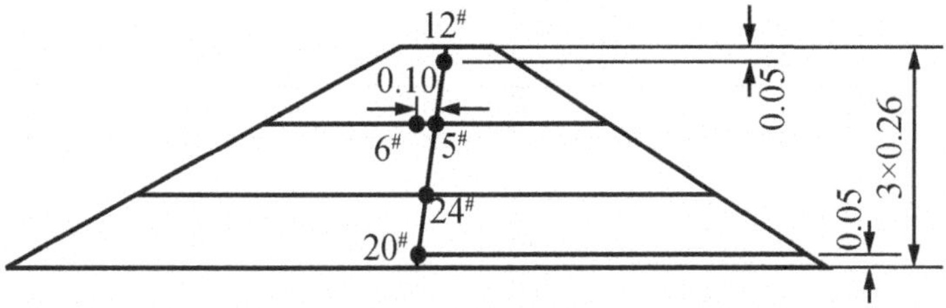

Fig. 3.31 Acceleration monitoring point arrangement plan on 50° surface

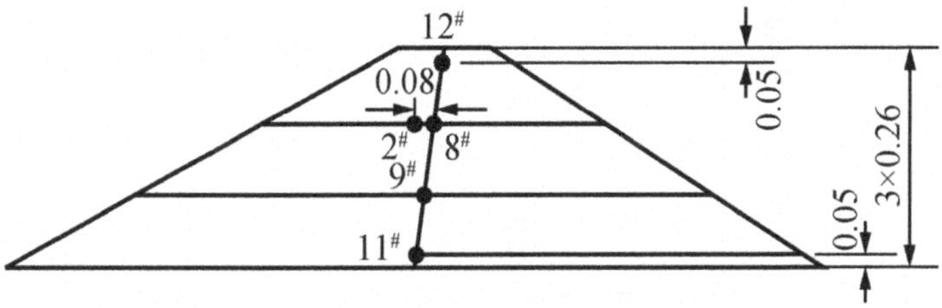

Fig. 3.32 Acceleration monitoring point arrangement plan on 60° surface

based on the principle to strengthen major factors, this experiment simplified the geologic body. In accordance with the model test purpose, this experiment designed slopes of different angles as characteristics to study the slope angle influence on seismic dynamic response rules.

For the two major kinds of rocks discovered in the Wenchuan earthquake region, i.e. hard rocks and soft rocks, this experiment adopts hard rocks with the representative of limestone and soft rocks with the representative of mudstone. Soft rocks simulate the slopes, while hard rocks simulate the bedrocks. In addition, this experiment adopts homogeneous structure to simplify influencing factors. See Figs. 3.33 and 3.34 for detailed characteristic information of the model. In line with the similarity relations, the length scale of prototype and model is 10:1, and the major parameters of the model are as follows:

① The total height of the model is 1.8 m, of which the height of the slope is 0.8 m, that of overlying soft rocks 0.5 m, and that of hard rocks 0.5 m. The length and width of the model are, respectively, 4.4 m.
② The length and width of the bottom line are, respectively, 3.53 and 1.55 m and that of the top surface are, respectively, 0.78 and 0.1 m.
③ The model contains four slopes of different angles: 30°, 45°, 50°, and 60°.
④ The material composition of the model: The rock slopes are composed of soft rocks with physical parameters of mud rocks as simulation; the bedrocks are composed of hard rocks with physical parameters of limestones as simulation.
⑤ The slope structures: The slopes inside the model container are composed of rocks of homogeneous materials.

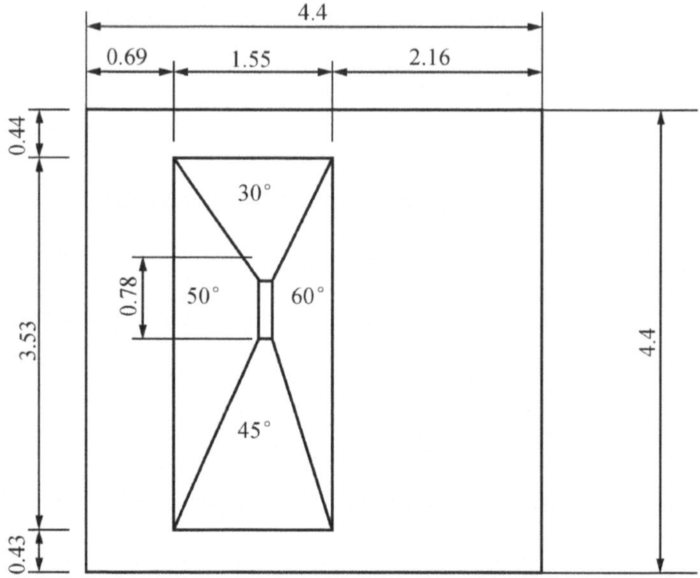

Fig. 3.33 Top view of shaking table test models

Fig. 3.34 Axonometric view of the finished shaking model test models

① Production and transportation of shaking model test materials and location of the model containers

Fig. 3.35 Smashing of the clay

Fig. 3.36 Bagging of the smashed clay

This experiment used homogeneous materials and in situ casting to build test models. When the models are built, we firstly prepared all the needed materials, of which river sand, barite powder, and plaster were all directly purchased from the manufactures, while clay was taken on-site by our research group members. Then, we dealt with them with the process of smashing, grinding, and sifting. This experiment needs 20 tons of clay. See Figs. 3.35 and 3.36 for detailed dealing process (Figs. 3.37 and 3.38).

② Model filling of Shaking Table Test

As the materials were of large amount, this experiment used mixer for blending for 10 min, and when the materials are mixed well, gantry crane is used to fill them into the model container. See Fig. 3.39. The filling was by the measure of from top to bottom and layer by layer. After filling 20 cm each time, we conducted a density test and compacted it with vibrating by both machine and manpower, i.e. using manual work to vibrate at the boundary of the model and machine at the central parts so as

Fig. 3.37 Loading and unloading of clay

Fig. 3.38 Location of the model containers

to ensure the compaction quality. See Figs. 3.40 and 3.41. In order to prevent some sensors breaking down or being damaged in the vibrating, the embedding number should be over the designed number.

To avoid vibration wave reflecting at the model container border, it should eliminate most of the "boundary effect" influence on the experiment. Wave-absorbing materials of 30-mm thickness were added to the end walls in the vibrating direction, and foam was used as wave-absorbing materials (see Fig. 3.42).

3.2.1.7 Embedding of Shaking Table Test Sensors

The correct installation and embedding of the test instrument play a key role in the accuracy of the test data. Before the embedding of the instruments, we need to use apparatus like multimeter to detect and rectify them to ensure normal operation

Fig. 3.39 Machine blending of the model materials

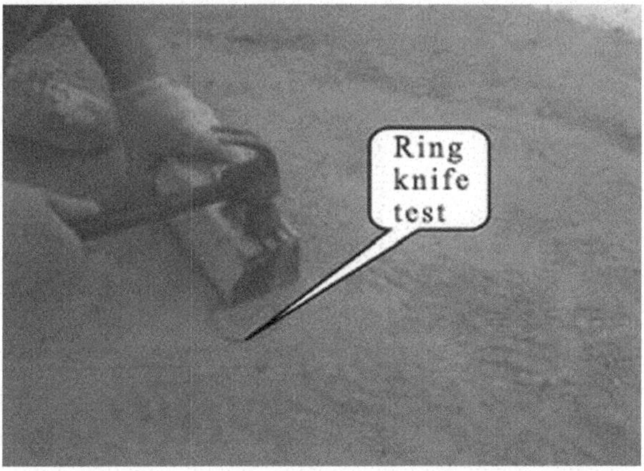

Fig. 3.40 Cutting ring test for density

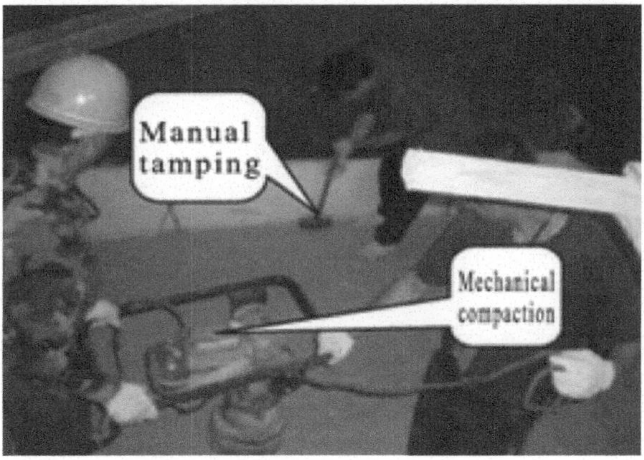

Fig. 3.41 Vibrating of the model materials

Fig. 3.42 Boundary of the test model

Fig. 3.43 Rectification of three-direction acceleration sensors

Fig. 3.44 Seal of three-direction acceleration sensors

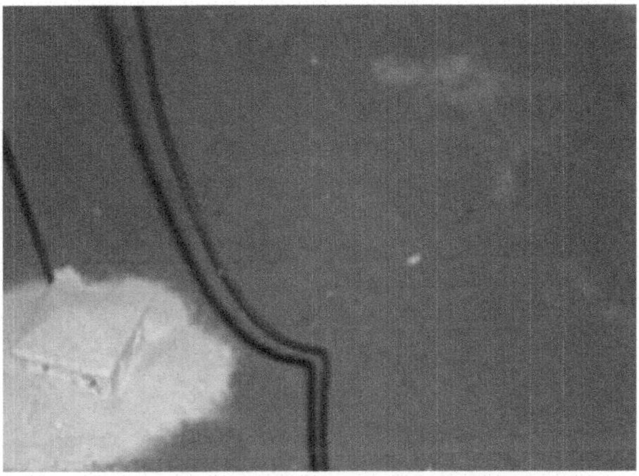

Fig. 3.45 Single-direction acceleration sensors

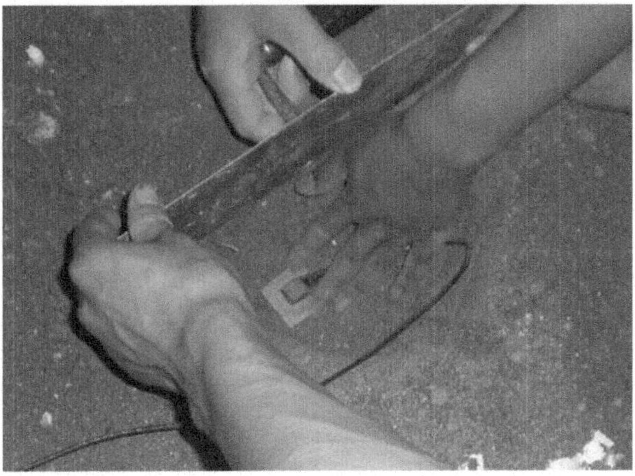

Fig. 3.46 Three-direction acceleration sensors

(see Fig. 3.43). In addition, as the model was built by blending and filling, water must be included in the similarity materials. In order to keep the normal operation of the sensors, they needed waterproofing work first. A layer of acid silicone glass sealant was coated on the sensors (see Fig. 3.44). Before embedding the single-direction acceleration sensors, they needed to be put inside the insulated boxes, and the weight of the acceleration sensors and the boxes must be kept the same with that of the same volume of soil. The acceleration sensors and the boxes were connected with screws (see Fig. 3.45). Due to high demands on security direction by the three-direction acceleration sensors, so in order to prevent larger rotation of acceleration in the vibrating process, this experiment bonded a light sheet steel of high stiffness with size of 5 cm × 5 cm × 1 mm on the surface of the acceleration sensors. The embedding of the acceleration sensors adopted the measure of modeling and embedding at the same time (see Fig. 3.46). Displacement sensors were installed directly after the model was built. Before the installation, the brackets of the sensors needed to be fixed,

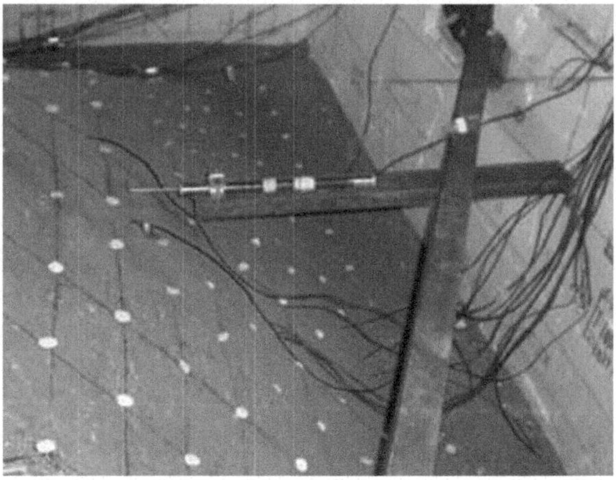

Fig. 3.47 Setup of differential displacement sensors

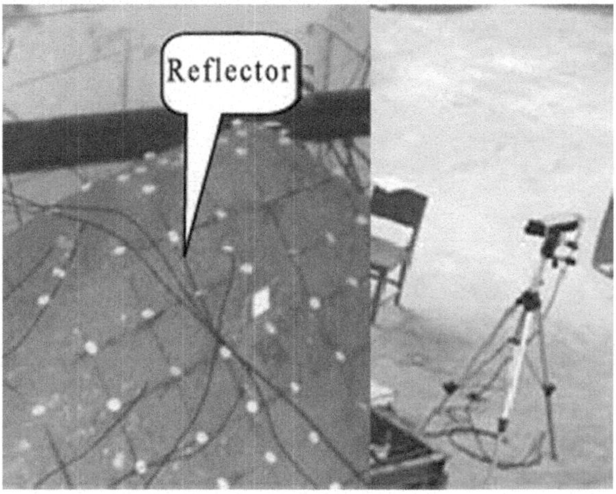

Fig. 3.48 Fix of laser displacement sensors

and the brackets were made of i beam and were connected to the model containers. Holes were dug in the middle of the brackets, and the displacement sensors were fixed on the brackets through the holes with nuts. The end of the displacement sensors bar needed to be buried inside the slopes to ensure the needles would not slip (see Fig. 3.47). The laser displacement sensors were fixed three meters away from the shaking table, and baffle board was embedded on the slopes to reflect light beam of the laser displacement sensors (see Fig. 3.48). The high-frequency digital photography system was fixed on the crane of the gantry crane at the top of the laboratory so as to reduce interference from the shaking table and increase test accuracy. At the same time, a large amount of reflex attachments were set up evenly on the surface of the model surface, which are basically of the 0.2 m × 0.2 m network distribution (see Figs. 3.49 and 3.50).

Fig. 3.49 High-frequency digital photography system

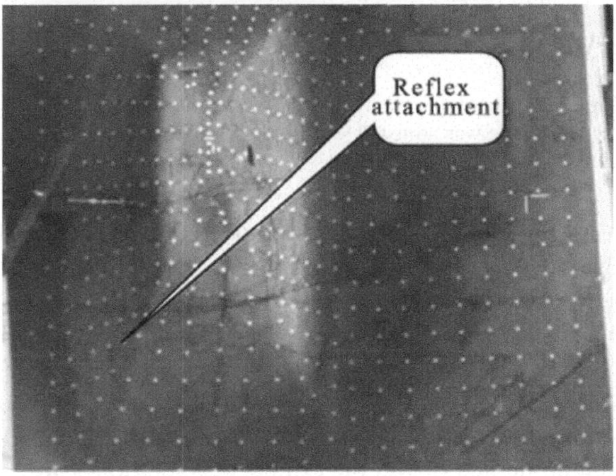

Fig. 3.50 Reflex attachments of high-frequency digital photography system

3.2.1.8 Loading Scheme of Shaking Table Test

This experiment used input accelerogram to achieve earthquake wave simulation. According to the similarity rule, the seismic input accelerogram waveform of the model and the prototype is exactly the same, we only modified by the time similarity ratio. This experiment aimed to study on the influence of seismic wave type, frequency, and amplitude on the dynamic response rules of rock slopes, so we began by normalization processing of input seismic waves and then exerted intensity by adjusting their amplitude to 7°, 8°, and 9° (modus earthquake, design earthquake, and rare earthquake), the three-direction earthquake time history. The major seismic types were Wenchuan wolong seismic waves, El Centro seismic waves, and Kobe seismic waves. Figure 3.51 shows the three-direction seismic dynamic accelerogram of Wenchuan wolong seismic waves after the amplitude was adjusted to 0.1 g, and acceleration on the EW, NS, and UP directions were, respectively, input through the

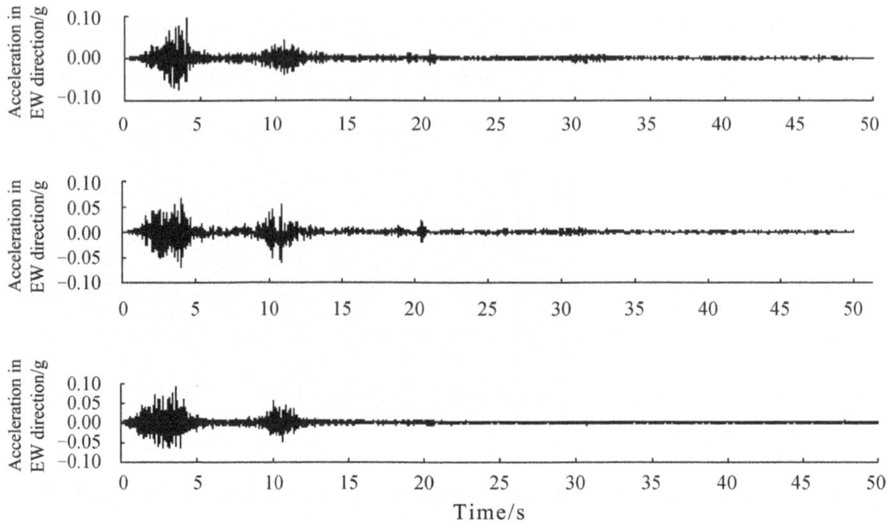

Fig. 3.51 Three-direction seismic dynamic accelerogram of Wenchuan wolong seismic waves after the amplitude was adjusted to 0.1 g

X, *Y*, and *Z* directions of the model (Fig. 3.34 specified this). Figure 3.52 shows the three-direction seismic dynamic accelerogram of El Centro seismic waves, and acceleration on the EW, NS, and UP directions, was, respectively input through the *X*, *Y*, and *Z* directions. Figure 3.53 shows the three-direction seismic dynamic accelerogram of Kobe seismic waves, and acceleration on the EW, NS, and UP directions was, respectively, input through the *X*, *Y*, and *Z* directions. In addition, in order to study on the dynamic response rules under different seismic dynamic acceleration amplitude and obtain the critical acceleration value when the slopes showed plastic deformation, the experiment planned to exert the acceleration load of different amplitudes from small to large on the model.

White noise scanning is the key measure to test the dynamic characteristics of model structure. In order to know the dynamic characteristic changes of the model before and after vibration, this experiment used white noise scanning in different stages before and after vibration. It is initially decided that the first step was to load Kobe, El Centro, and Wenchuan wolong seismic waves under seven-degree modus earthquake, and then added amplitude gradually. See Table 3.10 for detailed loading scheme.

3.2.1.9 Data Collecting System of Shaking Table Test

This experiment adopts two data collecting systems simultaneously. The NPIC adopts BBM data collecting system with 128 channels, and its maximum fiducial error is less than or equal to 0.5%. Sensor signal conditioner was connected to electric

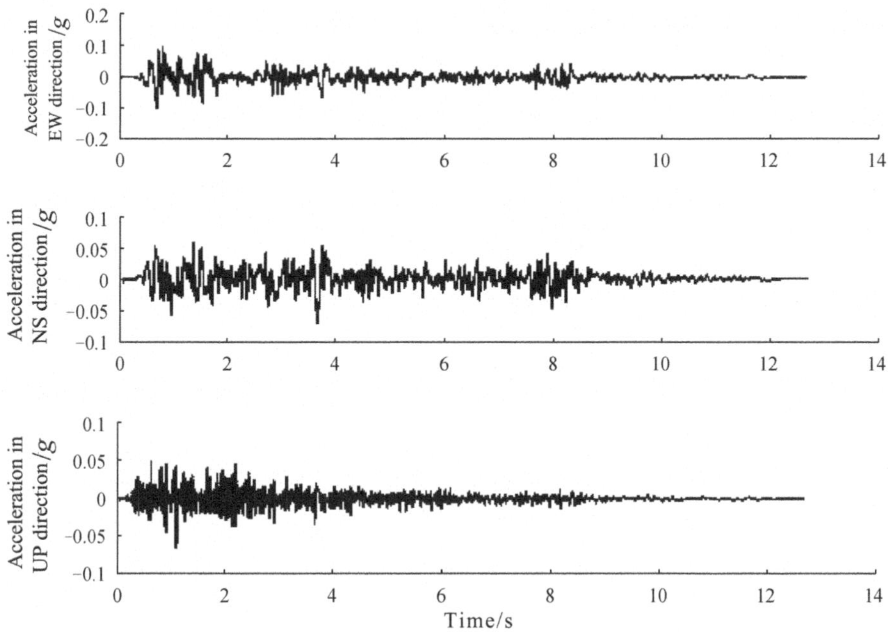

Fig. 3.52 Three-direction seismic dynamic accelerogram of El Centro seismic waves after the amplitude was adjusted to 0.1 g

charge transfer to convert the voltage signal, and its maximum fiducial error is less than or equal to 1%. Data collecting, signal monitoring, and online analyzing were conducted at the same time (see Fig. 3.54). Southwest Jiaotong University adopted Donghua dynamic data acquisition instrument DH5923 with 32 channels, and the maximum sampling frequency was 5 kHz. It used 1394 or PCI interface, DMA measure, and computer communication to control, collect, store, and analyze in real time (see Fig. 3.55). In addition, it adopts high-frequency digital photography system to monitor the displacement field changes of the whole model site in real time. When model making, brackets setup, and sensor embedding are finished, we connect the sensors with the data collecting system and collect electric signals like acceleration and displacement measured by the sensors (see Figs. 3.56 and 3.57).

The measurement indexes and observing contents are acceleration, displacement, and stress response value of different parts of the slopes as well as the deformation characteristics of the slopes. After the experiment, analyze the seismic response principle of slopes through comparison of acceleration, displacement, and deformation characteristics of different parts of the slopes.

Table 3.10 Loading scheme of shaking table test

Loading order: firstly 0.05 g white noise, then load one by one					
Loading scheme	PGA		Loading scheme	PGA	
1	7° modus earthquake	Kobe seismic wave	17	9° modus earthquake	Kobe seismic wave
2		El Centro seismic wave	18		El Centro seismic wave
3		Wenchuan wolong seismic waves	19		Wenchuan wolong seismic waves
4	0.05 g	White noise	20	0.05 g	White noise
5	7° design earthquake	Kobe seismic wave	21	9° design earthquake	Kobe seismic wave
6		El Centro seismic wave	22		El Centro seismic wave
7		Wenchuan wolong seismic waves	23		Wenchuan wolong seismic waves
8	0.05 g	White noise	24	0.05 g	White noise
9	8° modus earthquake	Kobe seismic wave	25	7° rare earthquake	Kobe seismic wave
10		El Centro seismic wave	26		El Centro seismic wave
11		Wenchuan wolong seismic waves	27		Wenchuan wolong seismic waves
12	0.05 g	White noise	28	0.05 g	White noise
13	8° design earthquake	Kobe seismic wave	29	8° rare earthquake	Kobe seismic wave
14		El Centro seismic wave	30		El Centro seismic wave
15		Wenchuan wolong seismic waves	31		Wenchuan wolong seismic waves
16	0.05 g	White noise	32	0.05 g	White noise
33	White noise 0.05 g				

3.2.1.10 Result Record of Shaking Table Test

The experiment result record was composed of three major parts: the initial states of the test instrument models, experiment data, and the deformation of the test models.

(1) Before vibration test

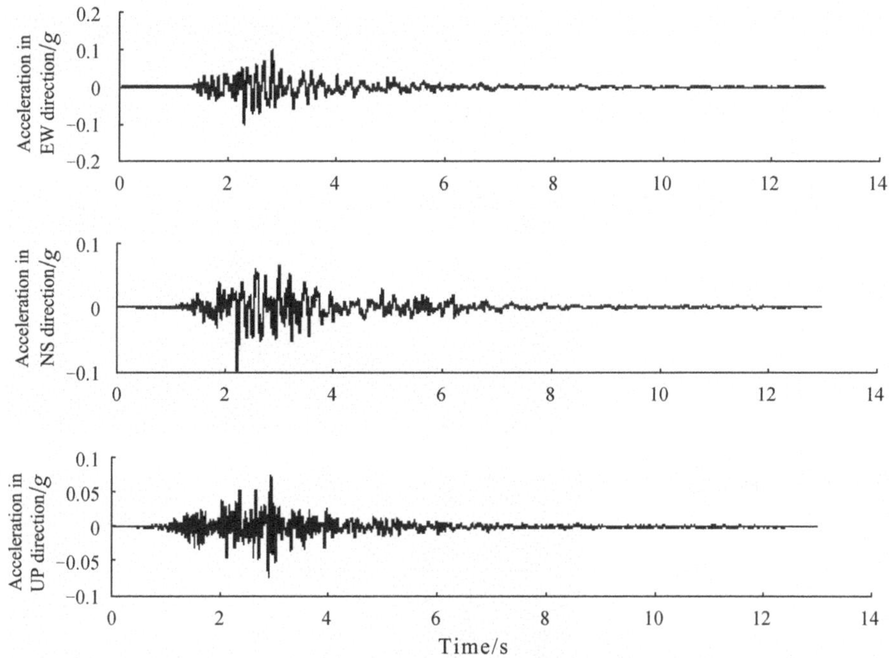

Fig. 3.53 Three-direction seismic dynamic accelerogram of Kobe seismic waves after the amplitude was adjusted to 0.1 g

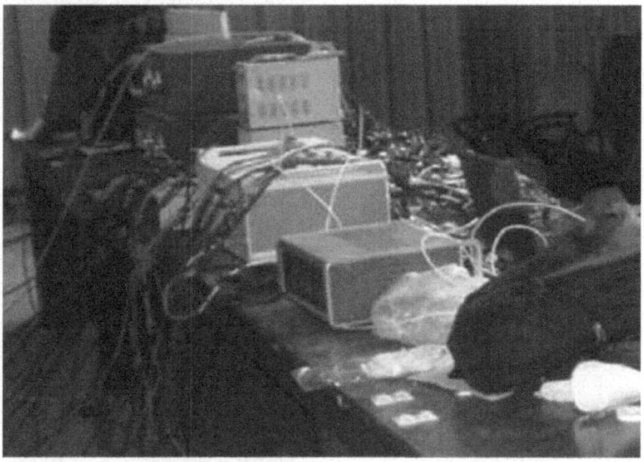

Fig. 3.54 BBM data collecting system

① Check the initial states of the models and the preparation of all the test instruments, and debug the whole system.
② Use high-frequency digital photography system to record the initial states of different parts of the model before vibration test.

(2) During vibration test

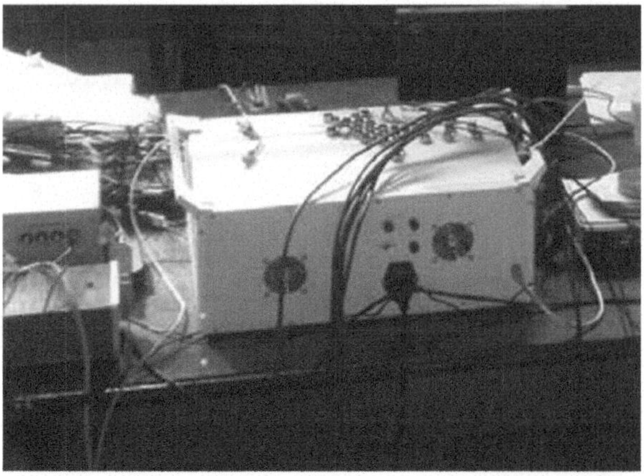

Fig. 3.55 DH5923 data collecting system

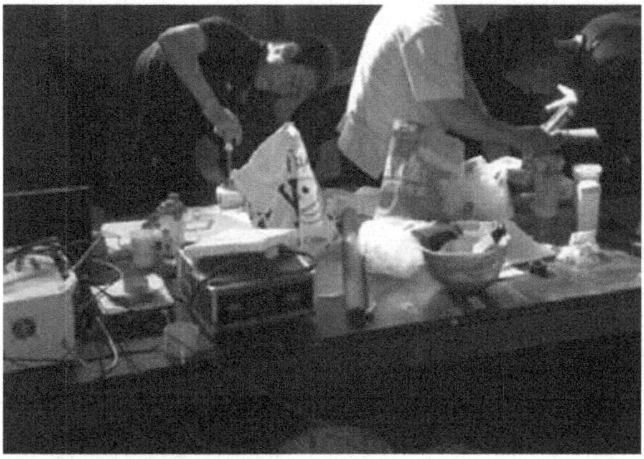

Fig. 3.56 Connection of sensors with data collecting systems

Fig. 3.57 Field debugging of shaking table test systems

① Set high-frequency digital photography system up above the shaking table test model to record the whole test process.
② After input of white noise and loading scheme of seismic wave of different levels, the computer data collecting system would automatically collect data about all the monitoring instruments during the vibration process.
③ Check the test model conditions and the working states of monitoring instruments.

(3) After vibration test

① Use steel ruler to measure the surface deformation and displacement of the model.
② Use camera to record the slope deformation and the cracks developments.

3.2.2 *Result Analysis of Shaking Table Test*

For test data of slope dynamic response rules experiment, acceleration is intuitive and easy to monitor and analyze, so many researchers see acceleration data as the main study object in the slope dynamic response rules experiment. This experiment arranged 19 three-direction acceleration test points inside the slopes, four of which are evenly set up from top down on the 30°, 45°, 50°, and 60° slope surfaces, and free-field acceleration test points were set up on both sides of the model. Then, through the input of different seismic wave parameters, monitor the acceleration response values of different slope parts of different angles under earthquakes, analyze and monitor acceleration data, and study on the influence of slope angles and ground motion parameters of seismic waves exerting on the slopes on the acceleration response rules.

For an easy study on the acceleration response rules, the acceleration peak value was abbreviated as PGA, and acceleration amplification coefficient was defined as the ratio of acceleration peak value inside the slopes with that in the free field. Explain with acceleration on X-axis for example: Suppose that the acceleration peak value on the X-axis of point C inside the slope is AXC and the acceleration peak value on the X-axis of point D in the free field is AXD (see Fig. 3.58), then the acceleration amplification coefficient δ_X on the X-axis of this point could be represented as $\delta_X = ACX/ADX$.

3.2.2.1 Research of Influence Factors of Rock Slope Acceleration Peak Value Elevation Amplification Effect

Difference of slope surface angle, seismic density, and seismic ground motion input wave parameters have significant influence on the rock slope seismic ground motion characteristics. Therefore, this experiment built test model of bevel angles of 30°, 45°, 50°, and 60° and input Wenchuan wolong, El Centro, and Kobe seismic waves,

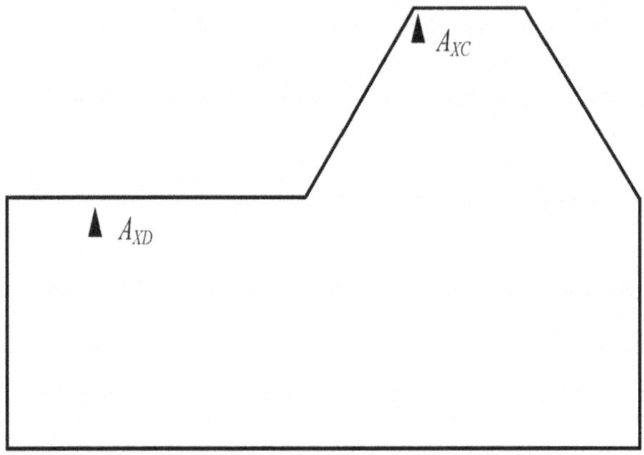

Fig. 3.58 Model schematic

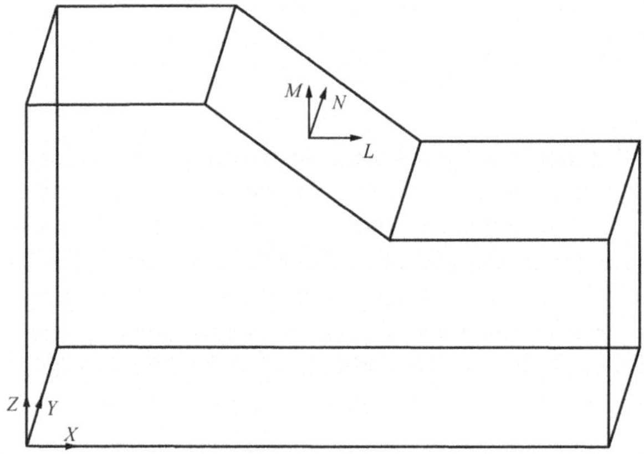

Fig. 3.59 Diagram of coordinate system changes

which belong to modus earthquake, design earthquake, and rare earthquake with respective peak value of seven, eight, and nine degrees. Further, we researched on the influence of the above three conditions on the rock slope acceleration peak value elevation amplification effect.

① Effects of Bevel Angles on the Rock Slope Acceleration Peak Value Elevation Amplification Effect.

At present, a large number of research results show that bevel angles have a significant influence on the seismic ground motion acceleration peak value elevation amplification effect. To clearly explain the influence of bevel angles on the seismic ground motion acceleration peak value elevation amplification effect, this chapter only selects peak values of test points on *X*, *Y*, and *Z* directions on slope surface of bevel angles of 30°, 45°, 50°, and 60° under the effect of Wenchuan wolong seismic

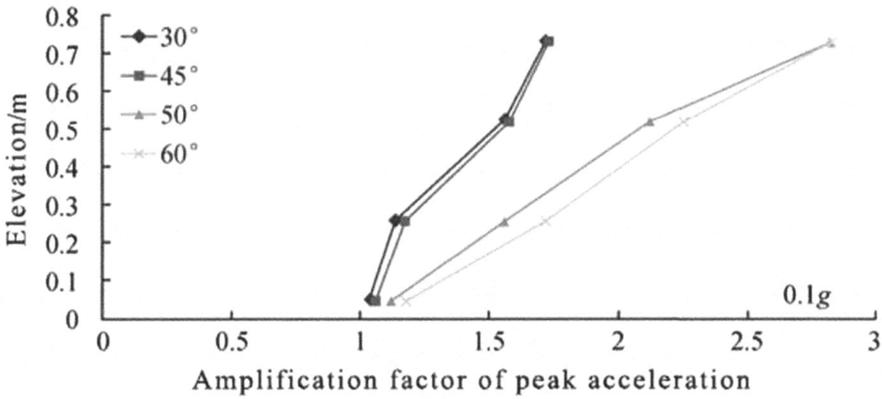

Fig. 3.60 PGA amplification coefficient in *X* direction

waves with acceleration peak values of 0.1, 0.2, and 0.4 g. It is worth noticing that the model used in this experiment contains four surfaces, and if we analyze with the traditional method, i.e. using global coordinate system with *X*, *Y*, and *Z* directions, it will lead to that the free face of 30° and 45° surfaces face the direction of 50° and 60° surfaces, so it will be difficult to study on the influence of bevel angles on the seismic ground motion acceleration peak value elevation amplification effect.

Thus, this chapter selects two kinds of coordinate systems to study the acceleration peak value elevation amplification effect. One is the widely used global coordinate system with *X*, *Y*, and *Z* directions, which is used to simply explain the above problems, and the other is local coordinate system, whose three directions are the free face direction of the slopes (*L*), the strike of the slopes (*M*), and the vertical direction (*N*). The detailed information of these two coordinate systems is shown in Fig. 3.59. See Figs. 3.60 and 3.61 for analysis results under global coordinate system and Figs. 3.62 and 3.63 for that under local coordinate system. It is noteworthy that the free face direction (*L*) in the local coordinate system is not the outer normal direction of the slopes, but the horizontal direction pointing to the outside of the slopes. The vertical direction *N* is the same with the *Z* direction in the global coordinate system, and the slope strike is the extension direction of slope in the space.

We can see from Figs. 3.60 and 3.61 that, under the effect of Wenchuan wolong seismic waves with PGA = 0.1 g, the PGA of *X* and *Y* directions on the surfaces of bevel angles of 30°, 45°, 50°, and 60° is all amplified to different extent along elevation. For acceleration on the *X*-axil, the acceleration peak value amplification rules of different slope surfaces are as follows: 60° surface > 50° surface > 45° surface > 30° surface. For acceleration on the *Y*-axil, the acceleration peak value amplification rules of different slope surfaces are as follows: 45° surface > 30° surface > 60° surface > 50° surface, while as Figs. 3.62 and 3.63 show that under local coordinate system (*L*, *M*, *N*), for acceleration on the *L*-axil, the acceleration peak value amplification rules of different slope surfaces are as follows: 60° surface > 50° surface > 45° surface > 30°

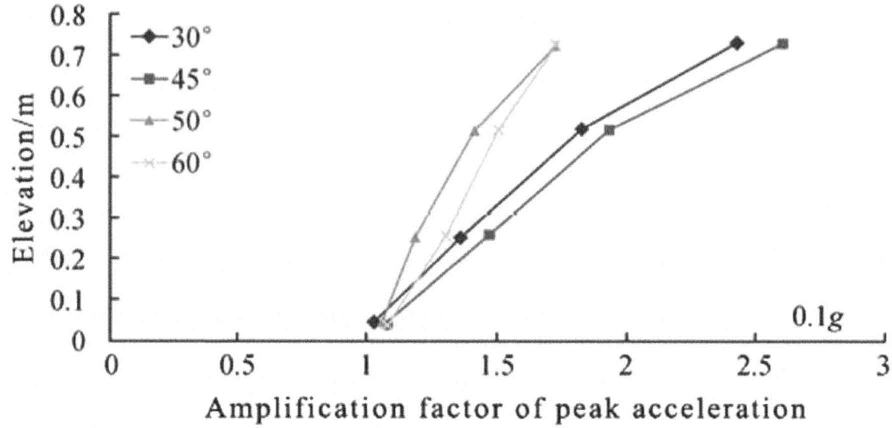

Fig. 3.61 PGA amplification coefficient in Y direction

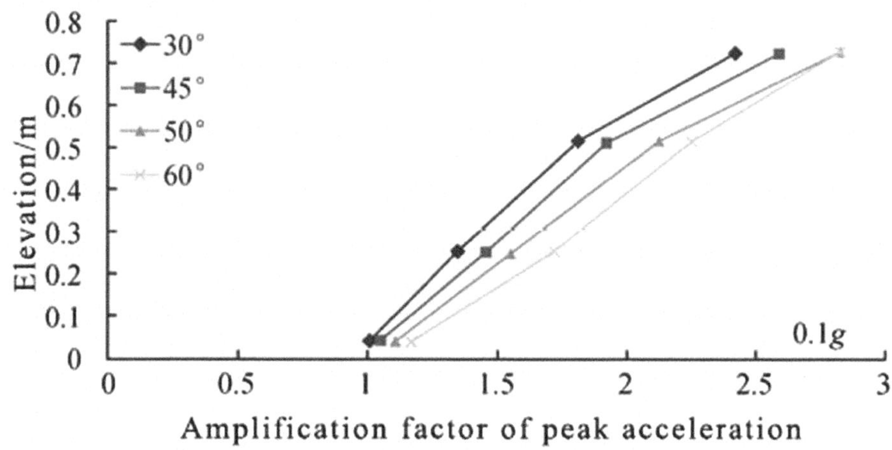

Fig. 3.62 PGA elevation amplification coefficient in L direction

surface. For acceleration on the M-axil, the acceleration peak value amplification rules of different slope surfaces are as follows: 60° surface, 50° surface, 45° surface, and 30° surface are basically the same. It can thus be seen that the results under global coordinate system and the local coordinate system are inconsistent. This may be because the $X(Y)$ direction is the free face direction for 50° and 60° surfaces (Y direction is the slope strike) and the slope strike for 30° and 45° surfaces (Y is the free face direction). However, it is unilateral to compare test results of slope free face surface with that of the slope strike to research the acceleration peak value elevation amplification effect as it is difficult to clearly study on the influence of bevel angles in the acceleration peak value elevation amplification effect. Therefore, this disserta-

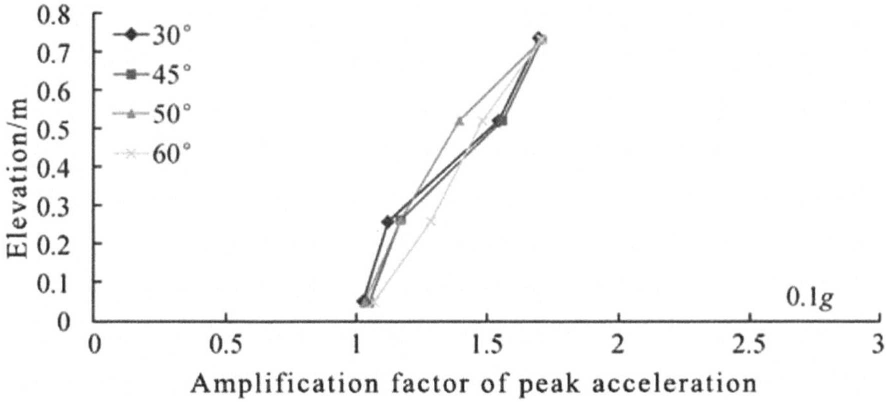

Fig. 3.63 PGA elevation amplification coefficient in *M* direction

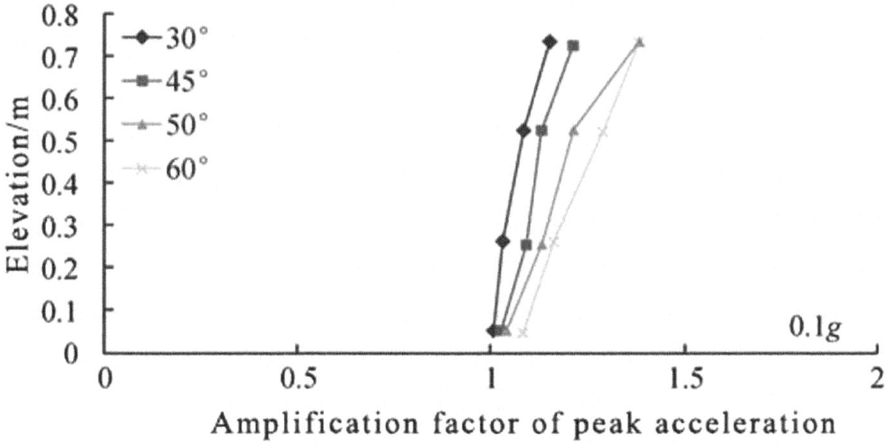

Fig. 3.64 Acceleration peak value elevation amplification coefficient in vertical direction

tion used local coordinate system to elaborate the acceleration peak value elevation amplification effect.

We can see from Figs. 3.62, 3.63, 3.64, 3.65, 3.66, 3.67, 3.68, 3.69, and 3.70 that, under the effect of Wenchuan wolong seismic waves with PGA = 0.1, 0.2, and 0.4 g, the PGA of free face direction (*L*), slope strike (*M*), and vertical direction (*N*) on the surfaces of bevel angles of 30°, 45°, 50°, and 60° is all amplified to different extent along elevation.

For acceleration on L direction, the acceleration amplification rules of all surfaces are as follows: 60° surface > 50° surface > 45° surface > 30° surface. For acceleration on *M* direction, the acceleration amplification rules of all surfaces are as follows: The amplification effects of 60° surface, 50° surface, 45° surface, and 30° surface are basically the same. For acceleration on *N* direction, the acceleration amplifica-

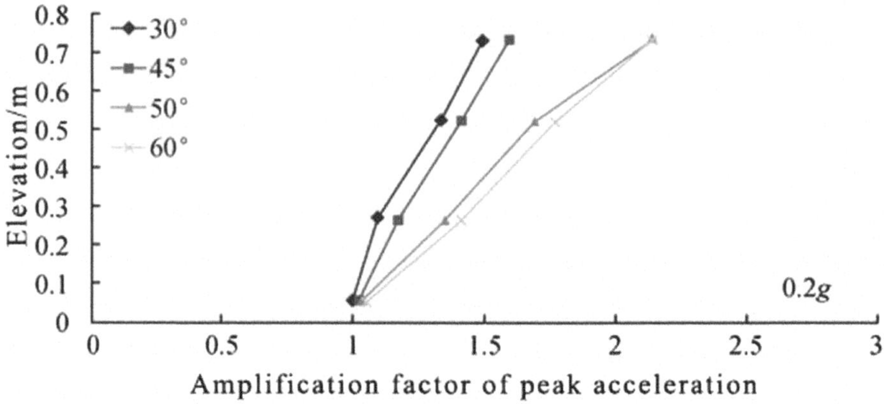

Fig. 3.65 Acceleration peak value elevation amplification coefficient in free face direction

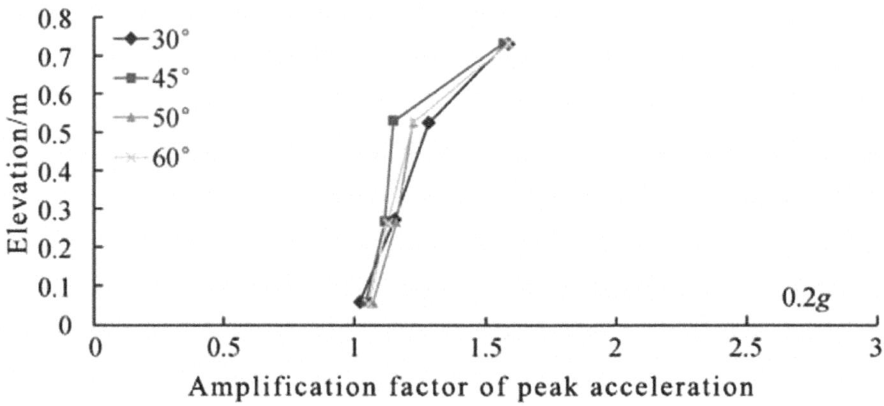

Fig. 3.66 Acceleration peak value elevation amplification coefficient in slope strike

tion rules of all surfaces are as follows: 60° surface > 50° surface > 45° surface > 30° surface.

While as Figs. 3.70 and 3.41 show that under local coordinate system (L, M, N), for acceleration on the L-axil, the acceleration peak value amplification rules of different slope surfaces are as follows: 60° surface > 50° surface > 45° surface > 30° surface, for acceleration on the M-axil, the acceleration peak value amplification rules of different slope surfaces are as follows: 60° surface, 50° surface, 45° surface, and 30° surface are basically the same. Therefore, with the increase of the bevel angles, the PGA elevation amplification effect in free face direction and vertical direction will strengthen gradually, while that in slope strike would basically stay the same.

For an easy explanation for the above phenomenon, this chapter will simplify the problem to plane strain problem and explain form the angle of seismic wave trans-

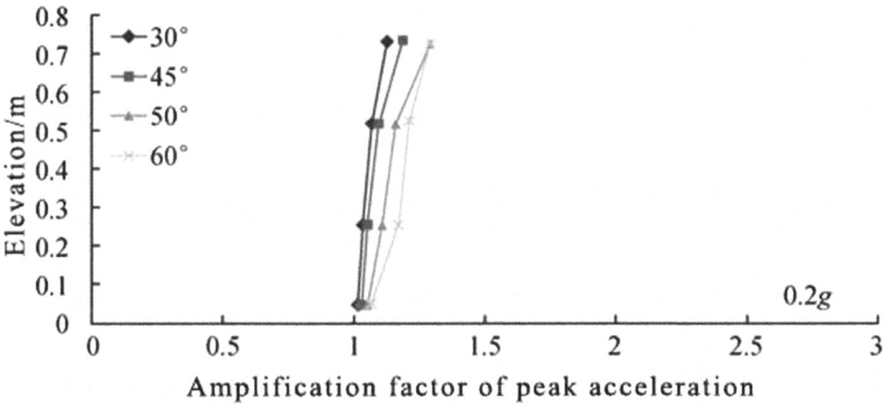

Fig. 3.67 Acceleration peak value elevation amplification coefficient in vertical direction

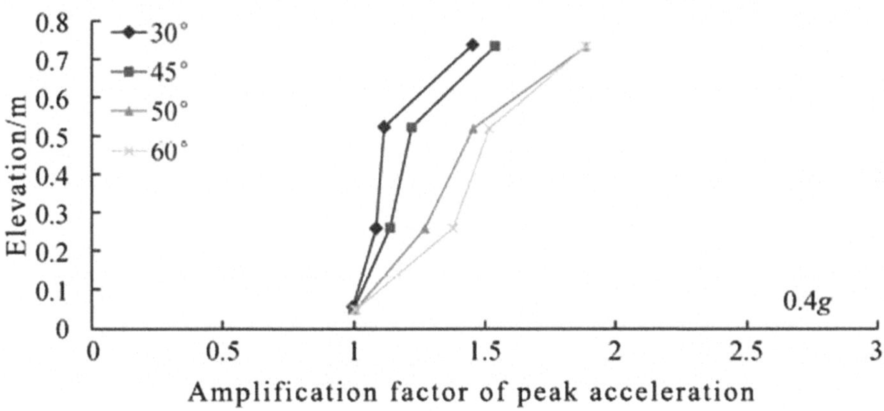

Fig. 3.68 Acceleration peak value elevation amplification coefficient in free face direction

mission characteristics. It can be known from elastic wave theory that the vertically incidenting SV wave will come across wave field separation when it reached the slope surface, and it will separate into SV reflected wave of the same type and *P* reflected wave of new type (see Fig. 3.71).

Potential function of SV incident wave, SV reflected wave, and *P* reflected wave is as follows: Potential function of SV incident wave: $\varphi_i = E_s e^{i(wt-k_1x+k_{2s}x_3)}$; potential function of SV incident wave: $\varphi_r = F_p e^{i(wt-k_1x_1-k_{2p}x_3)}$; that of SV reflected wave: $\varphi = E_s e^{i(wt-k_1x_1+k_{2S}x_3)} + F_s e^{i(wt-k_1x_1-k_{2S}x_3)}$; ϕ is the incident angle.

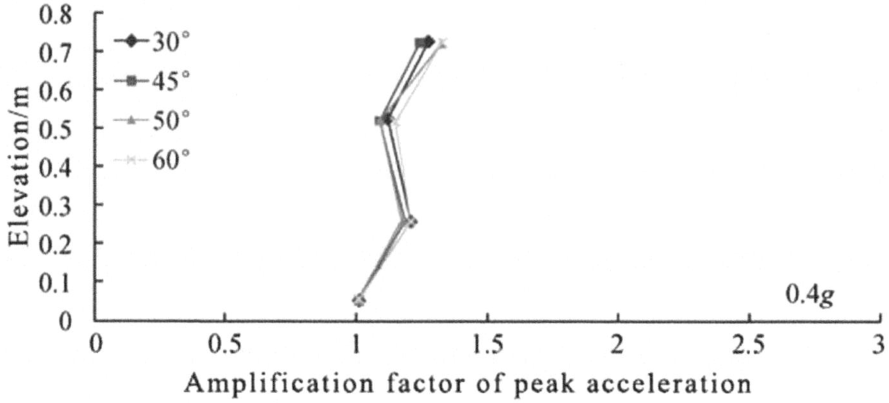

Fig. 3.69 Acceleration peak value elevation amplification coefficient in slope strike

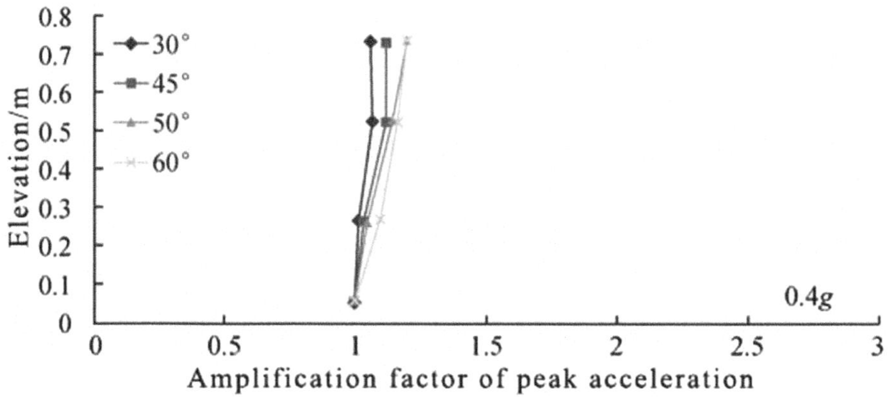

Fig. 3.70 Acceleration peak value elevation amplification coefficient in vertical direction

Fig. 3.71 Wave field separating phenomenon of SV incident wave

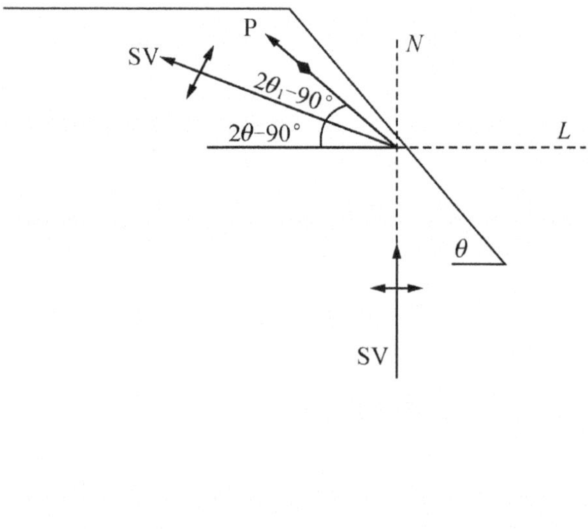

$$k_1 = \frac{\omega \sin \phi}{V_s} \quad k_{2s} = \begin{cases} \sqrt{\left(\frac{w}{V_s}\right)^2 - k_{1s}^2} \\ -i\sqrt{\left(\frac{w}{V_s}\right)^2 - k_{1s}^2} \end{cases} \quad k_{2p} = \begin{cases} \sqrt{\left(\frac{w}{V_p}\right)^2 - k_{1p}^2} & k_{1p} \le \frac{w}{V_p} \\ -i\sqrt{\left(\frac{w}{V_p}\right)^2 - k_{1p}^2} & k_{1p} > \frac{w}{V_p} \end{cases}$$

$$\frac{F_P}{E_S} = \frac{-2V_p^2 \sin 2\phi \cos 2\phi}{V_s^2 \sin 2\theta \sin 2\phi + V_p^2 \cos 2\phi} \quad \frac{F_s}{E_S} = \frac{V_s^2 \sin 2\theta \sin 2\phi - V_p^2 \cos 2\phi}{V_s^2 \sin 2\theta \sin 2\phi + V_p^2 \cos 2\phi}$$

Based on the above function, bring the relevant parameters of 30°, 45°, 50°, and 60° surface into it to solve, the FP/ES ratio get is negative, and the FS/ES ratio is positive, which means the phase position of SV incident wave is the same with SV reflected wave, but opposite to P reflected wave. The further explanation is that the vibration components of SV incident wave and SV reflected wave are the same in *L* and *N* directions, while vibration components of SV incident wave and *P* reflected wave are opposite in *L* and *N* directions. In addition, the vibration components of SV incident wave are Es and 0 in *L* and *N* directions, the vibration components of SV reflected wave are (ES + FS) sin (2θ − 90°) and (ES + FS) cos (2θ − 90°) in *L* and *N* directions, and the vibration components of *P* reflected wave in *L* and *N* directions are Fp cos ($2\theta 1$–90°) and Fp sin ($2\theta 1$–90°), so the total vibration in *L* and *N* directions is Es + (ES + FS) sin (2θ − 90°) + Fp cos ($2\theta 1$–90°) and (ES + FS) cos (2θ − 90°) + Fp sin ($2\theta 1$–90°). Besides, with the increase of incident angle θ, the reflection angle of SV wave increase gradually, and its vibration direction is gradually close to *L* direction, while the vibration component of *P*-wave in *L* direction decreases with its reflection angle $\theta 1$ increasing; i.e. (ES + FS) sin (2θ − 90°) + Fp cos ($2\theta 1$–90°) and (ES + FS) cos (2θ − 90°) + Fp sin ($2\theta 1$–90°) increase gradually. Therefore, with the increase of the bevel angles, the PGA elevation amplification effect in free face direction and vertical direction will strengthen gradually. In line with the above thoughts, the PGA elevation amplification effect in slope strike will come to the same conclusion.

In order to reveal more clearly the influence of bevel angles on PGA in L, M, and N directions, this chapter selects test point of *H*/4 to the slope top as research object, and detailed research results are shown in Figs. 3.72, 3.73, and 3.74.

It can be known from Figs. 3.72, 3.73 and 3.74 that with the increase of bevel angles, the acceleration amplification coefficient in free face direction, slope strike, and vertical direction increases gradually and are distributed in "step form"; i.e. the amplification coefficient of 30°–45° and 50°–60° increased slightly, while that of 45°–50° increased suddenly.

In addition, the changes of PGA amplification coefficient with step heights are as follows: free face direction > vertical direction > slope strike, and with the effects of Wenchuan wolong seismic waves with peak value of 0.1, 0.2, 0.4, and 0.7 g, the PGA amplification coefficient of 50° surface is, respectively, 42.3, 19.7, 19.7, and 13.8% higher of that of 45° surface. The above analysis results illustrate that with the increase of bevel angles, the PGA amplification coefficient did not increase all the same, but with two turning points; i.e. 45° surface is the sudden increase turning

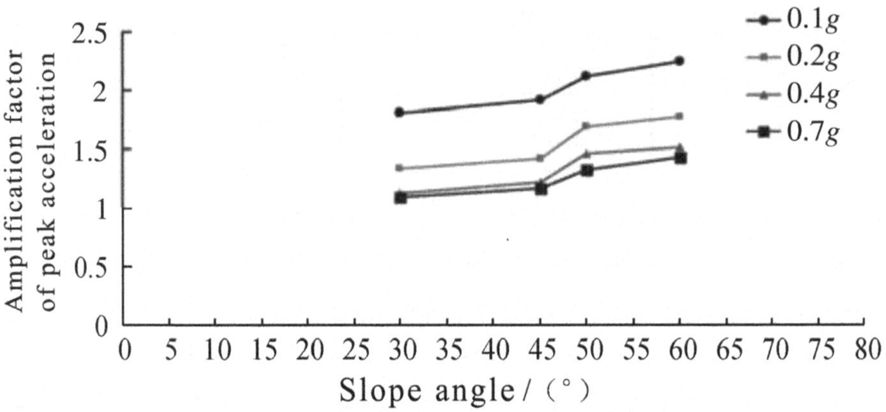

Fig. 3.72 Acceleration peak value elevation amplification coefficient in free face direction

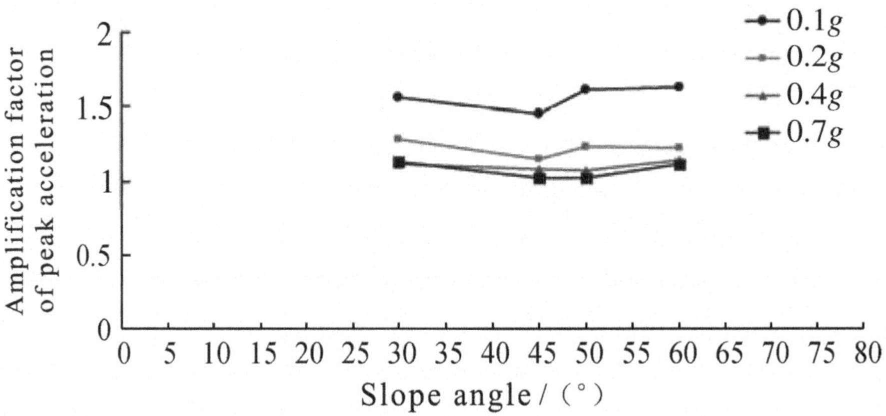

Fig. 3.73 Acceleration peak value elevation amplification coefficient in slope strike

point of PGA amplification coefficient, and 50° surface is the leveling off turning point of PGA amplification coefficient.

Therefore, the seismic dynamic response of rock slopes with bevel angles above 45° will be apparently higher of that below 45°, which is identical with the results of the Wenchuan earthquake damage survey of "*The Landslide Disaster Occurs Mainly in Slopes with Bevel Angles above 40°*". In addition, the above test results will be elaborated from the perspective of theoretical calculation in the fourth chapter "theoretical explanation of time and frequency analysis of two-sided rock slope elevation amplification effects."

From all the above analyses, it can be known that, with the increase of bevel angles, the PGA elevation amplification effects in free face direction and vertical direction will rise gradually, while at the same time, there exist a sudden increase turning point at bevel angles of 45° and a leveling off turning point at 50°. However,

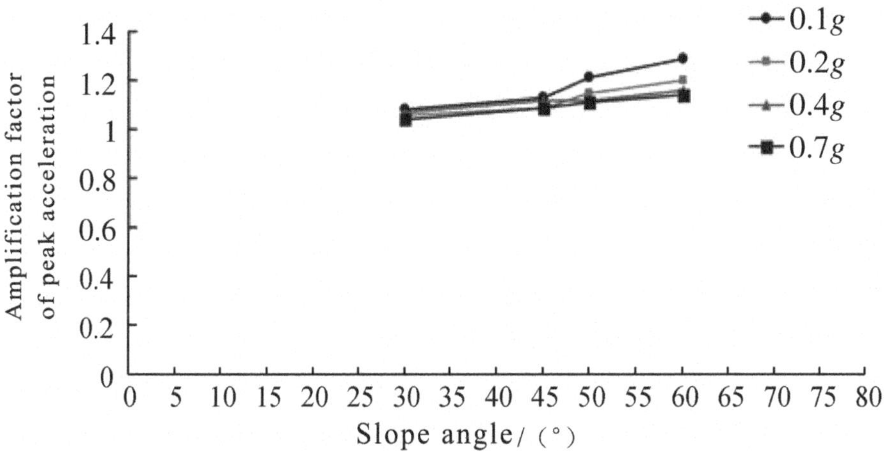

Fig. 3.74 PGA amplification coefficient in vertical direction

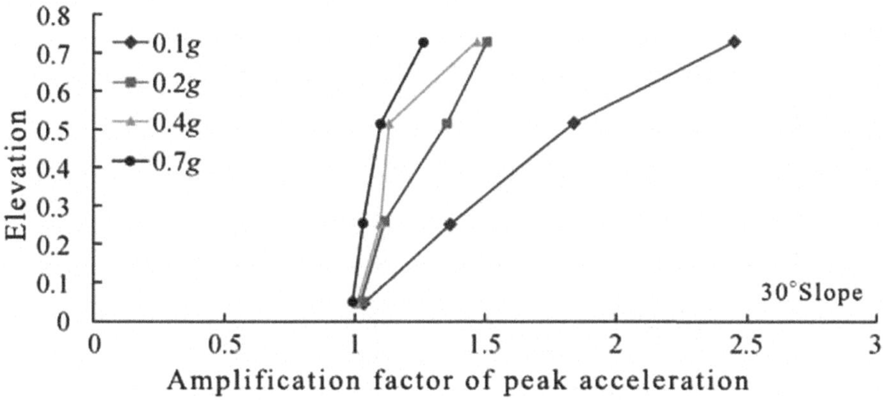

Fig. 3.75 Acceleration peak value elevation amplification coefficient in free face direction

the PGA elevation amplification effects in slope strikes stay unchanged with the increase of bevel angles, and its steps are comparatively gentle.

② Effects of seismic density on the rock slope acceleration peak value elevation amplification effect.

To clearly explain the influence of seismic density on the seismic ground motion acceleration peak value elevation amplification effect, this chapter only selects peak values of test points on *L*, *M*, and *N* directions on slope surface of bevel angles of 30°, 45°, 50°, and 60° under the effect of Wenchuan wolong seismic waves with acceleration peak values of 0.1, 0.2, 0.4, and 0.7 g. See Figs. 3.75, 3.76, 3.77, 3.78, 3.79, 3.80, 3.81, 3.82, 3.83, 3.84, 3.85, and 3.86 for detailed calculation results.

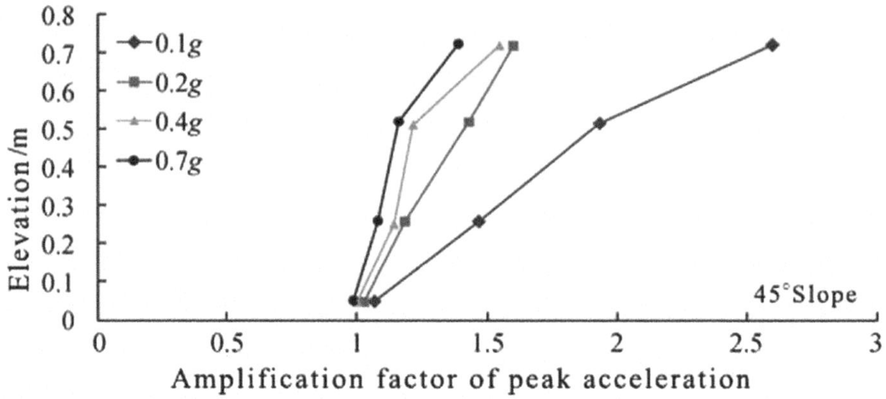

Fig. 3.76 Acceleration peak value elevation amplification coefficient in free face direction

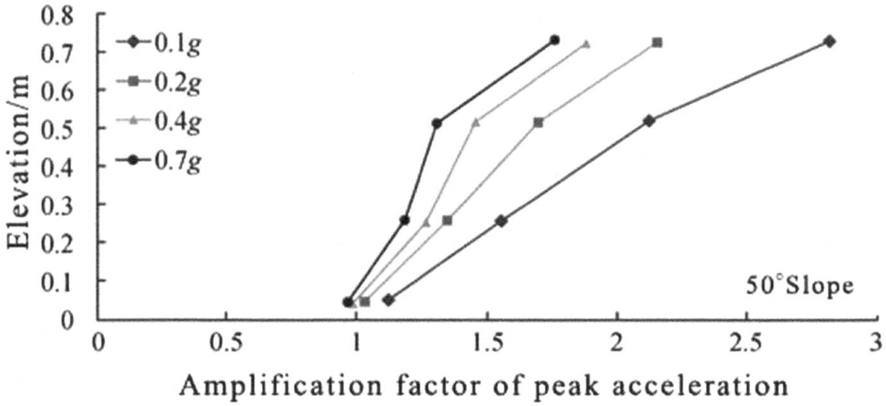

Fig. 3.77 Acceleration peak value elevation amplification coefficient in free face direction

We can see from Figs. 3.75, 3.76, 3.77, 3.78, 3.79, 3.80, 3.81, 3.82, 3.83, 3.84, 3.85, and 3.86 that, under the effect of Wenchuan wolong seismic waves with PGA = 0.1, 0.2, and 0.7 g, the PGA amplification effects of free face direction (*L*), slope strike (*M*), and vertical direction (*N*) on the surfaces of bevel angles of 30°, 45°, 50°, and 60° are identical to the changes of seismic wave peak values, i.e. PGA = 0.1 g > PGA = 0.2 g > PGA = 0.4 g > PGA = 0.7 g. Therefore, with the increase of PGA input, the PGA elevation amplification effect in free face direction, vertical direction, and slope strike will decrease gradually. In order to reveal more clearly the influence of bevel angles on PGA in *L*, *M*, and *N* directions, this chapter selects test point of *H*/4 to the slope top as research object, and detailed research results are shown in Figs. 3.87, 3.88, and 3.89.

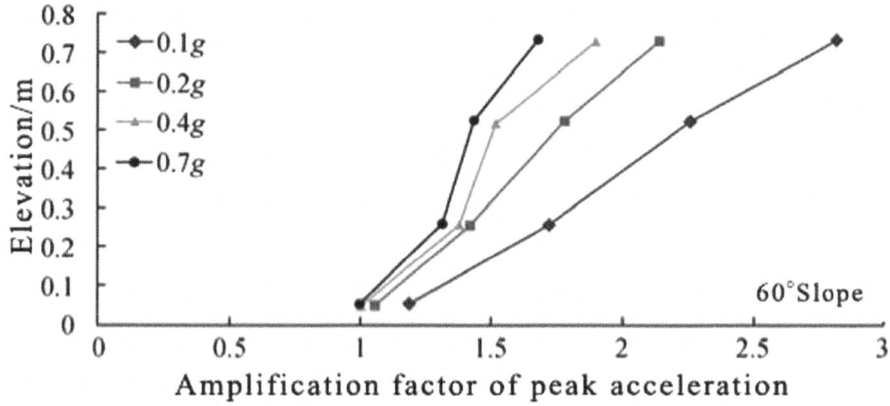

Fig. 3.78 Acceleration peak value elevation amplification coefficient in free face direction

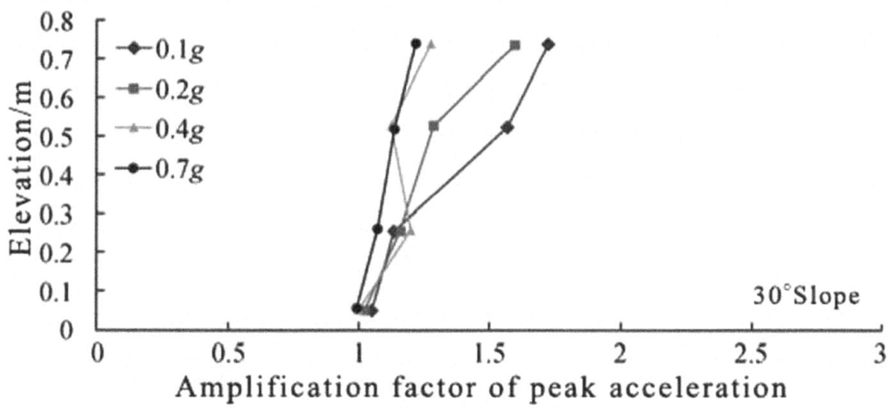

Fig. 3.79 Acceleration peak value elevation amplification coefficient in slope strike

It can be known from Figs. 3.87, 3.88, and 3.89 that, with an increase of input ground motion PGA, the acceleration elevation amplification effect decreases gradually and shows characteristics of magnitude saturation.

Magnitude saturation characteristics mainly refer to that with the input ground motion PGA, the dynamic shear strength and modulus of soils declined, damping ratio of soils increases, nonlinear characteristics strengthen gradually, and energy consumption of seismic waves enlarges, which causes the attenuation of acceleration amplification effect, and the increase of rock and earth mass shock insulation and shock absorption effects.

From the above analysis, we can know that, with an increase of input ground motion PGA, the acceleration elevation amplification effect decreases gradually and shows characteristics of magnitude saturation.

③ Effects of Seismic Wave Types on the PGA Elevation Amplification Effect

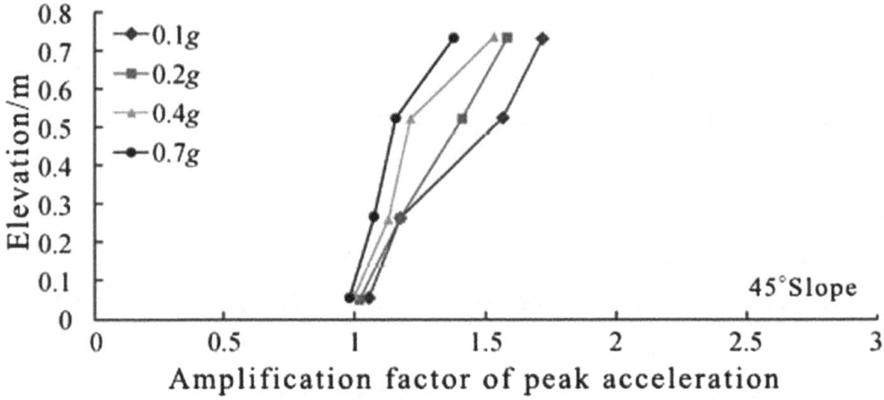

Fig. 3.80 Acceleration peak value elevation amplification coefficient in slope strike

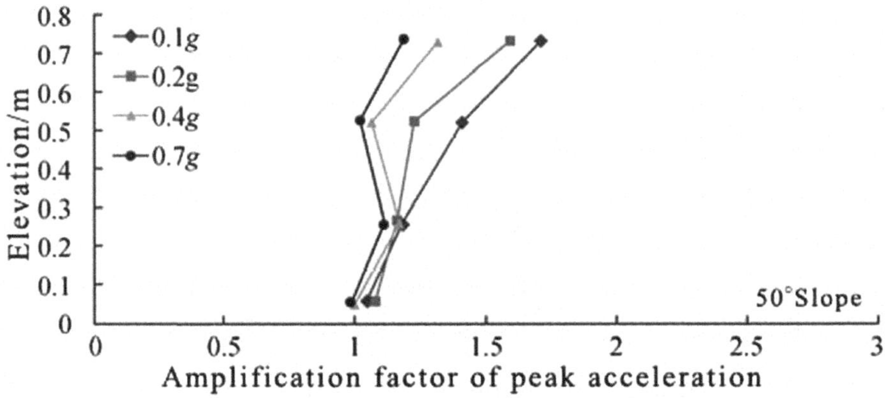

Fig. 3.81 Acceleration peak value elevation amplification coefficient in slope strike

To clearly explain the influence of seismic wave types on the seismic ground motion PGA elevation amplification effect, this chapter only selects peak values of test points on three directions on slope surface of bevel angles of 30°, 45°, 50°, and 60° under the effect of Wenchuan wolong, El Centro, and Kobe seismic waves with acceleration peak values of 0.1 g. See Figs. 3.90, 3.91, 3.92, 3.93, 3.94, 3.95, 3.96, 3.97, 3.98, 3.99, 3.100, and 3.101 for detailed results.

We can see from Figs. 3.90, 3.91, 3.92, 3.93, 3.94, 3.95, 3.96, 3.97, 3.98, 3.99, 3.100, and 3.101 that, under the effect of Wenchuan wolong, El Centro, and Kobe seismic waves with PGA = 0.2 g, the PGA amplification effects of free face direction (*L*), slope strike (*M*), and vertical direction (*N*) on the surfaces of bevel angles of 30°, 45°, 50°, and 60° are identical to the changes of input seismic wave types; i.e. the concrete rules are as follows: El Centro seismic wave > Wenchuan wolong seismic wave > Kobe seismic wave. It also shows that PGA elevation amplification coefficient in vertical direction changes little related to the seismic wave-type changes. The

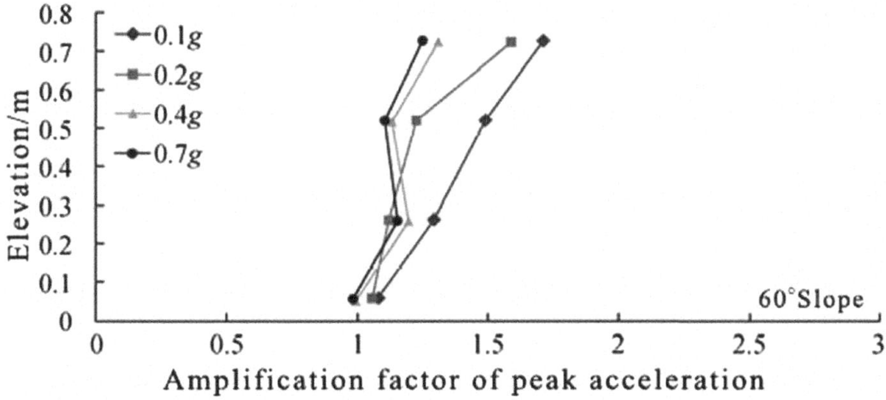

Fig. 3.82 Acceleration peak value elevation amplification coefficient in slope strike

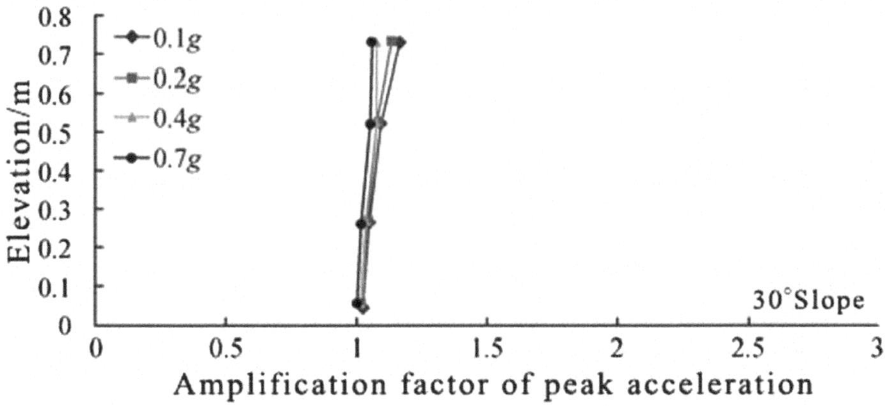

Fig. 3.83 Acceleration peak value elevation amplification coefficient in vertical direction

above results may be due to that the dominant frequencies of El Centro seismic wave in three directions are, respectively, 4.6 Hz (*X* direction), 4.49 Hz (*Y* direction), and 6.5 Hz (*Z* direction), that of Kobe seismic wave are, respectively, 6.84 Hz (*X* direction), 6.60 Hz (*Y* direction), and 8.09 Hz (*Z* direction), and that of Wenchuan wolong seismic wave are, respectively, 4.24 Hz (*X* direction), 3.56 Hz (*Y* direction), and 6.23 Hz (*Z* direction). Using mode analysis method, we can get that the first natural frequency of the test model is 4.5 Hz, and the dominant frequency of El Centro seismic wave on *X* and *Y* directions is the closest to it, so it is easy to cause resonance. The second close to it is Wenchuan wolong seismic wave, and the last one is Kobe seismic wave. In addition, the dominant frequency of all three kinds of seismic waves in vertical direction differentiates a lot to the first natural frequency of the test model. Therefore, the seismic wave types have little influence on the PGA amplification effect in vertical direction.

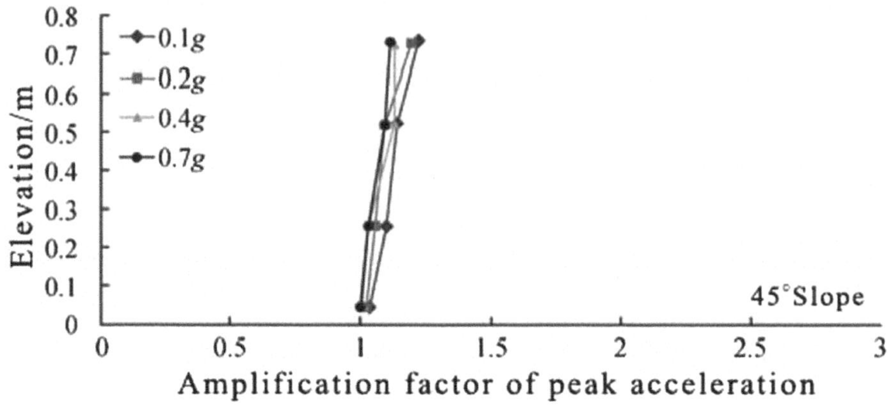

Fig. 3.84 Acceleration peak value elevation amplification coefficient in vertical direction

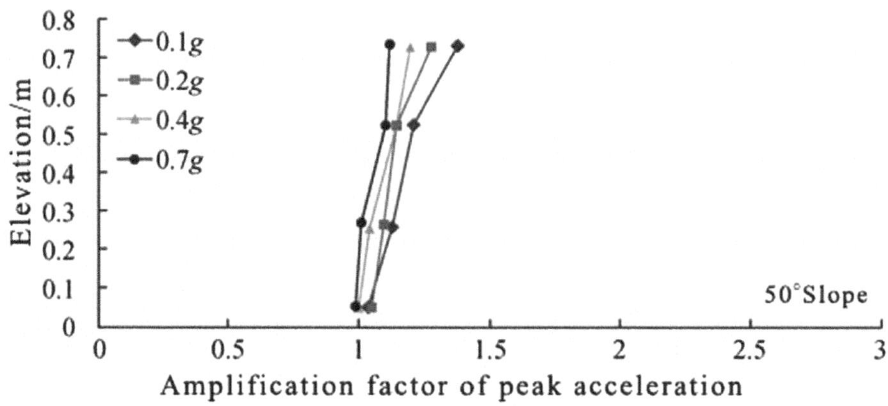

Fig. 3.85 Acceleration peak value elevation amplification coefficient in vertical direction

While it is noteworthy that the El Centro, Wenchuan wolong, and Kobe seismic wave used in this chapter are all acceleration versus time data of same-elevation free-field test points at the bottom of the slopes, but not the input seismic waves at the bottom of shaking model test model, which is mainly because the spectral components of seismic ground motion get changed when it transmitted from the model container bottom through the soil layer of the slope bottom field upward, and what has real influence on slope dynamic response should be seismic waves directly input through the slope bottom, but not seismic ground motion input through the bottom of model container. Besides, this experiment did not set up monitoring points at the bottom of slope bottoms, but at the free field of the same elevation. Therefore, this chapter selects acceleration versus time data of same-elevation free-field test points at the bottom of the slopes to analyze.

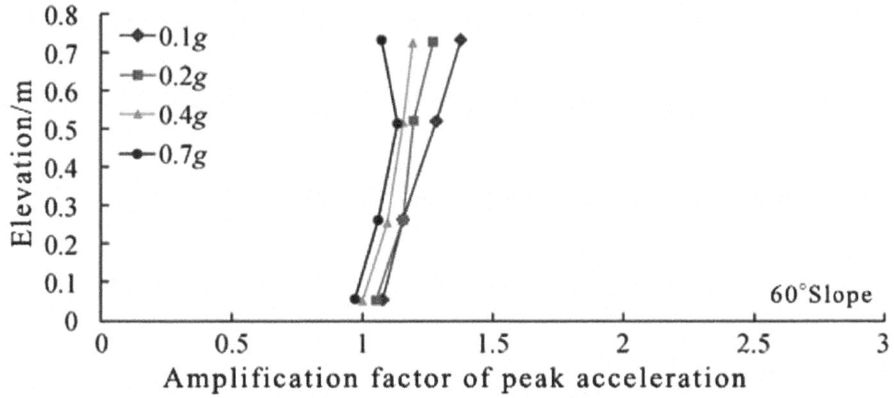

Fig. 3.86 Acceleration peak value elevation amplification coefficient in vertical direction

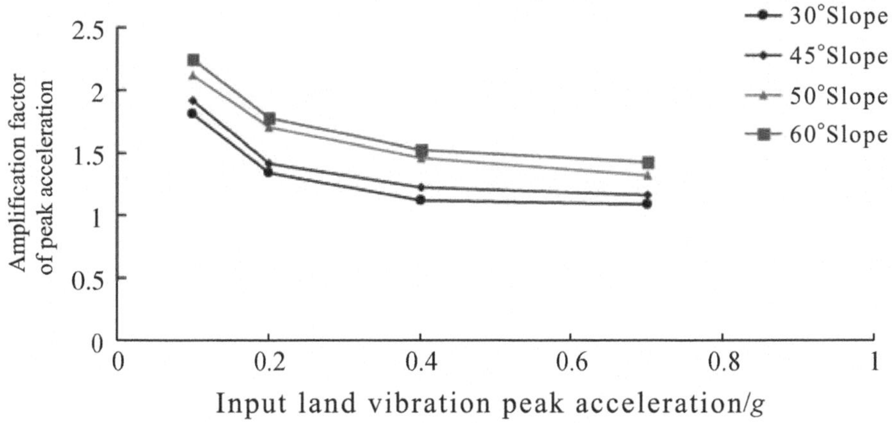

Fig. 3.87 Acceleration peak value elevation amplification coefficient in free face direction

Based on all the above analysis, the seismic wave types have significant importance on rock slope acceleration elevation amplification effects, which is mainly exemplified in the relation between the dominant frequency of seismic waves and the natural frequency of slopes.

④ Influence of Seismic Ground Motion Response Direction PGA Elevation Amplification Effect.

To clearly explain the influence of seismic ground motion direction on the PGA elevation amplification effect, this chapter only selects peak values of test points on *L*, *M*, and *N* directions on slope surface of bevel angles of 30°, 45°, 50°, and 60° under the effect of Wenchuan wolong seismic waves with acceleration peak values of 0.1, 0.2, and 0.4 g. See Figs. 3.102, 3.103, 3.104, 3.105, 3.106, 3.107, 3.108, 3.109, 3.110, 3.111, 3.112, and 3.113 for detailed results.

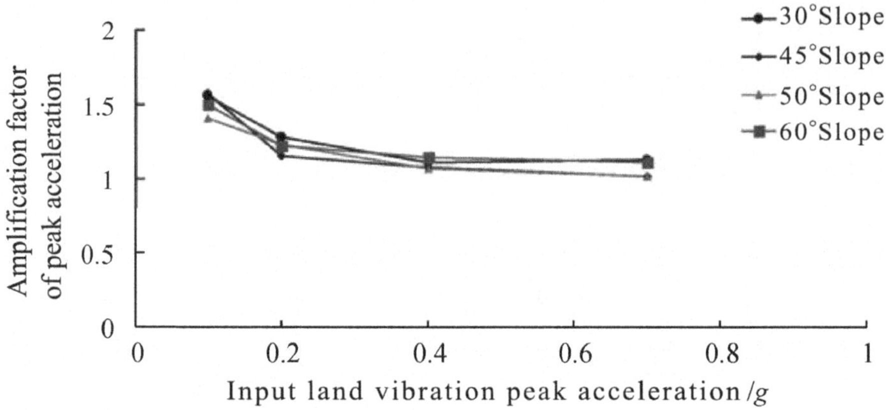

Fig. 3.88 Acceleration peak value elevation amplification coefficient in slope strike

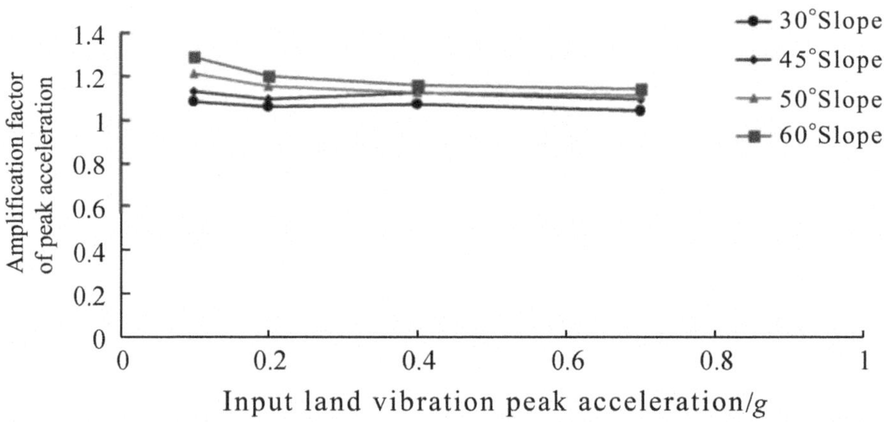

Fig. 3.89 Acceleration peak value elevation amplification coefficient in vertical direction

We can see from Figs. 3.102, 3.103, 3.104, 3.105, 3.106, 3.107, 3.108, 3.109, 3.110, 3.111, 3.112, and 3.113 that, under the effect of Wenchuan wolong seismic waves with PGA = 0.1, 0.2, and 0.4 g, the PGA amplification effects of free face direction (*L*), slope strike (*M*), and vertical direction (*N*) on the surfaces of bevel angles of 30°, 45°, 50°, and 60° are basically identical, i.e. free face direction > slope strike > vertical direction, which shows apparent "free face direction amplification effects." The reason for the above results may be that the seismic waves came across reflection and refraction and formed complex wave field which caused larger acceleration.

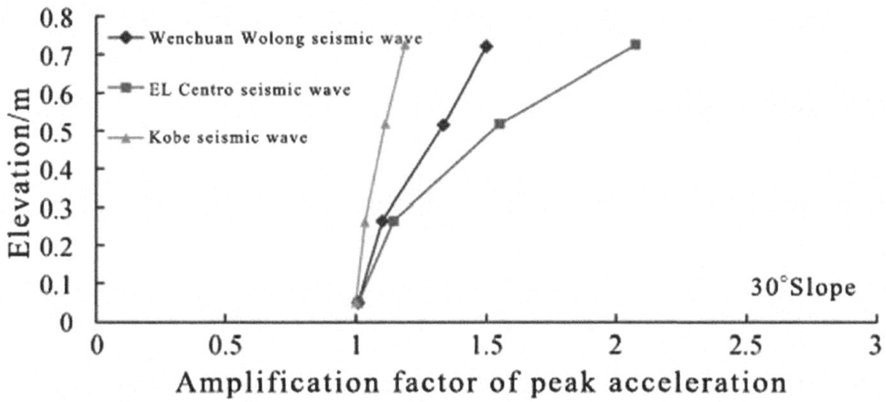

Fig. 3.90 Acceleration peak value elevation amplification coefficient in free face direction

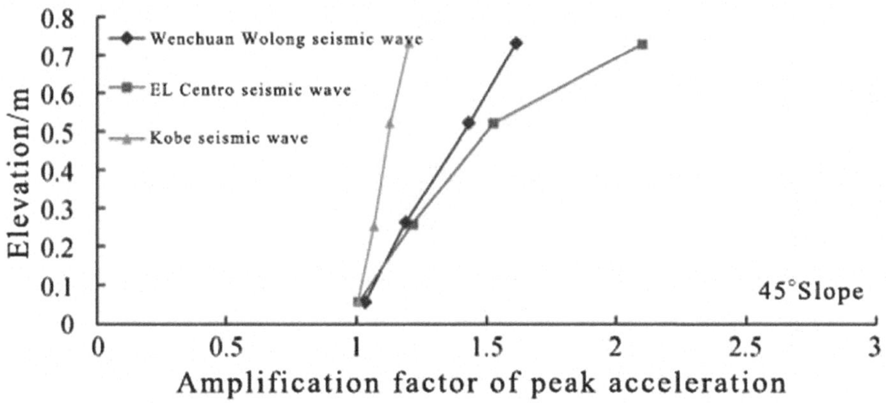

Fig. 3.91 Acceleration peak value elevation amplification coefficient in free face direction

3.2.2.2 Rock Slope Acceleration Fourier Spectrum and Response Spectrum Change Rules Along Elevation

According to analysis and summaries of macroscopic seismic damage experience and apparatus observing data, the main characteristics of seismic ground motion, for engineering seismology, also can be described by frequency spectrum and duration besides amplitude. The curves that show the relation between amplitude and frequency or period in an earthquake are collectively known as frequency spectrum, which mainly includes Fourier spectrum and response spectrum. Lots of research results demonstrate that the spectrum components of rock slope seismic ground motion acceleration change with the slope height elevation [90–100]. Therefore, this shaking table test arranged large amount of monitoring points to study on the frequency spectrum characteristic change rules of rock slopes along elevation.

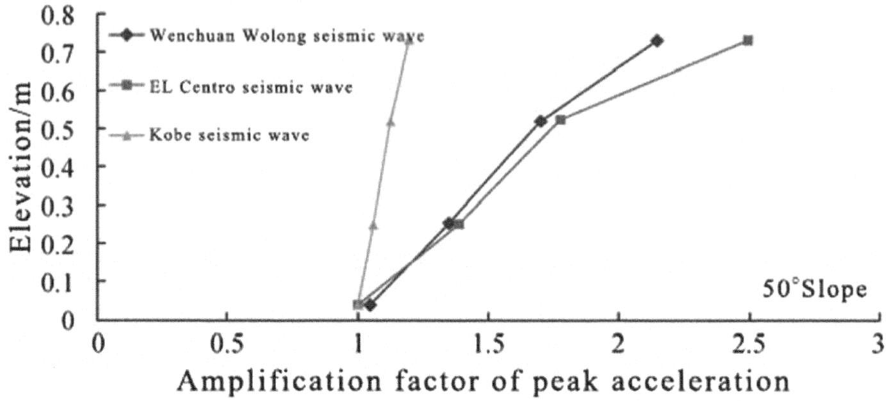

Fig. 3.92 Acceleration peak value elevation amplification coefficient in free face direction

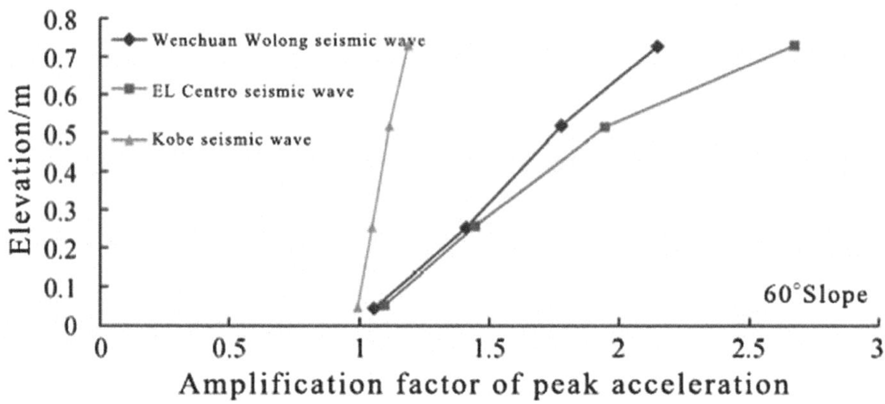

Fig. 3.93 Acceleration peak value elevation amplification coefficient in free face direction

① Rock Slope Acceleration Fourier Spectrum Change Rules along Elevation

Due to lack of space, this chapter only selects test points on 45° surface in L, M, and N directions under effects of Wenchuan wolong seismic waves with PGA = 0.2 g to analyze. There set up four test points on the 45° surface with respective number of 4#, 10#, 16#, and 22#, and their vertical distance to slope foot are, respectively, 0.05, 0.26, 0.52, and 0.73 m. See Figs. 3.114, 3.115, 3.116, 3.117, 3.118, 3.119, 3.120, 3.121, 3.122, 3.123, 3.124, 3.125, and 3.126 for Fourier spectrum calculation results of all test points.

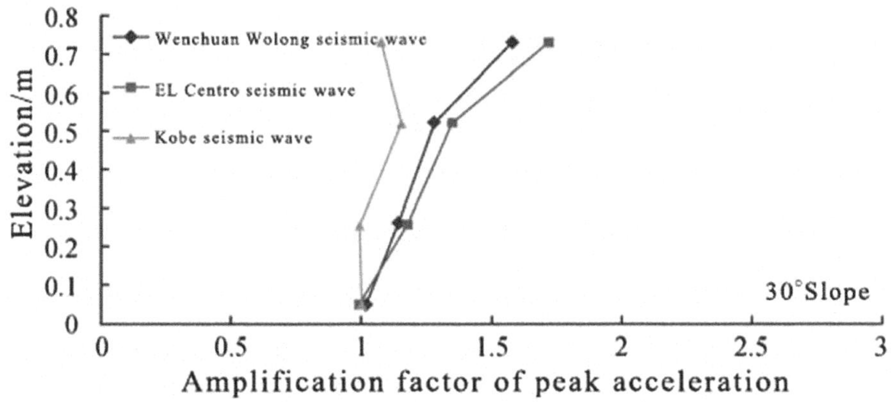

Fig. 3.94 PGA elevation amplification coefficient in slope strike

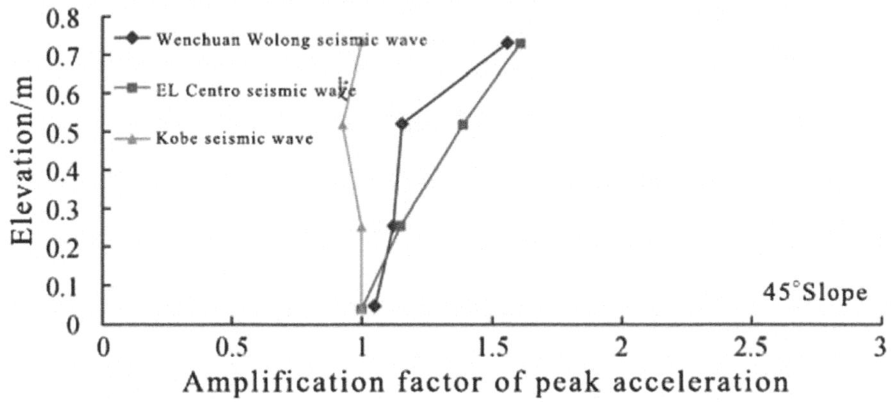

Fig. 3.95 PGA elevation amplification coefficient in slope strike

It can be known from Figs. 3.114, 3.115, 3.116, and 3.117 that the dominant frequencies of Test Points 4#, 10#, 16#, and 22# on 45° surface under effects of Wenchuan wolong seismic waves with PGA = 0.2 g concentrate in the range of 0.0–1.0 Hz and 3.69–5.48 Hz, and with the increase of elevation, the former gradually weakened and the latter gradually strengthened. Figures 3.118, 3.119, 3.120, and 3.121 show that the dominant frequencies of Test Points 4#, 10#, 16#, and 22# concentrate in the range 3.3–3.6 Hz, and with the increase of elevation, other frequency components gradually weakened. Figures 3.122, 3.123, 3.124, and 3.125 show that the dominant frequencies of Test Points 4#, 10#, 16#, and 22# in vertical direction concentrate in the range of 4.8–5.0 Hz and 6.1–6.4 Hz, and with the increase of elevation, the former gradually strengthened and the latter gradually weakened. The reason for the above phenomenon lies in that the first natural frequency of the model test model is 4.5 Hz, which enlarges the surrounding frequency components. This

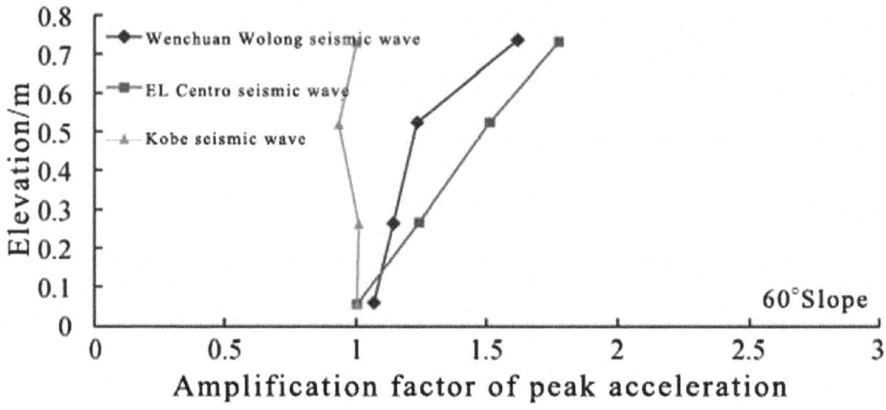

Fig. 3.96 PGA elevation amplification coefficient in slope strike

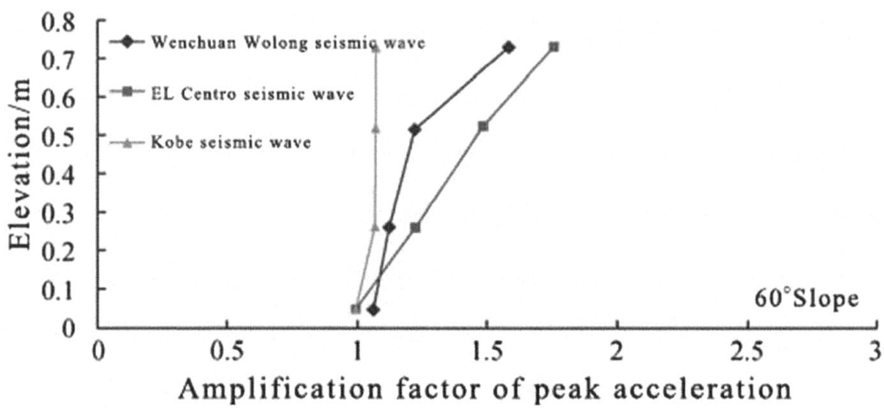

Fig. 3.97 PGA elevation amplification coefficient in slope strike

is identical with the conclusion of the normally said "the slope soil has filter effects of the high-frequency components of seismic waves and has amplification effects for the low-frequency components" [147]. This is because, for normal slope soil, the natural frequency f is small, which has amplification effects for frequency range around $f \pm \Delta f$ in seismic waves and has filter effects for other frequency ranges.

From the above analysis, we can know that the changes of Fourier spectrum frequency components of acceleration in free face direction (L), slope strike (M), and vertical direction (N) along elevation are basically identical; i.e. with the increase of elevation, the slope soil has significant amplification effects for frequency components around its natural frequency range and has filter effects for other frequency ranges.

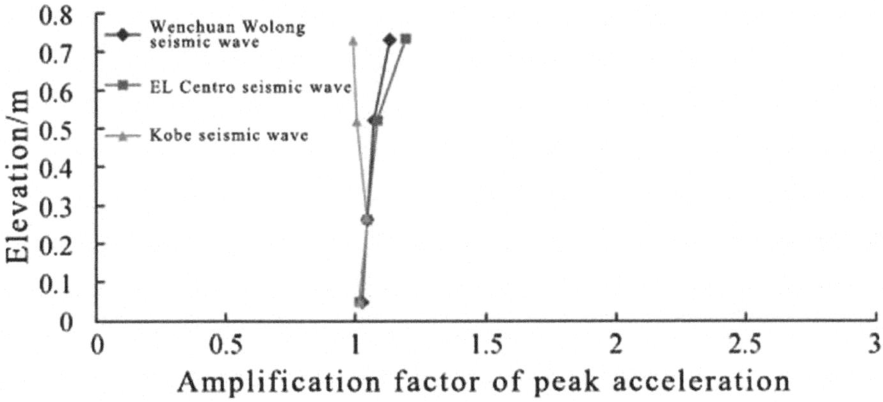

Fig. 3.98 PGA elevation amplification coefficient in vertical direction

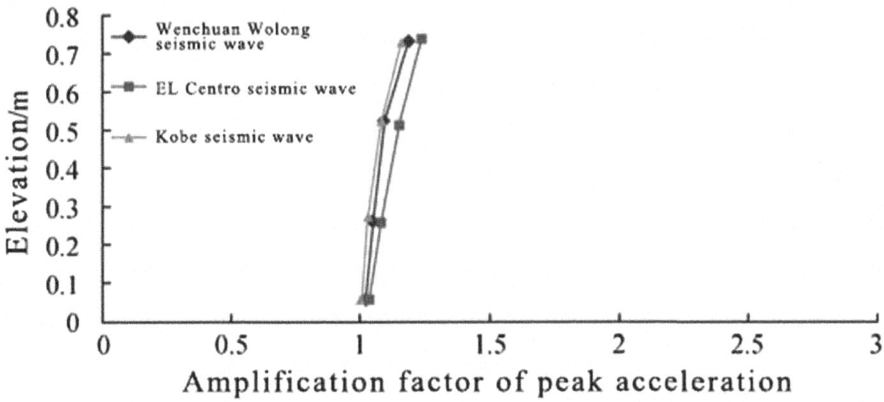

Fig. 3.99 PGA elevation amplification coefficient in vertical direction

3.2.2.3 Rock Slope Acceleration Response Spectrum Change Rules Along Elevation

The shape of rock slope acceleration response spectrum has an important reference value for recognition of rock slope dynamic response characteristics and for the design of rock slope control. In order to study on change rules of rock slope acceleration response spectrum along elevation, this chapter only selects test points on 45° surface in *L*, *M,* and *N* directions under effects of Wenchuan wolong seismic waves with PGA = 0.2 g to analyze. There set up four test points on the 45° surface with respective number of 4#, 10#, 16#, and 22#, and their vertical distance to slope foot are, respectively, 0.05, 0.26, 0.52, and 0.73 m. At the same time, in order to more clearly state the elevation amplification effects of different periodic spectrum values in the response spectrum, this chapter takes the three-direction response spectrum

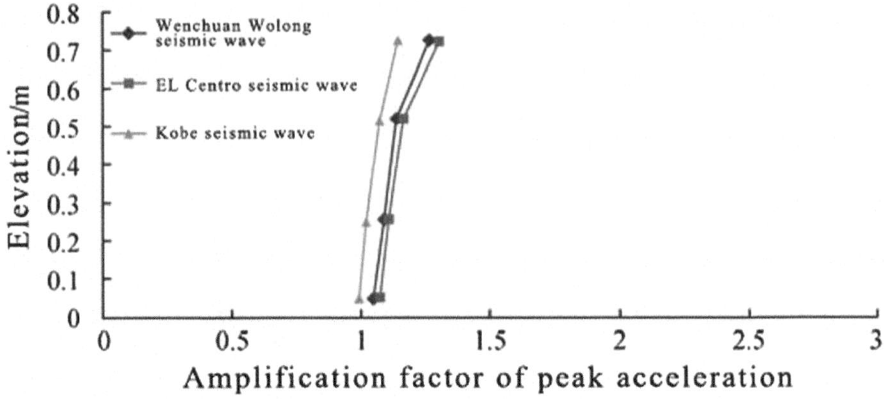

Fig. 3.100 PGA elevation amplification coefficient in vertical direction

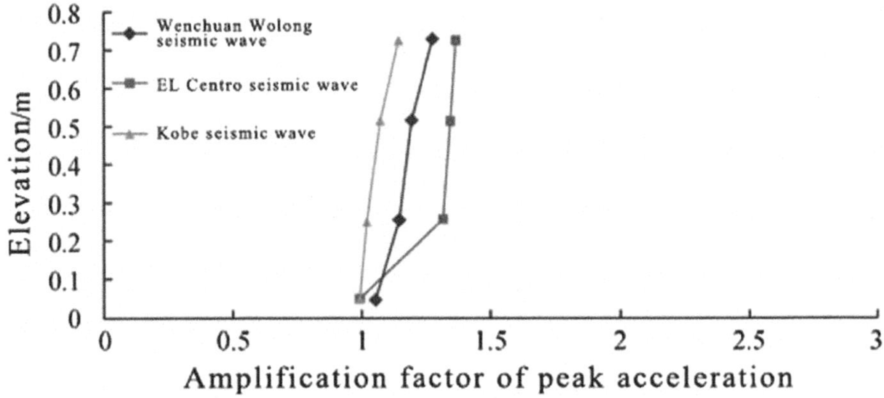

Fig. 3.101 PGA elevation amplification coefficient in vertical direction

of Test Point 4# at the bottom of the slopes as basis to work out the acceleration response spectrum amplification coefficient of Test Points 10#, 16#, and 22#. See Figs. 3.127, 3.128, 3.129, 3.130, 3.131, and 3.132 for detailed calculation results. It is noteworthy that this acceleration response spectrum calculation adopts damping ratio of 5% which is commonly used by engineering.

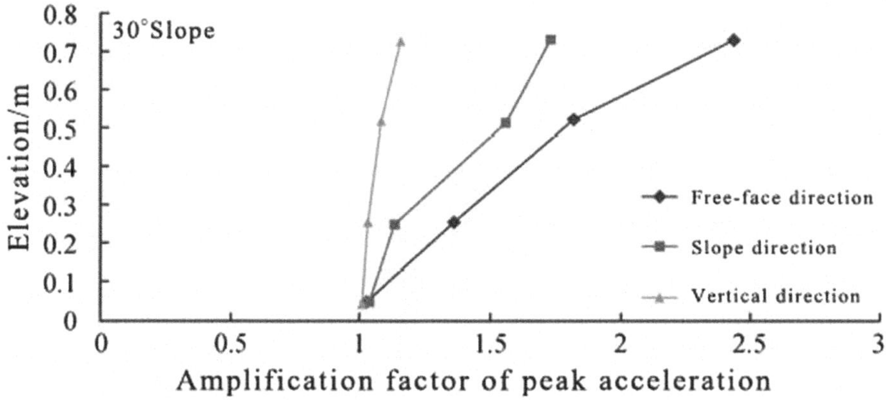

Fig. 3.102 PGA elevation amplification coefficient (PGA = 0.1 g)

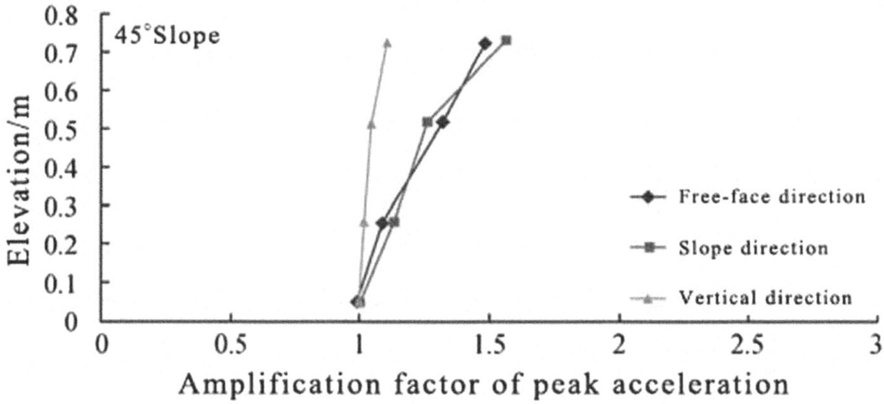

Fig. 3.103 PGA elevation amplification coefficient (PGA = 0.1 g)

It can be known from Figs. 3.127 and 3.128 that the shape of acceleration response spectrum along elevation in free face direction on 45° surface under the effects of Wehchuan wolong seismic waves with PGA = 0.2 g is basically consistent and has peak features. The predominant period concentrates at about $T = 0.22$ s (the corresponding frequency is $f = 4.54$ Hz). The response spectrum amplitude corresponding to short period ($T = 0.0$–0.5 s) has certain amplification effects with amplification coefficient between 1.0 and 4.0, while as for other T periods, especially at long period (low frequency), it has certain reduction effects with amplification coefficient between 0.93 and 1.18. Figures 3.129 and 3.130 show that the shape of acceleration response spectrum along elevation in slope strike direction is basically consistent, and its peak features are not obvious as in free face direction. The predominant period concentrates at about $T = 0.28$ s (the corresponding frequency is $f = 3.57$ Hz). The response spectrum amplitude corresponding to short period ($T = 0.0$–0.8 s) has cer-

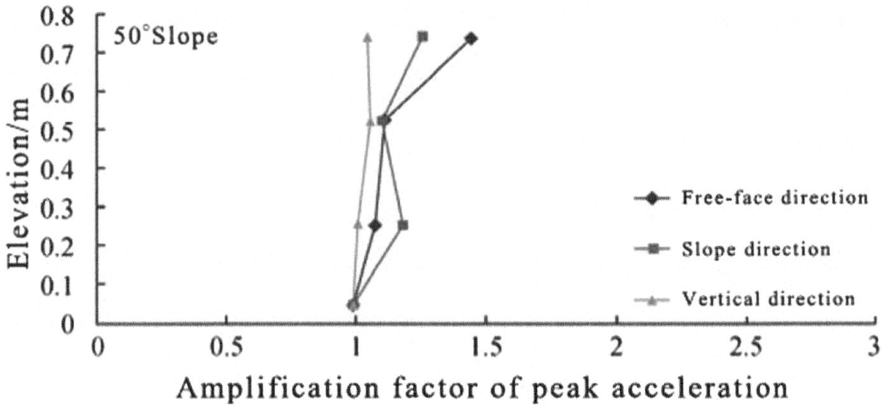

Fig. 3.104 PGA elevation amplification coefficient (PGA = 0.1 g)

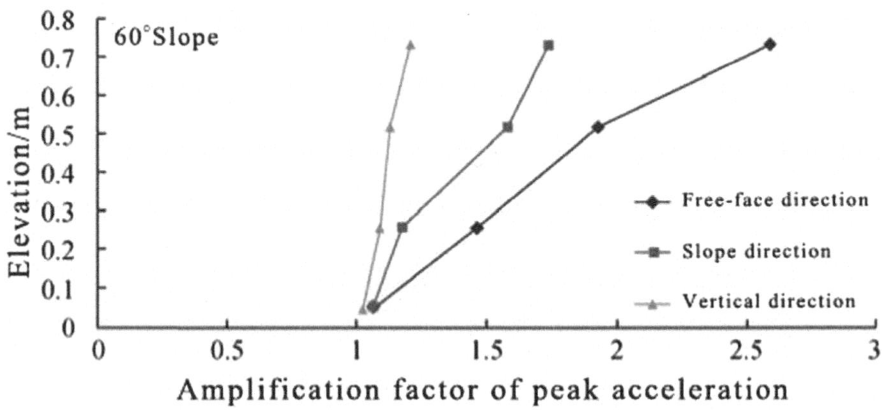

Fig. 3.105 PGA elevation amplification coefficient (PGA = 0.1 g)

tain amplification effects with amplification coefficient between 1.0 and 2.5, while as for other T periods, especially at long period (low frequency), it has certain reduction effects with amplification coefficient between 0.76 and 1.05. Figures 3.131 and 3.132 demonstrate that the shape of acceleration response spectrum along elevation in vertical direction has double peak phenomena. The predominant period concentrates at about $T = 0.22$ s (the corresponding frequency is $f = 4.54$ Hz). The response spectrum amplitude corresponding to short period ($T = 0.0$–0.3 s) has slight amplification effects, while as for other T periods, especially at long period (low frequency), the amplification coefficient fluctuates at about 1. The reason for the above phenomenon may lie in that the first natural frequency of the model test model is 4.5 Hz.

It can be known from the above analysis that for the short period, especially around predominant period, the amplification response spectrum in free face direction is stronger than that in slope strike direction, and the amplification response spectrum

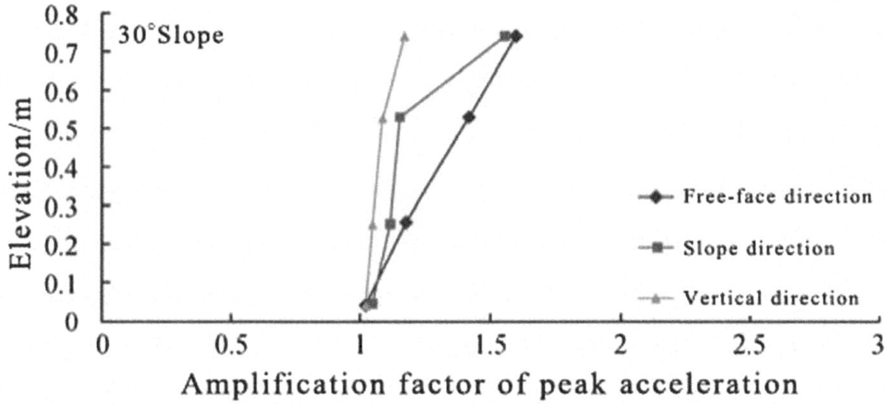

Fig. 3.106 PGA elevation amplification coefficient (PGA = 0.2 g)

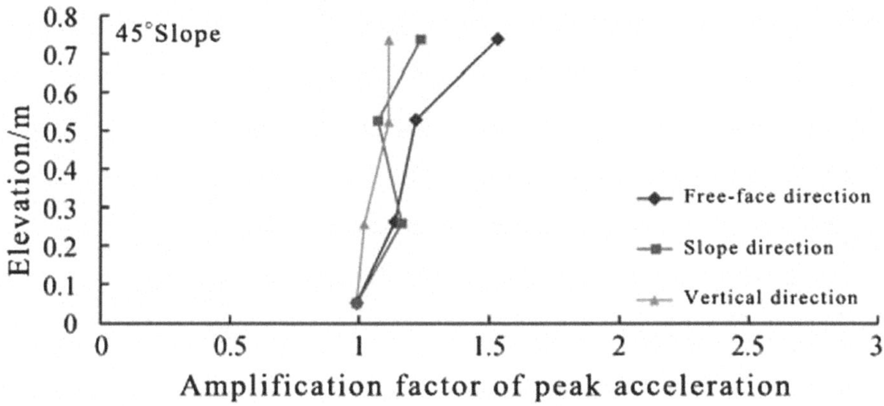

Fig. 3.107 PGA elevation amplification coefficient (PGA = 0.2 g)

in vertical direction is the weakest. For the long period, the amplification response spectrum along elevation has certain reduction effects in all three directions.

3.2.3 *Brief Summary*

This chapter analyzed the monitored acceleration data under different seismic working conditions, discussed the influence rules of bevel angles, seismic intensity, and seismic ground motion wave types on rock slope acceleration response from the perspective of PGA of slopes, Fourier spectrum, and response spectrum, and came to the following conclusions:

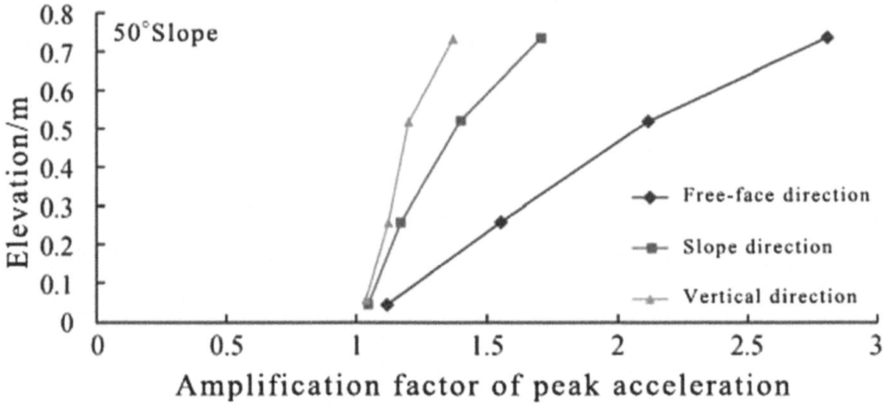

Fig. 3.108 PGA elevation amplification coefficient (PGA = 0.2 g)

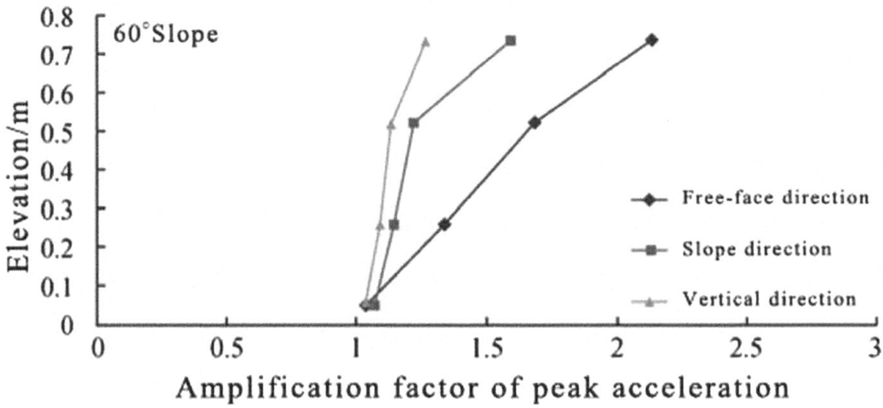

Fig. 3.109 PGA elevation amplification coefficient (PGA = 0.2 g)

(1) For mountains containing four slope surfaces, if we analyze with traditional method, i.e. using *X*, *Y*, and *Z* three-direction coordinate system to analyze PGA elevation amplification effects, the results may be unilateral; i.e. *X* direction is the free face direction of one slope surface and slope strike direction of another one, which cannot clearly reflect the influence of bevel angles on PGA elevation amplification effects on each direction. Therefore, when studying on the slope dynamic problems, it is suggested to use the three-dimensional local coordinate system, i.e. the free face direction, slope strike direction, and vertical direction of slopes.

(2) With the increase of bevel angles, the PGA elevation amplification effects in vertical direction will rise gradually, while at the same time, there exists a sudden increase turning point at bevel angles of 45° and a leveling off turning point at 50°. However, the PGA elevation amplification effects in slope strikes

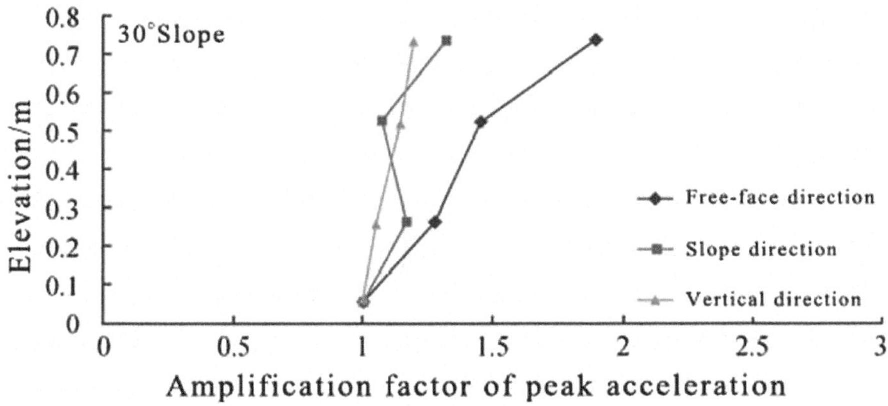

Fig. 3.110 PGA elevation amplification coefficient (PGA = 0.4 g)

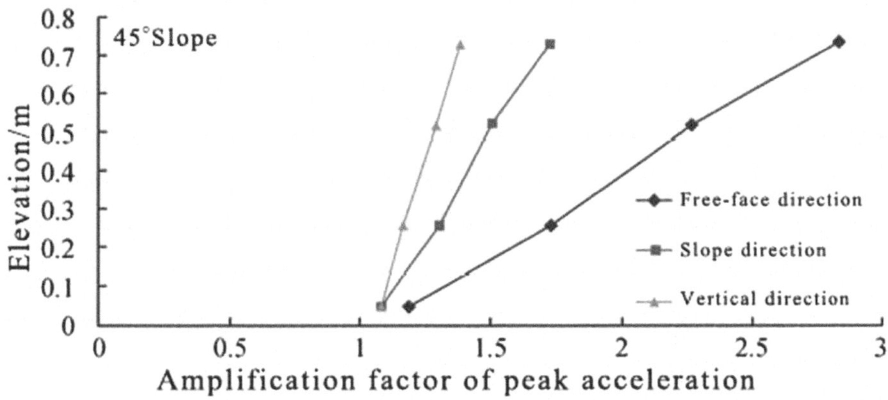

Fig. 3.111 PGA elevation amplification coefficient (PGA = 0.4 g)

stay unchanged with the increase of bevel angles, and its steps are comparatively gentle.

(3) With the increase of seismic ground motion PGA, the peak acceleration elevation amplification effect decreases gradually; slope strike direction and vertical direction reduce gradually and show characteristics of magnitude saturation. Under the effects of El Centro, Wenchuan wolong, and Kobe seismic waves, the rules of peak acceleration elevation amplification effect in free face direction, slope strike direction, and vertical direction are as follows: El Centro seismic wave > Wenchuan wolong seismic wave > Kobe seismic wave, and free face direction > slope strike direction > vertical direction.

(4) The changes of Fourier spectrum frequency components of acceleration in free face direction (*L*), slope strike (*M*), and vertical direction (*N*) along elevation have certain regularity; i.e. with the increase of elevation, the slope soil has

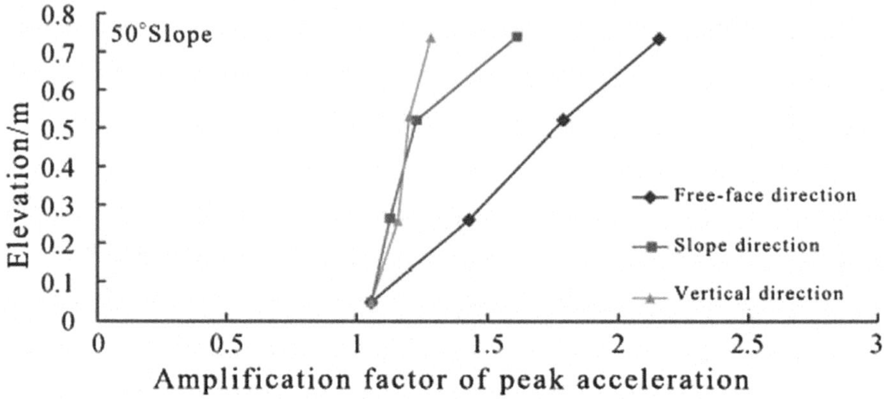

Fig. 3.112 PGA elevation amplification coefficient (PGA = 0.4 g)

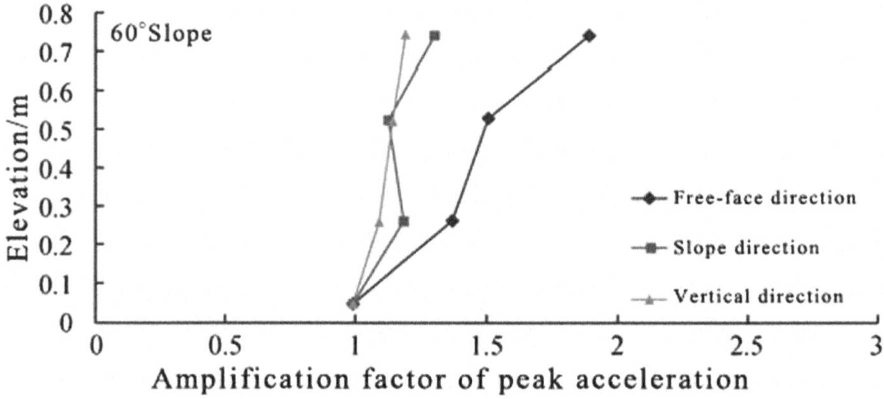

Fig. 3.113 PGA elevation amplification coefficient (PGA = 0.4 g)

significant amplification effects for frequency components around its natural frequency range f and has filter effects for other frequency ranges.

For the short period, especially around predominant period, the amplification response spectrum in free face direction is stronger than that in slope strike direction, and the amplification response spectrum in vertical direction is the weakest. For the long period, the amplification response spectrum along elevation has certain reduction effects in all three directions. In addition, it has obvious peak phenomena for free face direction, while for slope strike direction and vertical direction, the peak phenomena are less obvious.

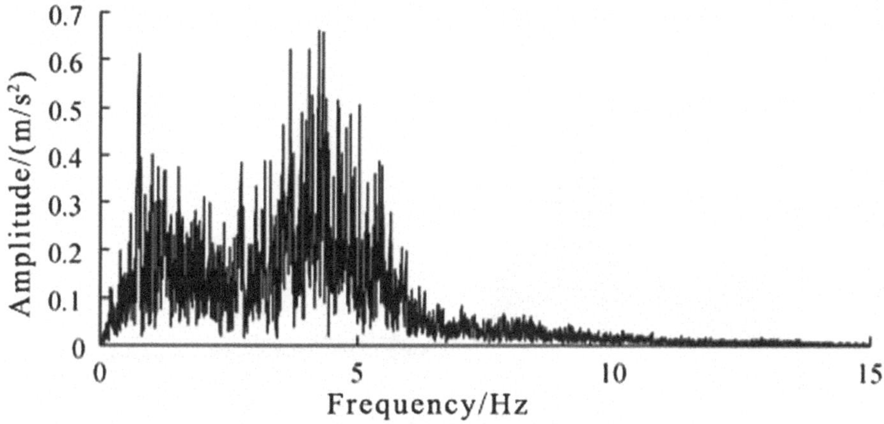

Fig. 3.114 Fourier spectrum of acceleration in free face direction (4#)

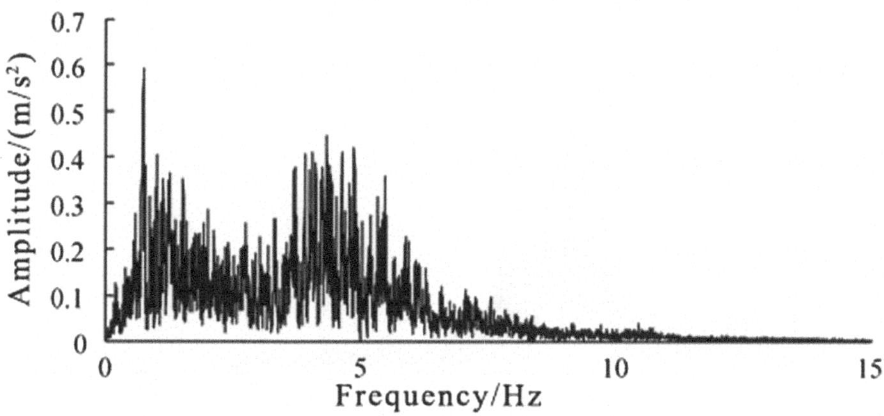

Fig. 3.115 Fourier spectrum of acceleration in free face direction (10#)

3.3 Numerical Analysis Research of Rock Slope Seismic Ground Motion Characteristics

The field monitoring results of rock slope seismic array and shaking table test results of rock slopes in Xishan Park, Zigong, reveal the existence of rock slope acceleration elevation amplification effects, and we come to some valuable conclusion through researches on the influence of different aspects of bevel angles, seismic density, seismic wave types, and seismic ground motion direction on rock slope seismic ground motion characteristics. While the slopes in Xishan Park, Zigong, and rock slopes used in the shaking table test are all the independent slopes containing four surfaces, in reality, there exists large amount of continuous slopes of dozens of meters, hundreds of meters, or even much longer, so there must be difference in the dynamic character-

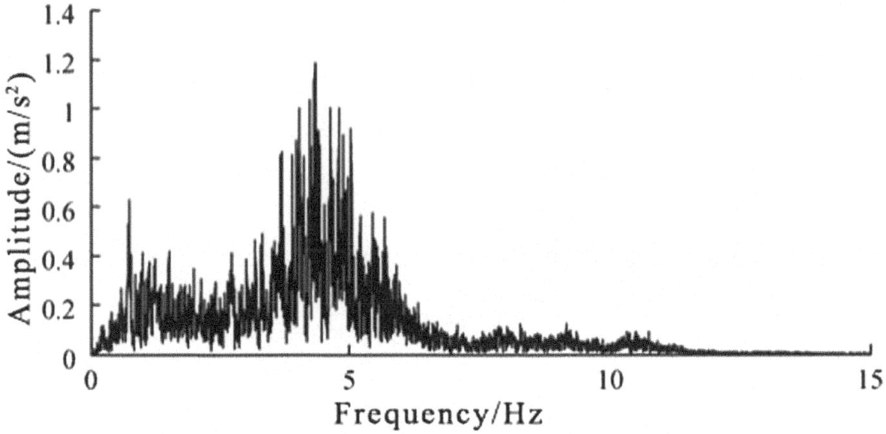

Fig. 3.116 Fourier spectrum of acceleration in free face direction (16#)

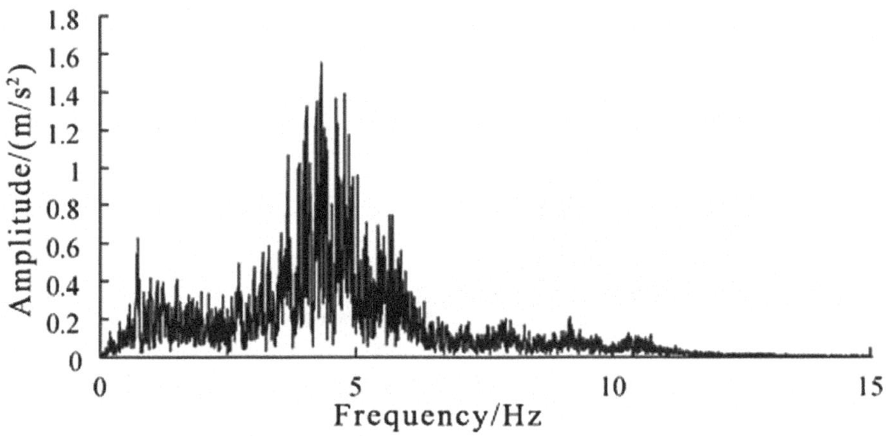

Fig. 3.117 Fourier spectrum of acceleration in free face direction (22#)

istics in reality and the seismic ground motion characteristics of independent slopes. Therefore, this chapter will, based on the previous research results, generally analyzes the model and adopts the new numerical calculation software—GDEM—to further analyze the seismic response rules of rock slopes focusing on the above slope types and to achieve systematic study of field array monitoring (macro-phenomena), shaking model test (further research on regularity), data simulation (micro-complement) so as to enrich research contents and expand research results.

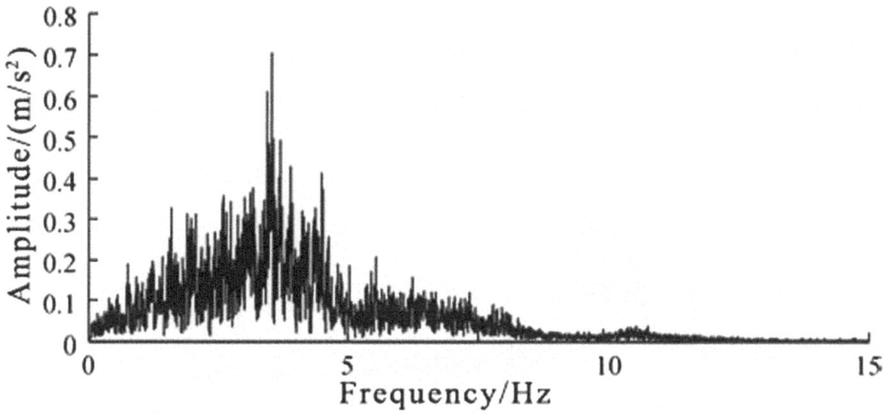

Fig. 3.118 Fourier spectrum of acceleration in slope strike direction (4#)

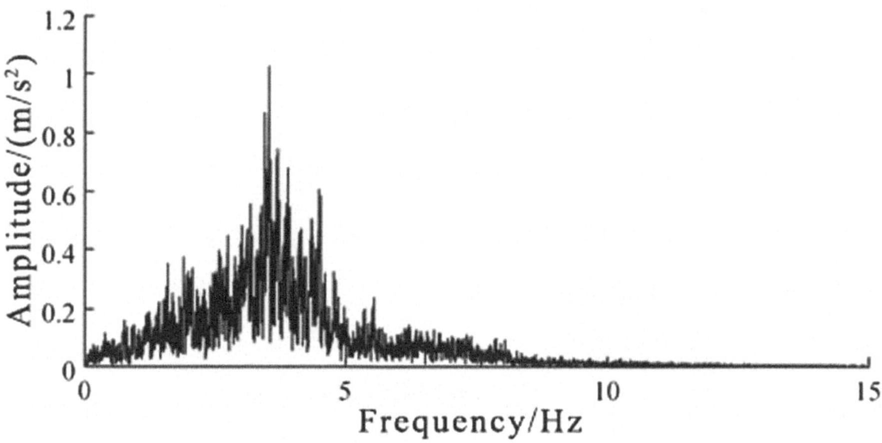

Fig. 3.119 Fourier spectrum of acceleration in slope strike direction (10#)

3.3.1 Brief Introduction of GDEM

CDEM is shorted for Continuum-based Discrete Element Method, which is a new numerical analysis method in recent years developed by Mechanical Institute of Chinese Academy of Sciences joint laboratory of non-continuum mechanics and engineering disaster. This method makes finite element method and discrete element method coupling together, i.e. using FEM inside the slope and DEM at the boundary of the slope. Through cracks of internal parts and boundary of slopes, it cannot only simulate the deformation and motion features of materials in continuous state and non-continuous state, but also can achieve the gradual damage course of materials from continuum to non-continuum. GPU is the abbreviation of graphics

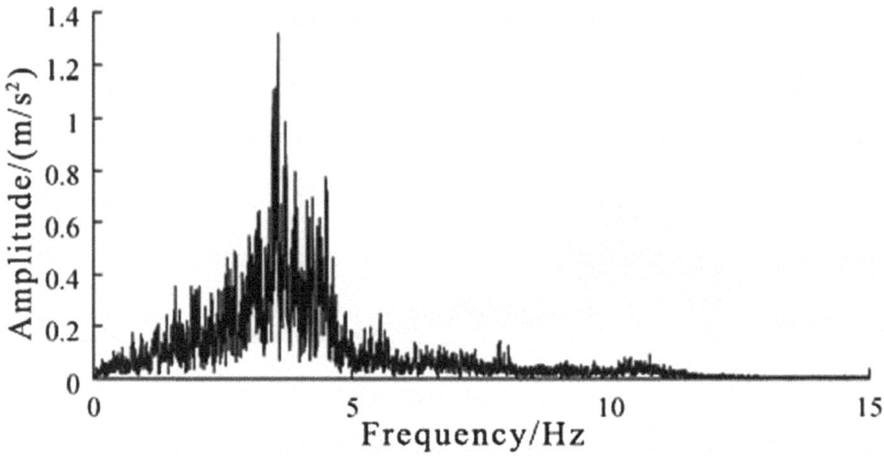

Fig. 3.120 Fourier spectrum of acceleration in slope strike direction (16#)

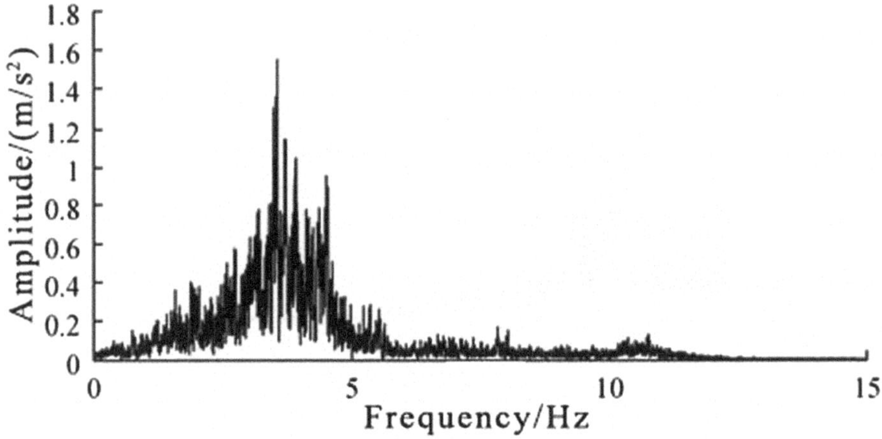

Fig. 3.121 Fourier spectrum of acceleration in slope strike direction (22#)

processing unit, which is the core part of computer graphics card and is the natural high-performance parallel device. Beijing GDEM Technology Ltd. developed the commercial software GDEM with independent intellectual property right together with Mechanical Institute of Chinese Academy of Sciences on the basis of CDEM and combining GPU technology, which improves the computing speed by two orders of magnitude and computing scale by one order of magnitude.

The GDEM software achieves interaction between kernel calculation and model display through hybrid programming of MFC with OPENGL and supports GUI and command stream dual operation. The main features of it are as follows.

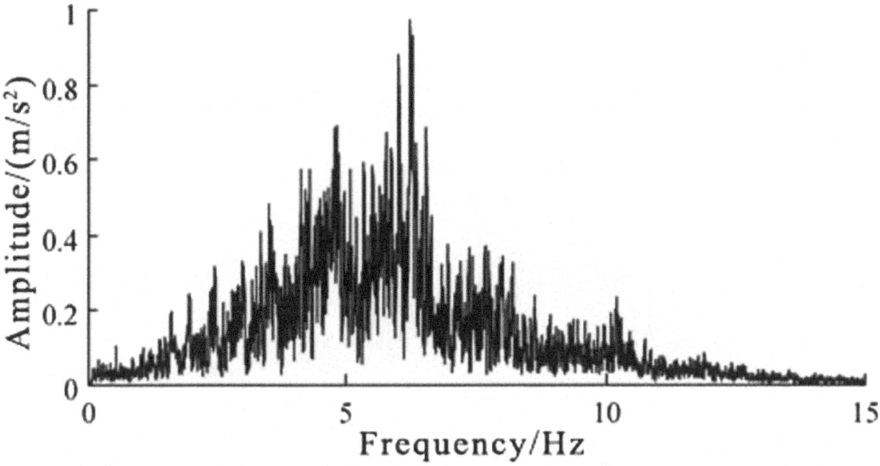

Fig. 3.122 Fourier spectrum of acceleration in vertical direction (4#)

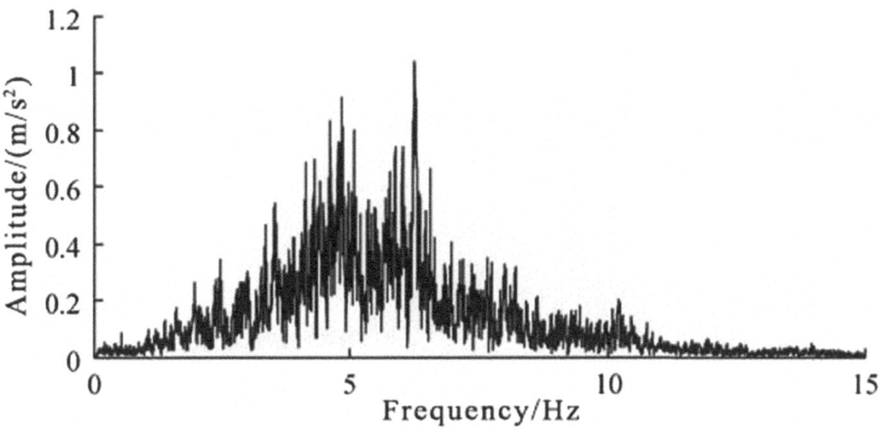

Fig. 3.123 Fourier spectrum of acceleration in vertical direction (10#)

First, using finite volume method to work out unit stress and strain and achieve large rotating calculation through real-time updating normal vector and unit area of all unit surfaces.

Second, using semi-spring model and semi-arris model to work out contact force among units, and converting judges on complex contacting state (angle–angle contact, angle–side contact, angle–surface contact, side–side contact, side–surface contact, and surface–surface contact) into judges on semi-spring and target surface, and semi-arris and target arris.

Third, using explicit calculation technology based on front inserting method, which could give slopes stress, strain features, and motion state at any time, and

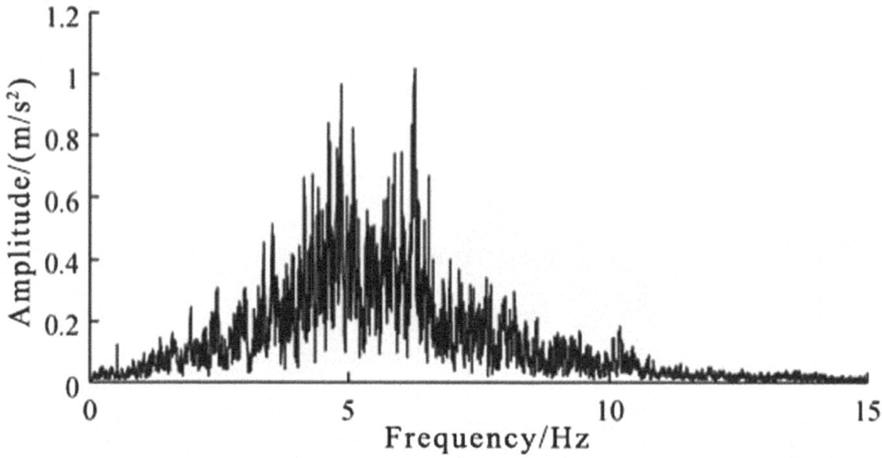

Fig. 3.124 Fourier spectrum of acceleration in vertical direction (16#)

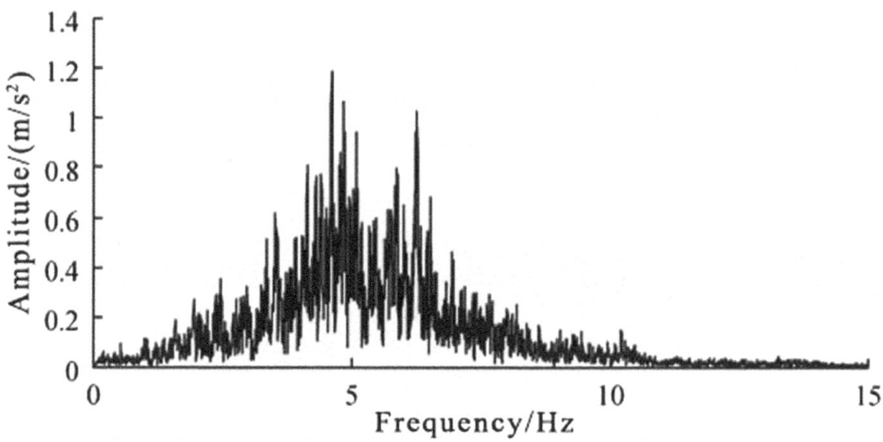

Fig. 3.125 Fourier spectrum of acceleration in vertical direction (22#)

make working out ill-condition and distortion as easy as working out the normal problems.

Currently, the GDEM software has been widely used in geotechnical engineering, mining engineering, tunnel engineering, hydraulic engineering, geological engineering, and structural engineering, and the gradual damage course simulation and global stability evaluation of geologic materials in all kinds of engineering.

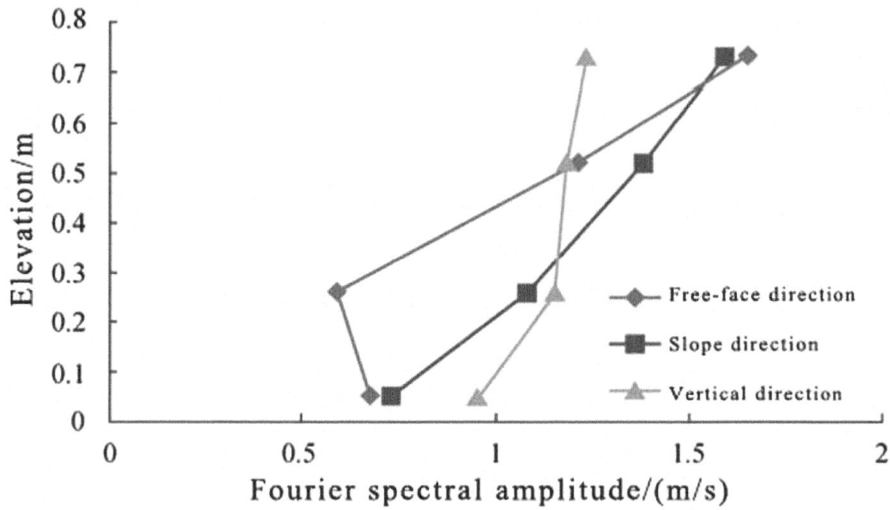

Fig. 3.126 Fourier spectral amplitude changes along the elevation

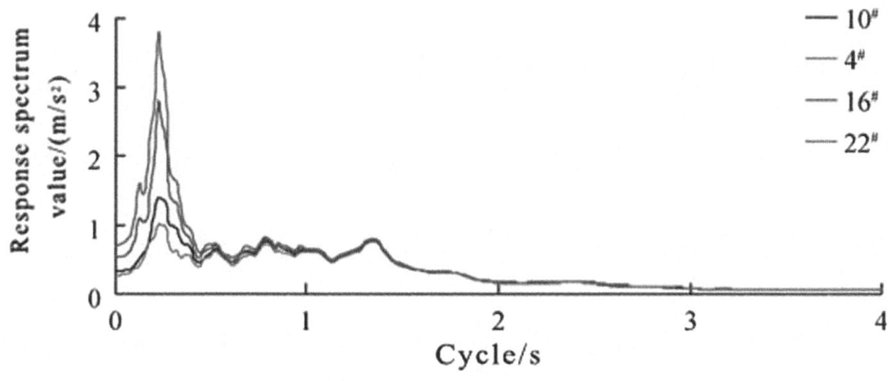

Fig. 3.127 Acceleration response spectrum in free face direction

3.3.1.1 Advantages of GDEM Software

Compared with the current numerical software, the advantages of GDEM can be summarized as follows: First, it is suitable to solve dynamic problems, nonlinear problems, and ill-conditioned problems, and there are explicit physical results after each time step, so there will not be problems of malfunction after stiffness matrix singularity turns out; second, it will not form the whole stiffness matrix which saves a lot of memory, so it is suitable to solve large-scale problems; third, GDEM combines three sets of software(FLAC, UDEC, PFC) together, so it can simulate the whole process of geologic calamity body from continuum to non-continuum; fourth, GDEM

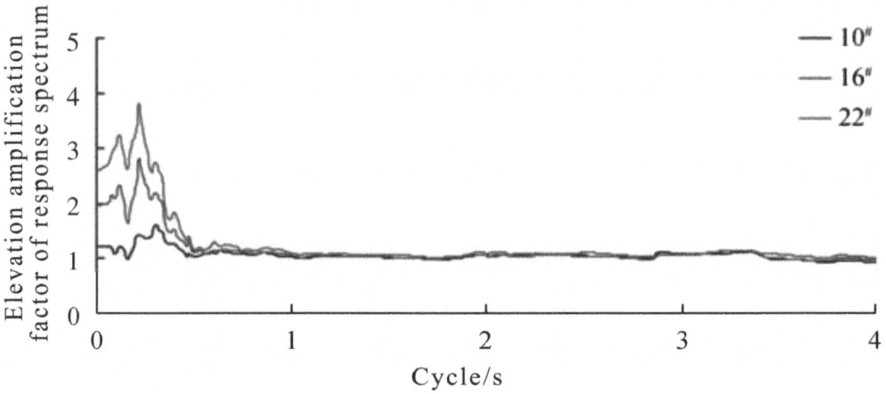

Fig. 3.128 Acceleration response spectrum amplification coefficient in free face direction

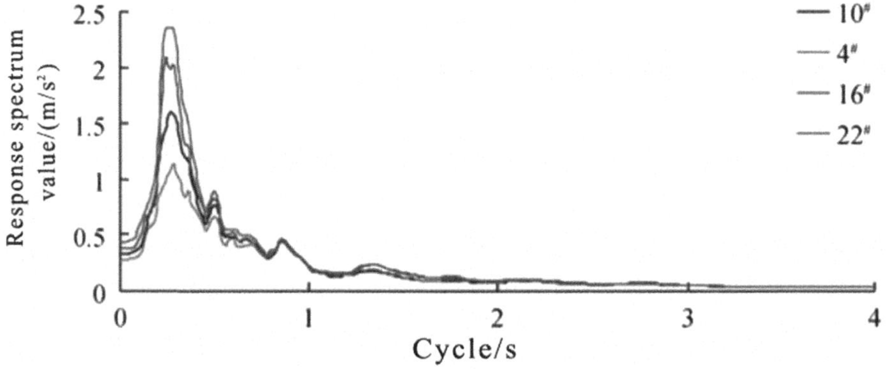

Fig. 3.129 Acceleration response spectrum in slope strike

uses GPU for acceleration, which is 100–150 times faster than solutions of the same type, and its computing scale could amount to millions of scale grid.

3.3.1.2 The Fundamental Principle of GDEM Computing Programming

Supposing that Ω is bounded closed domain in a space, and its boundary surface $\partial\Omega$ is composed of limited blocks of smooth or piecewise smooth surfaces, and suppose that function $f(x, y, z)$ has first-order continuous partial derivatives in Ω, so by Gauss divergence theorem we can get:

$$\oiint_{\partial\Omega+} f n_i \mathrm{d}S = \iiint_{\Omega} \frac{\partial f}{\partial x_i} \mathrm{d}\Omega \tag{3.17}$$

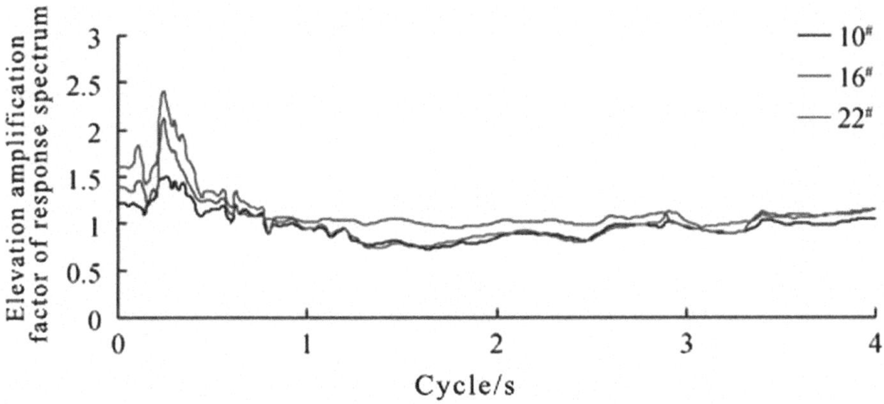

Fig. 3.130 Acceleration response spectrum amplification coefficient in slope strike

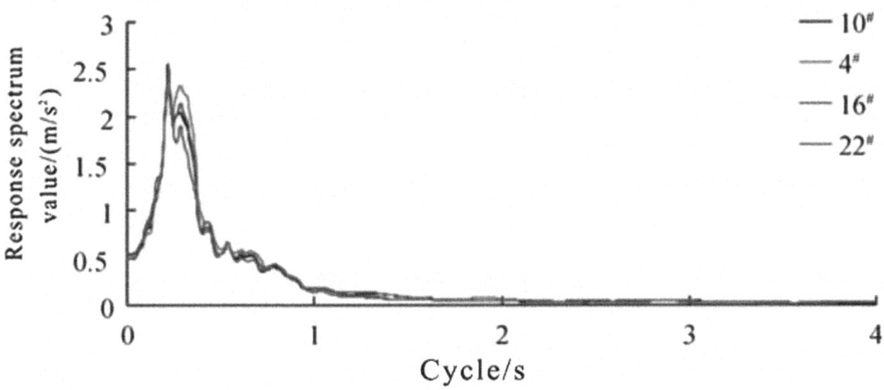

Fig. 3.131 Acceleration response spectrum in vertical direction

In which n_i is the unit vector in outside normal direction of boundary surface.

The average value $\left\langle \frac{\partial f}{\partial x_i} \right\rangle$ in bounded closed domain of $\frac{\partial f}{\partial x_i}$ can be expressed by Formula (3.18), and V is the total volume of bounded closed domain Ω. When $\Omega \to 0$, there is $\left\langle \frac{\partial f}{\partial x_i} \right\rangle = \frac{\partial f}{\partial x_i}$.

$$\left\langle \frac{\partial f}{\partial x_i} \right\rangle = \frac{1}{V} \iiint \frac{\partial f}{\partial x_i} \mathrm{d}V = \frac{1}{V} \oiint_{\partial\Omega+} f n_i \mathrm{d}S \tag{3.18}$$

Supposing that a polyhedron has N surfaces, then Formula (3.18) can be rewritten into Formula (3.19):

$$\left\langle \frac{\partial f}{\partial x_i} \right\rangle \approx \frac{1}{V} \sum_{k=1}^{N} \bar{f} n_i^k \Delta S^k \tag{3.19}$$

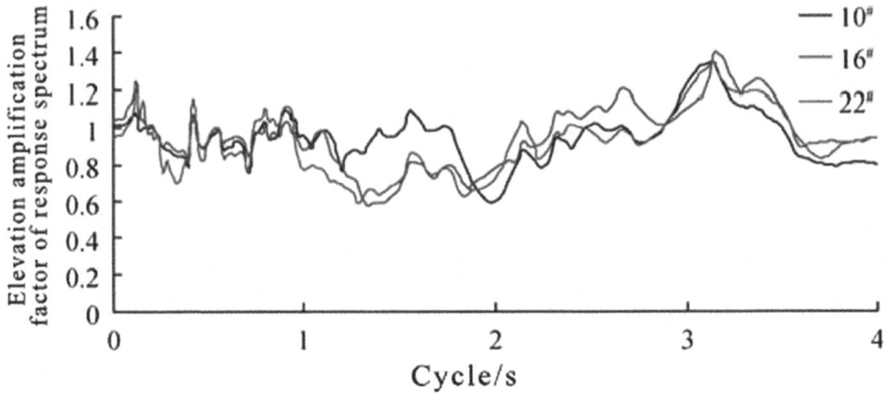

Fig. 3.132 Acceleration response spectrum amplification coefficient in vertical direction

ΔS^k is the area of the kth surface, the classification of outside normal direction of this surface in direction x_i is n_i, and $\bar{f}$ is the average value of f on the kth surface. Taking tetrahedron as example, if we want to work out stress increment inside the unit, we should firstly work out the velocity gradient average inside the tetrahedron, as Formula (3.20) shows:

$$\left\langle \frac{\partial v_i}{\partial x_j} \right\rangle = \frac{1}{V} \sum_{k=1}^{4} \overline{v_i} n_j^k \Delta S^k \tag{3.20}$$

The stress increment can be expressed as Formula (3.21):

$$\Delta \varepsilon_{ij} = \frac{1}{2} \left(\left\langle \frac{\partial v_i}{\partial x_j} \right\rangle + \left\langle \frac{\partial v_j}{\partial x_i} \right\rangle \right) \Delta t \tag{3.21}$$

The stress increment can be expressed as Formula (3.22):

$$\Delta \sigma_{ij} = C_{ijkl} \Delta \varepsilon_{kl} \tag{3.22}$$

The total component of stress can be expressed as Formula (3.23):

$$\sigma_{ij} = \sum \Delta \sigma_{ij} \tag{3.23}$$

The node stress component of all nodes in the tetrahedron is as Formula (3.24):

$$f_i^N = \sum_{k=1}^{3} \int_s \sigma_{ij} n_j^k \mathrm{d}s \approx \sigma_{ij} \sum_{k=1}^{3} \left(n_j^k \Delta S^k \right) \tag{3.24}$$

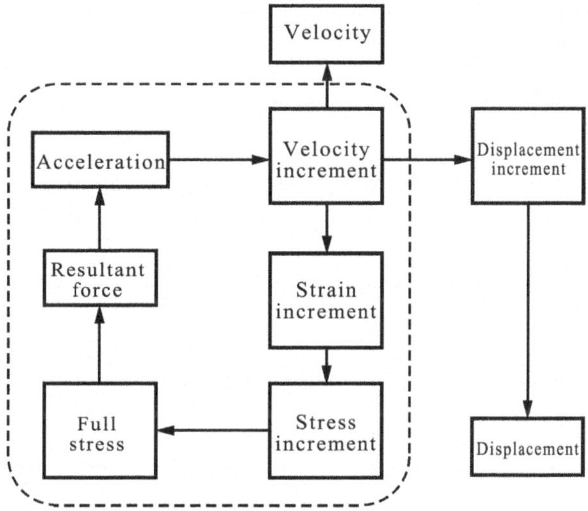

Fig. 3.133 Calculation flow of FVM

In there, N is representative of the Nth node, k is the serial number of the surface contains node N (each node inside the tetrahedron belongs to three coplanar vectors), n_j^k is the component of unit vector in outside normal direction of the surface, and ΔS^k is the node area of node N in the kth area.

If node N is shared by M nodes, the calculation formula of the total joint forces is as Formula (3.25):

$$F_i^N = \sum_{l=1}^{M} \left(f_i^N\right) = \sum_{l=1}^{M} \left[\sigma_{ij} \sum_{k=1}^{3} \left(n_j^k \Delta S^k\right)\right] \tag{3.25}$$

3.3.1.3 Program Implementation and Example of Verification

GDEM uses FVM to work out unit strain and stress, which takes the basis of computing source program and adopts incremental mode to achieve the above process. Its calculation flow is as Fig. 3.133 shows. The dotted line outlined part shows the solution iteration part through FVM.

Taking the elastic displacement field calculation of homogeneous slope as example to check the correctness of program, the comparison with the calculating results of FLAC3D is shown in Figs. 3.134 and 3.135. Through comparison, it can be known that the calculating results of GEDM based on FVM are exactly the same with that of FLAC3D.

On the basis of incremental method, the classic ideal elastoplastic criterion—Drucker–Prager and Mohr–Coulomb criteria—is introduced. Through uniaxial compression test, the calculating results of GEDM software are completely the same with that of commercial FLAC3D software (see Figs. 3.136 and 3.137).

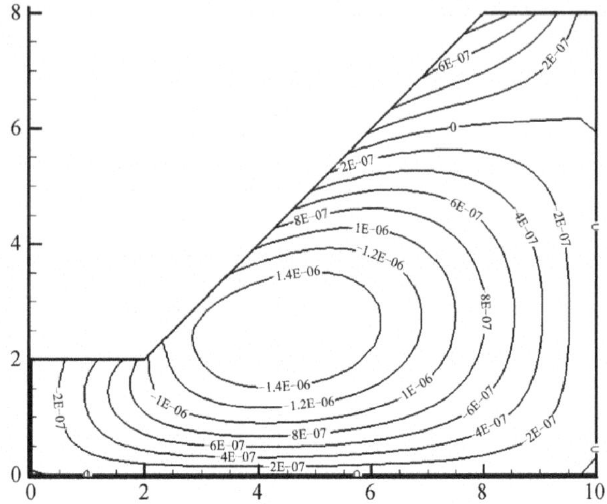

Fig. 3.134 Rock slope horizontal displacement field comparison diagram based on GEDM

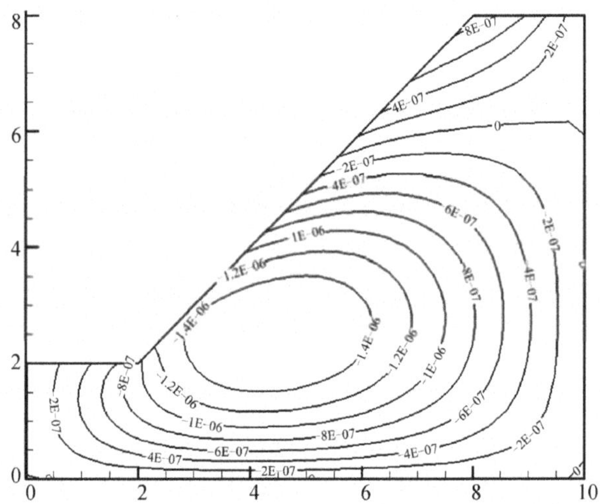

Fig. 3.135 Rock slope horizontal displacement field comparison diagram based on FLAC3D

3.3.1.4 Fundamental Function of GDEM Software

The fundamental function of GDEM mainly includes preprocessing function, main body function, and post-processing function. The main contents are as follows:

(1) Preprocessing function
GDEM could produce simple two-dimensional rectangle and homogeneous slope model network. The software supports mesh import of commercial software like ANSYS, Patran, and FLAC3D.
(2) Main body function

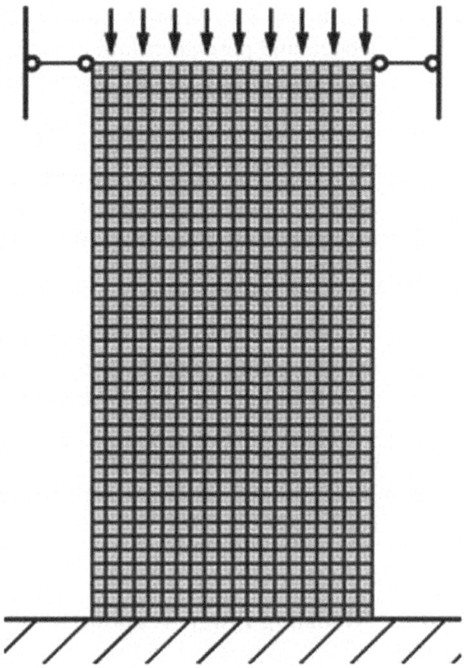

Fig. 3.136 Model diagram of uniaxial compression test

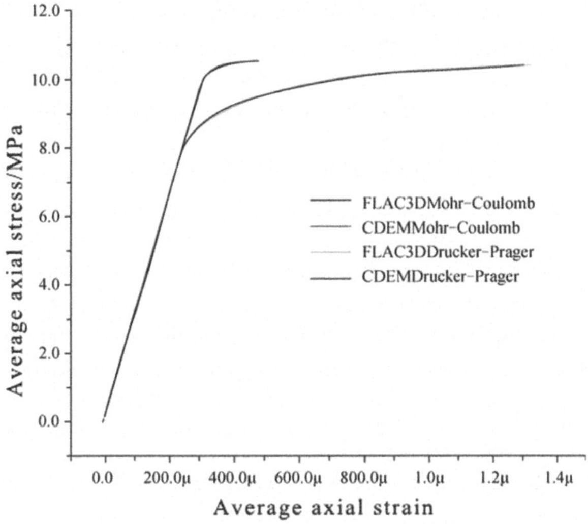

Fig. 3.137 Results of comparison diagram of uniaxial compression test

① Block models: linear elastic model, transversely isotropic model, ideal elastoplastic model (Drucker–Prager model and Mohr–Coulomb model), strain-softening model based on Mohr–Coulomb criterion, ubiquitous-joint model based on Mohr–Coulomb criterion, creep model (coupling model

of Burger with Mohr–Coulomb, Kelvin plasticity model), the excavation model, and pore percolation model.
② Ubiquitous-joint models: linear elastic model, brittle fracture model, strain-softening model, and ideal elastoplastic model.
③ Bar models: pile model, beam model, bolt model, and anchor cable model.
④ The basis of kinetics includes: explosion loading, seismic loading, the applied module to impact loading, non-reflecting boundary condition, and the applied module under free-field boundary conditions.

(3) Post-processing function

① It could check the acceleration, velocity, displacement, stress, strain, flow and flow velocity nephogram of rod model and the real units.
② It could check the time-history curve of monitoring points.
③ It could draw real unit velocity graph, displacement vector graph, and principal stress direction graph.
④ It could draw the unit displacement, force, and moment vector graphs of rod model.

3.3.1.5 Modeling Progress and Methods of GDEM

Before modeling, we should begin by clearly recognizing the subordinate system of GDEM and build two-dimensional or three-dimensional model according to your own needs. The default straight up direction by the software is the Y-axis, which is also the positive direction, and the opposite is the negative direction. We used to name the due north direction Z-axis of the software, which is also the positive direction, and the opposite is the negative direction. We have the tendency to conform to the geographical location with due north at 0°, and dip angle is identical with the recognized principle in geology field. X-axis takes the due east as positive direction and the opposite as negative direction, which suits geographical orientation.

We have mentioned previously that, in preprocessing function, GDEM could generate simple two-dimensional model on its own, but for complex models, it needs to import from other software. GDEM can operate with both GUI and command stream, and it provides rich commands for simple and easy modeling and adjusting process. It mainly adopts the following three modes:

(1) Using GUI operation;
(2) Using single command input method;
(3) For complex modeling needing large amounts of commands, it can use text input method and directly modify and save on the texts.

A complete joint model needs six steps in sequence to finish: (1) The software could produce simple two-dimensional rectangle and homogeneous slope model network or support mesh import of commercial software like ANSYS, Patran, and FLAC3D; (2) input parameters like dip angle, inclination, initial point, quantity, spacing, and continuity of joints; (3) set up material parameters for the model as well as the mechanic parameters of the joint its own; (4) set up monitoring curve or stress nephogram of displacement, velocity, stress, and strain according to your own needs; (5) set up boundary conditions of the model; and (6) work out the results and output it.

3.3.2 Key Problems in Numerical Analysis

In general case, continuous slopes of dozens of meters, hundreds of meters, or even much longer can be simplified into plane strain problems. Regarding influence factors like rock and earth mass nature and seismic wave types, the above-mentioned types of rock slopes can be generalized into two major kinds: single-side rock slope and dual-side rock slope. And neither field seismic array monitoring nor shaking table test have conducted researches on slopes of such types. Therefore, this chapter will, focusing on the above two types of rock slopes, select classic hazard point in "5.12" Wenchuan earthquake and use methods of shaking table test combining with numerical simulation to study seismic ground motion characteristics of rock slopes under the effects of earthquakes from the general rule so as to enlarge research results. What is important to note is that the shaking table test in this chapter is completed by both our research group and research group of Researcher Cui Peng from Chengdu Institute of geological hazards and Environment of Ministry of Water Resources of the Chinese Academy of Sciences, and most of the research results have been used by the latter. In addition, because of the limits of external conditions, limited monitoring points have been set up in the shaking table test, so it could not comprehensively reflect the seismic ground motion characteristics of rock slopes. Therefore, this chapter only builds numerical analysis model based on the shaking table test model and selects parts of the shaking table test data to verify the correctness of the numerical analysis results.

3.3.2.1 Numerical Analysis Model

The numerical analysis model in this chapter is built by geometric similarity ratio of 1:1 according to the shaking table test model. While the experiment model takes the prototype of two classic landslide hazard points by the left side of National Highway No. G213, it is built by the geometric similarity ratio of 1:600. Its geometric features are as follows: The single side high and steep model is 1807-mm high, 3500-mm wide, and its bevel angle is in the range of 50°–60°; the river valley at the slope foot is 173 mm, and the riverbed is slightly inclined; the bevel angle of the river valley right bank slope is in the range of 30°–40°, and the slope top is partly covered with seriously weathered layer. See Fig. 3.138 for concrete experiment model. The dual side high and steep model is 1810-mm high and 3500-mm wide, which include steep slope topography and gentle slope topography with bevel angles, respectively, in the range of 50°–60° and 42°–50°. At the same time, in order to more truly reflect the

Fig. 3.138 Shaking table test model of single side rock slope

Fig. 3.139 Shaking table test model of dual side rock slope

practical situations of the above-mentioned slopes, the soil layer and weathered layer on the surface of the slopes are generalized into slope wash in slope top, mid slope, and slope foot according to field survey results. See Fig. 3.139 for detailed shaking table test model. In addition, to thoroughly study on the seismic ground motion characteristics of high and steep rock slopes, we set up large amounts of monitoring points inside and on the surface of the numerical model to monitor the dynamic response time history changes of slope acceleration, stress, and strain under seismic effects. Among them, we select monitoring points K8, K11, K14 and B6, B8, B4 as comparison monitoring points of single side rock slope and dual side rock slope to verify the correctness of numerical results. See Figs. 4.8 and 4.9 for numerical analysis model and monitoring points distribution of single side rock slope and dual side rock slope (Figs. 3.140 and 3.141).

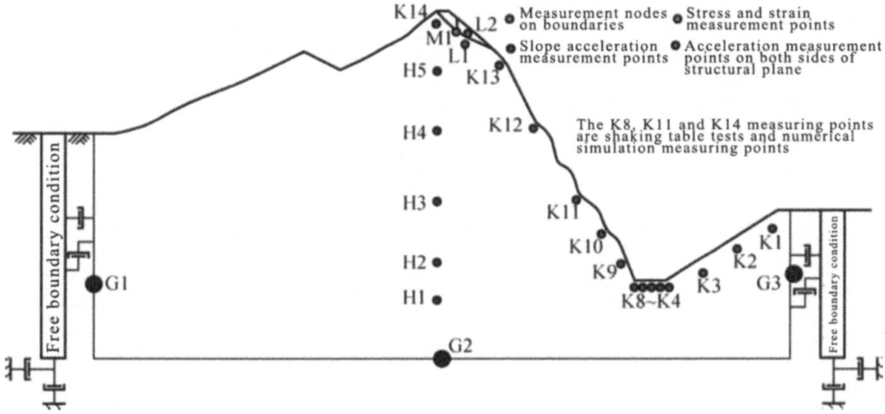

Fig. 3.140 Numerical model of single side rock slope

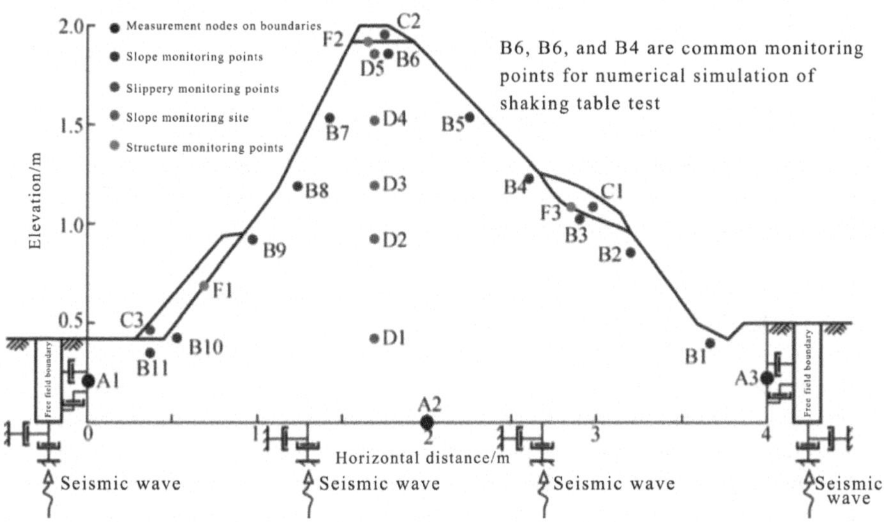

Fig. 3.141 Numerical model of dual side rock slope

3.3.2.2 Determination of Rock and Soil Mass Material Parameters and Constitutive Model

Based on the physico-mechanical parameters of slope materials of the prototype, the above-mentioned shaking table test used quartz sand, barite powder, clay, plaster or fly ash and water with different ratios to build the model so as to simulate dolomitic limestone of single and dual side high and steep slopes and surface weathering rock mass materials of cover layer. The numerical simulation of this part adopted exactly the same materials with the shaking table test, and the concrete physico-mechanical

Table 3.11 Physico-mechanical parameters of the shaking table test model materials

Model name	Unit weight (kN/m^3)	Cohesive force (kPa)	Internal friction angle (°)	Elastoplastic model (MPa)	Poisson ratio
Single side slope	23	13.7	39.4	100	0.2
Dual side slope	22	15.6	37.9	75	0.25
Cover layer	21	8.3	28.8	10	0.35
	Structural position	Normal stiffness (MPa)	Shear stiffness (MPa)	Internal friction angle (°)	Cohesive force (kPa)
Single- and double-sided slope	Inside of cover layer	600	300	32	8.1
	Bed rock cover layer	4500	2300	32	2.3

parameters are shown in Table 3.11. The sliding bed and slip mass adopted elasto-plastic model and Mohr–Coulomb failure principle (M–C failure principle), while the structural surface and the particles inside the slip mass used brittle fracture constitutive model and M–C failure principle.

3.3.2.3 Selection of Boundary Conditions

When carrying out dynamic numerical simulation to single side high and steep slopes and dual side high and steep slopes, it is an important work to reasonably exert artificial boundaries. Boundaries are set up according to the following steps: ① Constrain the bottom boundary in the vertical direction and the bilateral boundaries in the horizontal direction. ② Change the original model displacement boundary conditions into free-field boundaries and the bottom boundary constraints in vertical directions into viscid boundaries so as to reduce energy loss due to seismic wave reflection.

3.3.2.4 Seismic Ground Motion Input Conditions

The seismic wave selected by the shaking table test is the adjusted Wenchuan wolong seismic wave by time simulation ratio of 1:4, and this chapter selects the same seismic waves to conduct the numerical analysis. The real seismic wave PGA of Wenchuan wolong seismic array is, respectively, 957 and 857 cm/s^2 in horizontal and vertical directions, and the duration is 163 s. After adjusting by the time simulation ratio, the seismic wave duration is 51 s and the amplitude stays unchanged. See Figs. 3.142 and 3.143 for detailed acceleration time history curve.

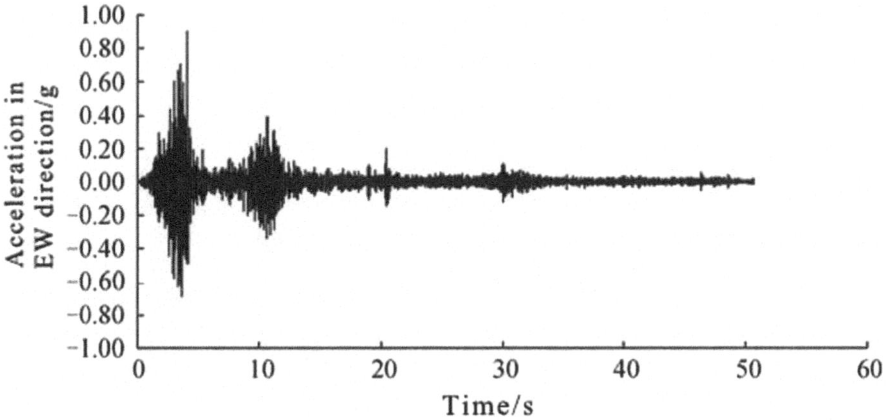

Fig. 3.142 Acceleration time history in EW direction (0.9 g)

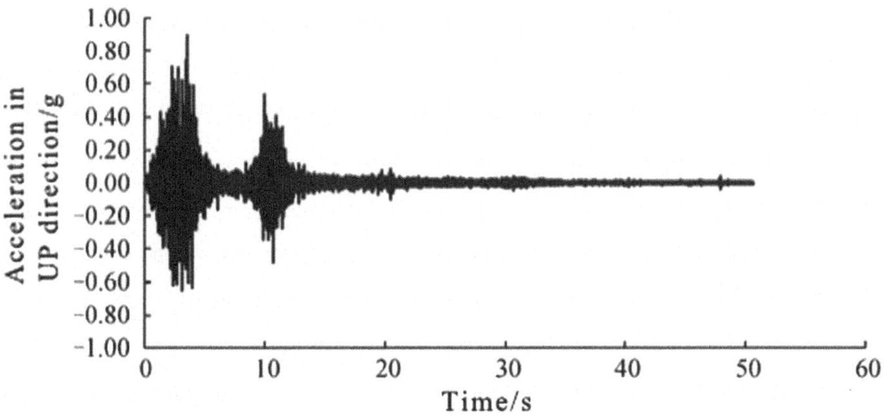

Fig. 3.143 Acceleration time history in UP direction (0.9 g)

During the numerical analysis process of GDEM, the dynamic loading represented by acceleration time-history curve cannot be directly exerted on the viscid boundaries, so it needed to be converted into stress curve before exerting on the bottom boundary. This chapter will use numerical integration to convert acceleration time-history curve into velocity time-history curve and then use formula (3.26)–(3.29) to convert velocity time-history curve into stress time-history curve and exert on the bottom boundary:

$$\sigma_n = -2(\rho C_p)\upsilon_n; \tag{3.26}$$

$$\sigma_s = -2(\rho C_s)\upsilon_s \tag{3.27}$$

$$C_P = \sqrt{\frac{K + 4G/3}{\rho}}; \tag{3.28}$$

$$C_s = \sqrt{G/\rho} \tag{3.29}$$

In the formula, σ_n is the exerted normal stress (MPa), σ_s is the exerted tangential stress (MPa), ρ is medium density (kg/m^3), C_p is the P-wave velocity of medium (m/s), C_s is the S-wave velocity of medium, υ_n is the vibration speed of input mass point in vertical direction, υ_s is the vibration speed of input mass point in horizontal direction, K is the medium bulk modulus, and G is the medium shear modulus.

According to suggestion by R. L. Kuhlemeyer and J. Lysme [149], the grid size is controlled by the shortest wavelength of input wave. Supposing the biggest size of grid is Δl, the shortest wavelength of input seismic wave is λ, then Δl must be less than (1/10–1/8)λ. Supposing that the highest frequency of input wave which does not distort wave shape is $f\text{max}$, then we can calculate according to Formula (3.30).

$$f\text{max} = \frac{C}{\lambda} = \frac{C}{10\Delta l} \tag{3.30}$$

Here, C is the small one of longitudinal wave velocity and transverse wave velocity.

3.3.3 Verification of the Correctness of Numerical Analysis Results

In order to verify the correctness of numerical calculation method, this dissertation sets up monitoring points K8, K11, K14 and B6, B8, B4 as comparison monitoring points in single side rock slope and dual side rock slope and compares the numerical calculation results with the shaking table test results. See Table 3.12 for concrete results. To more clearly explain the difference of shaking table test results and numerical analysis results in time history, this dissertation will list the horizontal and vertical time-history curves of Test Point B6 in T = 0–4 s (including PGA) to further explain (see Figs. 3.144 and 3.145).

Comprehensively analyzing Figs. 3.12, 3.144, and 3.145, we can know that the maximum error of horizontal acceleration and vertical acceleration of Test Points K8, K11, K14 and B4, B6, B8 is all less than 10.0%, in which the minimum error of horizontal acceleration could amount to −1.19%, while the minimum error of vertical acceleration could amount to 2.28%. Therefore, the numerical simulation results are basically identical with the shaking model test results, which fully show that the numerical model of this dissertation is correct and the calculation results are credible.

Table 3.12 PGA results comparison of shaking table test and numerical simulation

Monitoring points no.	GDEM calculation results (m/s/s)		Shaking table test results (m/s/s)		Error (%)	
	Horizontal acceleration	Vertical acceleration	Horizontal acceleration	Vertical acceleration	Horizontal acceleration	Vertical acceleration
K8	16.6	13.1	16.8	12.8	−1.19	2.34
K11	24.6	31.2	24.0	29.9	2.50	4.35
K14	19.3	15.6	20.5	14.3	−6.21	−8.3
B4	24.9	21.4	23.8	22.3	4.62	9.71
B6	27.5	32.5	27.0	30.5	1.75	6.54
B8	23.2	31.2	24.6	30.5	−5.69	2.28

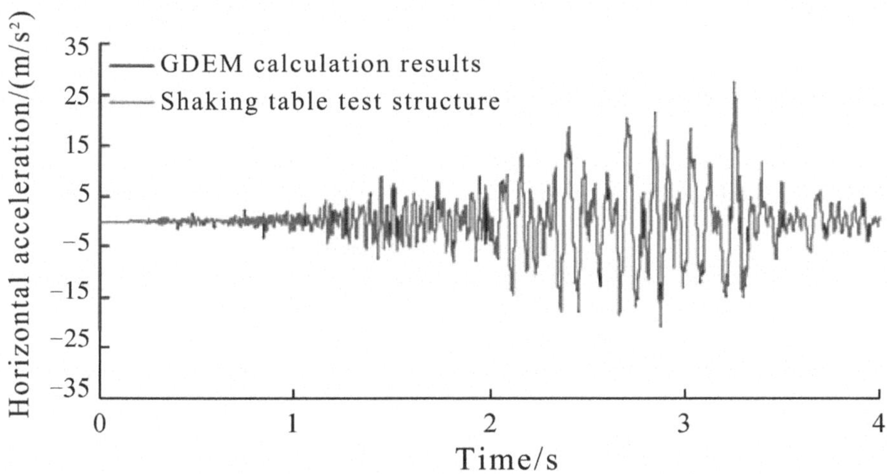

Fig. 3.144 Horizontal time-history curve of Test Point B6

3.3.4 *The General Rules of Height and Steep Rock Slope Seismic Dynamic Response Features*

Based on the numerical analysis results of single side and dual side high and steep slopes, this dissertation will reveal the general rules of high and steep rock slope dynamic response features from the perspective of time-domain and frequency.

3.3.4.1 The Elevation Amplification Effects of Slope Surface and Slope Mass PGA

In order to study on the propagation rule of acceleration along slope surface and slope mass elevation in single and dual side high and steep slopes under seismic effects, this dissertation sets up Test Points B1–B6 (except B3) in gentle slope topography

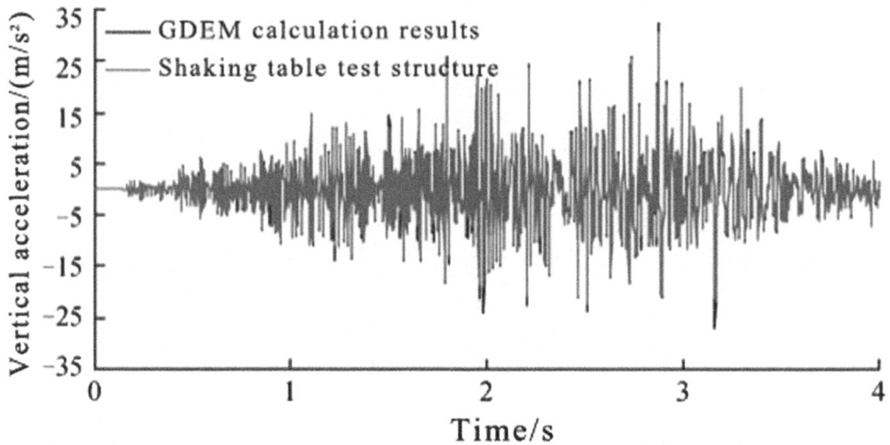

Fig. 3.145 Vertical time-history curve of Test Point B6

Table 3.13 Horizontal PGA amplification coefficient of single and dual side high and steep slopes

Height to the model bottom (m)	Dual side high and steep slope			Height to the model bottom (m)	Single-sided high and steep slope	
	Steep slope topography	Gentle slope topography	Inside slope mass		On the slope surface	Inside slope mass
1.75	1.91	1.91	1.93	1.78	1.84	1.84
1.35	1.81	1.73	1.65	1.53	2.23	2.15
1.05	1.73	1.61	1.51	1.21	2.72	2.05
0.75	1.62	1.52	1.32	0.84	2.35	1.87
0.25	1.52	1.34	1.15	0.32	1.57	1.17

of dual side high and steep slopes, B6–B10 in steep slope topography, and D1–D5 inside the slope mass. At the same time, we set up Test Points K14–K10 on the slope surfaces of single side high and steep slopes, and H2–H5 and K14 inside the slope mass. See Tables 3.13 and 3.14 for detailed results, and we draw Figs. 3.146 and 3.147 according to data from Tables 3.14 and 3.15.

Comprehensively analyzing Tables 3.13, 3.14 and Figs. 3.146, 3.147, we can know that for dual side high and steep slope, taking acceleration in the valley bottom as criterion, the horizontal acceleration coefficient on steep slope topography is between 1.52 and 1.81, and the vertical acceleration coefficient is between 1.10 and 1.36. While the horizontal acceleration coefficient on gentle slope topography is between 1.34 and 1.91, the vertical acceleration coefficient is between 0.68 and 1.36. The horizontal acceleration coefficient inside the slope mass is between 1.15 and 1.93, and the vertical acceleration coefficient is between 0.65 and 1.35. For single side high and steep slope, the horizontal acceleration coefficient on the slope surface

Table 3.14 Vertical PGA amplification coefficient of single and dual side high and steep slopes

Height to the model bottom (m)	Dual side high and steep slope			Height to the model bottom (m)	Single-sided high and steep slope	
	Steep slope topography	Gentle slope topography	Inside slope mass		On the slope surface	Inside slope mass
1.75	1.36	1.36	1.35	1.78	1.06	1.06
1.35	1.35	1.22	1.08	1.53	1.95	1.57
1.05	1.30	0.90	0.84	1.21	2.13	1.91
0.75	1.21	0.80	0.73	0.84	2.12	1.81
0.25	1.10	0.68	0.65	0.32	0.89	0.87

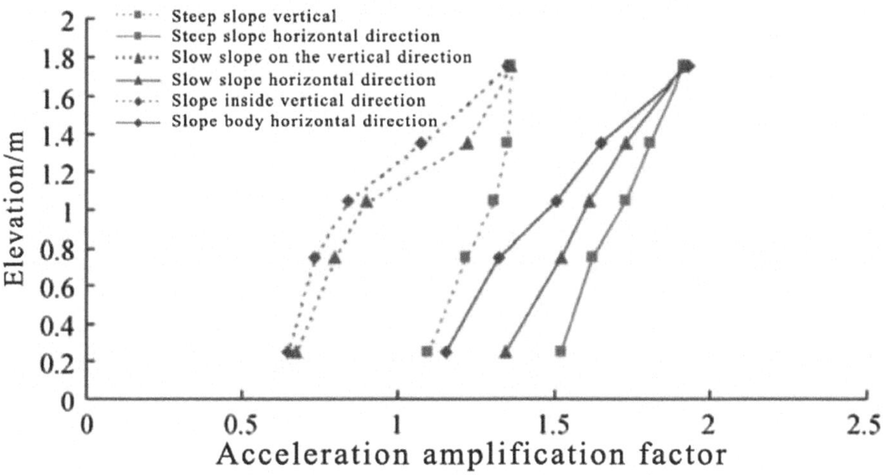

Fig. 3.146 Acceleration amplification effect distribution of double-sided high-steep slope

Table 3.15 Material parameters of rock and soil mass

Lithology	Elasticity modulus (E/Mpa)	Poisson's ratio (μ)	Cohesive force (c/kPa)	Internal friction angle (°)	Unit weight (γ/kN/m^3)
Rock mass	450	0.25	0.936	37.9	22.0

is between 1.57 and 2.72, and the vertical acceleration coefficient is between 0.89 and 2.13. The horizontal acceleration coefficient inside the slope mass is between 1.17 and 2.15, and the vertical acceleration coefficient is between 0.87 and 1.91. Therefore, no matter for single side high and steep slope or for dual side high and steep slope, the PGA of different positions along elevation amplifies to different extent, and it shows as follows: the amplification effects of dual side high and steep slope on acceleration > the amplification effects of single side high and steep slope on

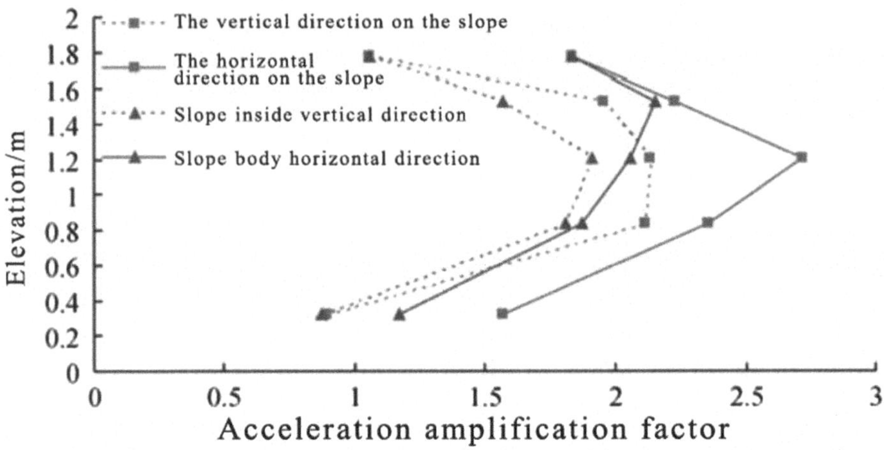

Fig. 3.147 Acceleration amplification effect distribution of single-sided high-steep slope

acceleration, which fully explains the hazard phenomenon in 5.12 Wenchuan earthquake that the damage of dual side slopes or independent slopes is more serious than that of single side slopes [150]; PGA amplification effects on slope surface > PGA amplification effects inside slope mass, which explains that the high and steep slopes have clear "amplification effect in free face direction"; the horizontal acceleration coefficient > the vertical acceleration coefficient *i*, and acceleration coefficient on steep slope topography > acceleration coefficient on gentle slope topography, which is the same with the shaking table test results and can explain landslide hazard distribution directivity in Wenchuan earthquake; i.e. the seismic damage on steep slope topography is more serious than that on gentle slope topography [151, 152].

3.3.4.2 Acceleration Fourier Spectrum Amplification Effects on Slope Surface Along Elevation

In order to study on distribution rules of surface acceleration Fourier spectrum along elevation on single and dual side high and steep slopes under seismic effects, this dissertation sets up Test Points B1, B4, B6 and B11, B8, B6, respectively, on gentle slope topography and steep slope topography on dual side high and steep slopes and, at the same time, sets up Test Points K8, K11, and K14 on single side high and steep slope surface. See Fig. 3.148 for detailed results. It is noteworthy that (a)–(i) in Fig. 3.148 represents the calculation results of horizontal acceleration Fourier spectrum along elevation and (A)–(I) represents the calculation results of vertical acceleration Fourier spectrum along elevation.

Through comprehensively analyzing Fig. 3.148, we can know that no matter single side or dual side high and steep slopes, the horizontal and vertical acceleration Fourier spectrum on slope surfaces along elevation shows the changing rule of low-

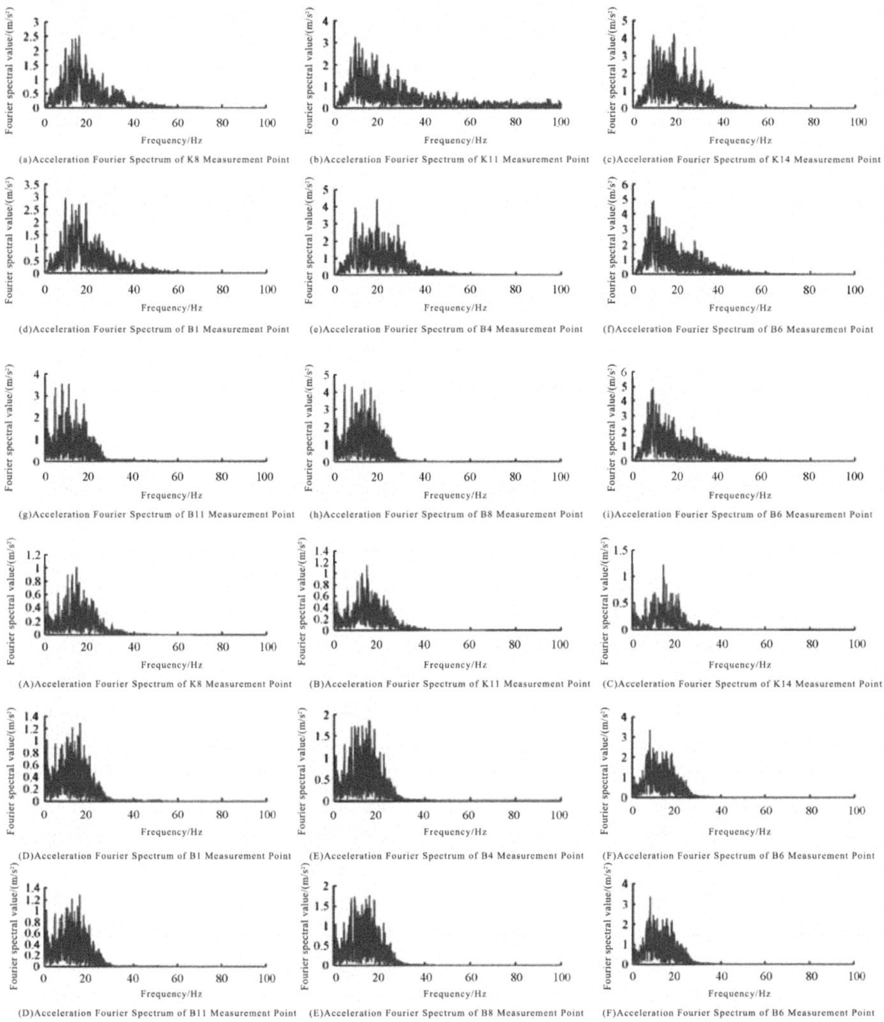

Fig. 3.148 Distribution graph of horizontal and vertical acceleration Fourier spectrum

frequency component amplification. This chapter will take the horizontal and vertical acceleration on single side high and steep slope surfaces as an example to explain. Figure 3.148 shows that the dominant frequencies of horizontal acceleration at slope foot, mid slope, and slope top of single side high and steep slopes concentrate in the range of 5–10 Hz, in which Test Points K8, K11, and K14 show the spectral value of horizontal acceleration dominant frequencies are, respectively, 2.51, 3.24, and 4.16, while that of vertical acceleration dominant frequencies are, respectively, 1.02, 1.14, and 1.25. Therefore, the horizontal and vertical acceleration Fourier spectrum

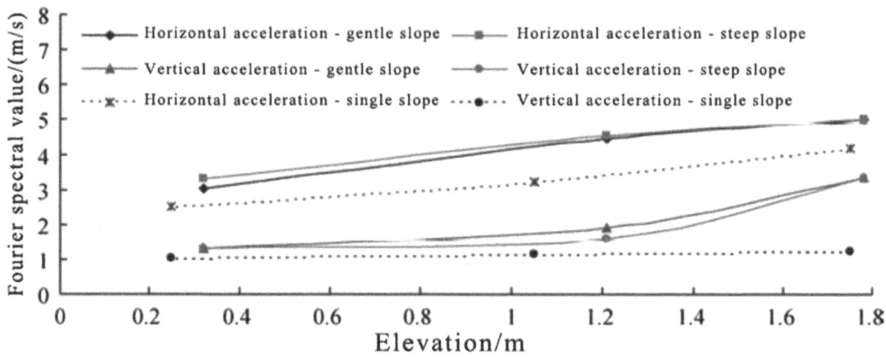

Fig. 3.149 PGA elevation amplification coefficient of low-frequency Fourier spectrum

on slope surfaces of single side high and steep slopes along elevation shows the changing rule of low-frequency component amplification.

The above phenomenon may be caused by that the material damping of the soil mass absorbs some energy of the seismic waves and has filter effects for seismic waves of high-frequency section and amplification effects for section around the first step natural frequency (about 6.06 Hz), which is consistent with the results of shaking table test results and field seismic array actual measurement. At the same time, to further study on the low-frequency section amplification rule of acceleration Fourier spectrum along surface elevation, this chapter selects the peak value of low-frequency section in acceleration Fourier spectrum of single and dual side high and steep slope surfaces and draws the peak value changes of acceleration Fourier spectrum along elevation in low-frequency section (see Fig. 3.149). The figure shows that, no matter for single or dual side high and steep slopes, the peak values of acceleration Fourier spectrum along elevation at different positions amplify to different extent, showing as the amplification effects of dual side high and steep slopes on acceleration Fourier spectrum > the amplification effects of single side high and steep slopes on acceleration Fourier spectrum, the horizontal amplification effects of acceleration Fourier spectrum > the vertical amplification effects of acceleration Fourier spectrum, the amplification effects of acceleration Fourier spectrum on steep slope surfaces (with bevel angles of 50°–60°) > that on gentle surfaces (with bevel angles of 42°–50°). Through the comprehensive analysis of Figs. 3.148 and 3.149, we can know that the slope topography and bevel angles have a significant influence on the changing rules of low-frequency components amplification.

3.3.4.3 The Elevation Amplification Effects of Slope Surface PGA

In order to study on distribution rules of surface acceleration response spectrum along elevation on single and dual side high and steep slopes under seismic effects, this dissertation sets up Test Points K8, K11, and K14 on single side high and steep slope surface and sets up Test Points B1, B4, B6 and B11, B8, B6, respectively, on gentle slope topography and steep slope topography on dual side high and steep slopes. See Fig. 3.150 for detailed results. It is noteworthy that the red frame shows the horizontal and vertical acceleration response spectrum and the blue frame shows the horizontal and vertical acceleration response spectrum value amplification coefficient along elevation.

It can be known from Fig. 3.150 that acceleration response spectrum and spectral ratio of single and dual side high and steep slope surface exist large difference. First of all, no matter horizontal acceleration or vertical acceleration, the acceleration response spectrum of single side high and steep slope along elevation is with single peak value, while the acceleration response spectrum of dual side high and steep slope along elevation is a process changing from single peak value to double peak value; i.e. the acceleration spectrum at slope foot is with single peak value, that at mid slope is with slight double peak value, and that at slope top is with obvious double peak value. The reason for the above phenomenon may be that the acceleration of dual side high and steep slope gradually amplifies along elevation under seismic effects, and the ground motion is more intense, so whiplash effect arises and double peak value phenomenon comes out. Secondly, the maximum horizontal acceleration response spectral value amplification coefficient of single and dual side high and steep slope occurs at period TH = 0.16 s, and the reason for the above phenomenon may be that TH = 0.16 s is close to the natural frequency period $T = 0.165$ s of single side high and steep slope and the natural frequency period $T = 0.156$ s of dual side high and steep slope, which results in resonance and caused intense seismic ground motion response on their surfaces. Lastly, around the short period ($T = 0.0$–0.8 s), the horizontal and vertical acceleration response spectral value amplification coefficients of single side high and steep slope are, respectively, in the range of 1.0–2.5 and 1.0–2.1, that of gentle slope surface are, respectively, in the range of 1.0–3.2 and 1.0–1.82, and that of steep slope surface are, respectively, in the range of 1.0–3.32 and 1.0–3.7. For the long period range ($T = 0.8$–4.0 s), the horizontal acceleration response spectral value amplification coefficient of single side high and steep slope reduces slightly, and it is in the range of 0.6–1.1, while the vertical acceleration response spectral value amplification coefficient stays at around 1.0. The horizontal and vertical acceleration response spectral value amplification coefficient of gentle and steep slope surface stays at around 1.0.

To sum up, no matter for single or dual side high and steep slopes, the horizontal and vertical acceleration response spectral value amplifies to different extent, showing as the amplification effects of dual side high and steep slopes on acceleration response spectrum > the amplification effects of single side high and steep slopes on acceleration response spectrum, the horizontal amplification effects of acceleration response spectrum > the vertical amplification effects of acceleration response spectrum, the amplification effects of acceleration response spectrum on steep slope surfaces > that on gentle surfaces. Therefore, the slope topography and bevel angles have significant influence on the changing rules of slope surface acceleration response spectrum along elevation.

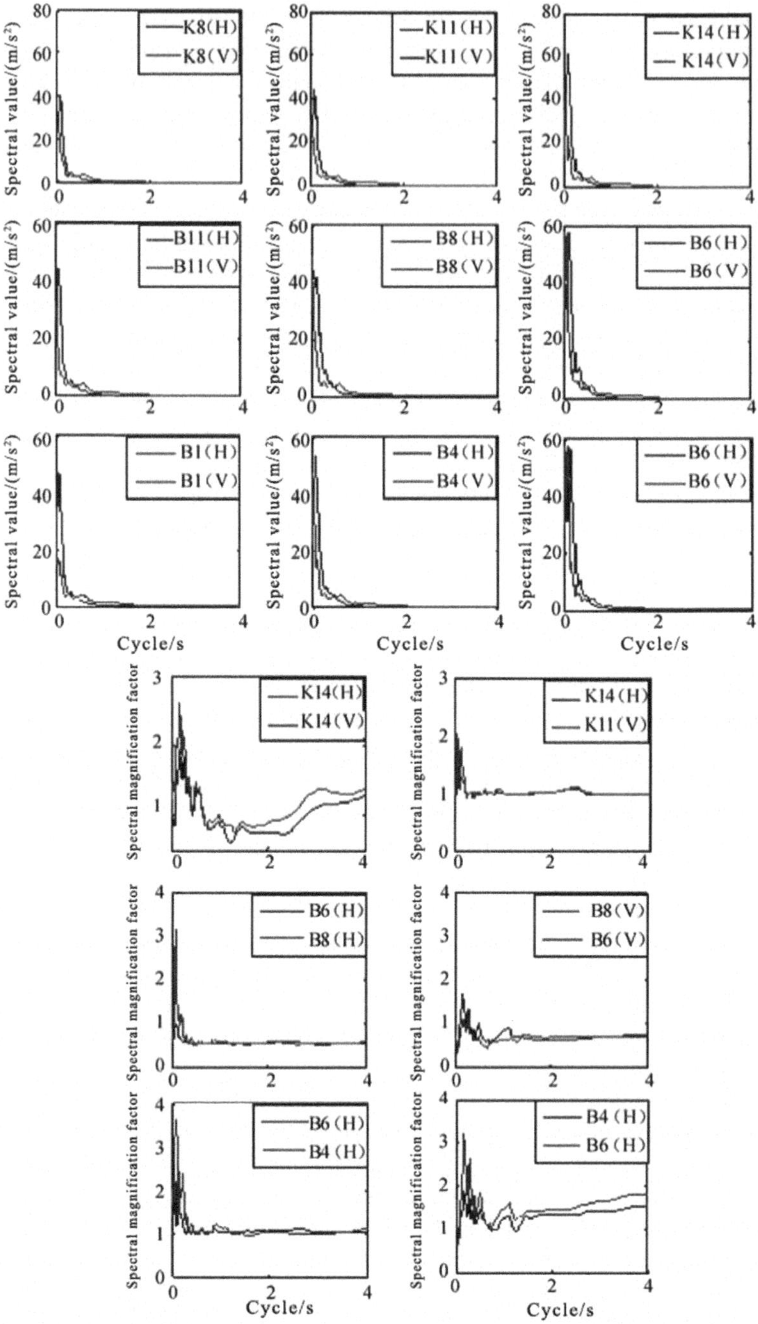

Fig. 3.150 Acceleration response spectrum and spectral ratio of single and dual side high and steep slope surface

3.3.5 Local Topography Effect of Rock Slope Seismic Dynamic Response Rules

For the study on rock slope seismic dynamic response rules, analysis from the perspective of the whole slope has been conducted by seismic array monitoring, shaking table test, and the above numerical analysis work on individual independent slope and continuous slopes of dozens of meters, hundreds of meters, or even much longer. While the field survey results of slope hazard for Wenchuan earthquake show that large amount of local topography like independent mountains, weak mountain ridge, and the turning parts of slope shape has intense dynamic response and serious hazards under strong seismic effects with obvious local topography effects, at the same time, the field monitoring results of slope seismic dynamic response also show that the seismic ground motion amplification effects of micro-topography with multiple free surfaces are also obvious. That indicates that local topography and microtopography perform important control functions in seismic dynamic response. Thus, this chapter will continue to focus on continuous slopes of dozens of meters, hundreds of meters, or even much longer, generally analyze the models, and study on the influence of slope tops with different forms on PGA elevation amplification effects so as to enrich research results of rock slope seismic dynamic characteristics from micro-perspective.

3.3.5.1 Key Problems in Numerical Analysis

This chapter will adopt slope models with different forms of slope tops (see Figs. 3.151, 3.152, 3.153, and 3.154). The slope height is about 80 m with bevel angles of about 45°, and the slope tops of different forms have the same or similar elevation. At the same time, test points of horizontal acceleration, horizontal velocity, and horizontal displacement are set up evenly on slope top position. The real seismic waves of Wenchuan wolong seismic array are selected simultaneously with horizontal and vertical PGA of 957 and 857 cm/s/s and the duration of 163 s. See Figs. 3.155 and 3.156 for detailed horizontal and vertical acceleration time history curve. In addition, the rock and soil materials select the materials of the prototype slope of the above single and dual side rock slope shaking table test. See Fig. 4.5 for concrete parameters.

3.3.5.2 Calculation Results Analysis

In order to further study the influence of slope surface forms on acceleration amplification effects along elevation, this chapter sorts out and analyzes the calculation results of Test Points 1, 2, and 3 on every model top and carries out research from the perspective of PGA, acceleration Fourier spectrum, and acceleration response spectrum.

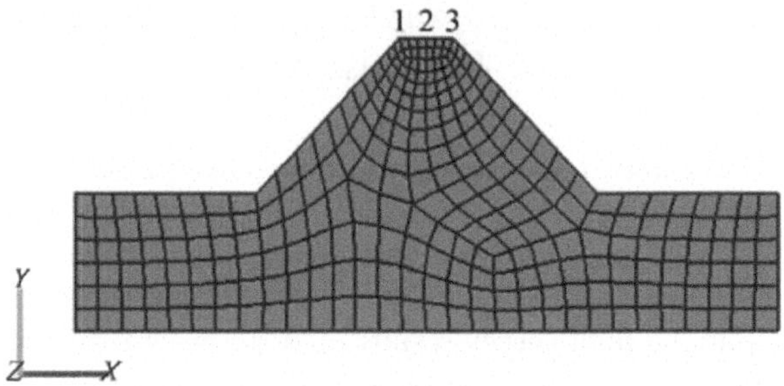

Fig. 3.151 Numerical analysis model of flattop slopes

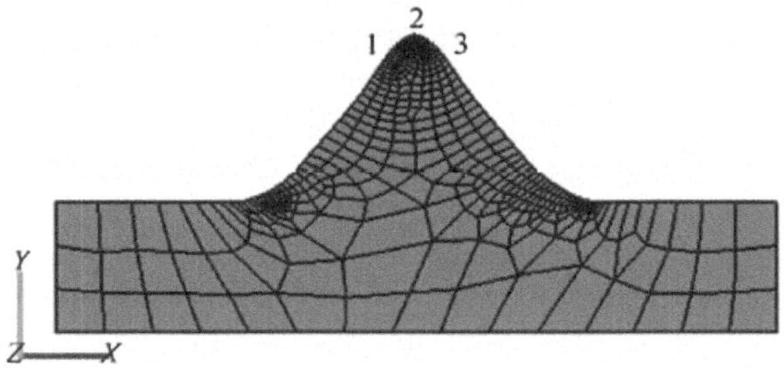

Fig. 3.152 Numerical analysis model of single peak slopes

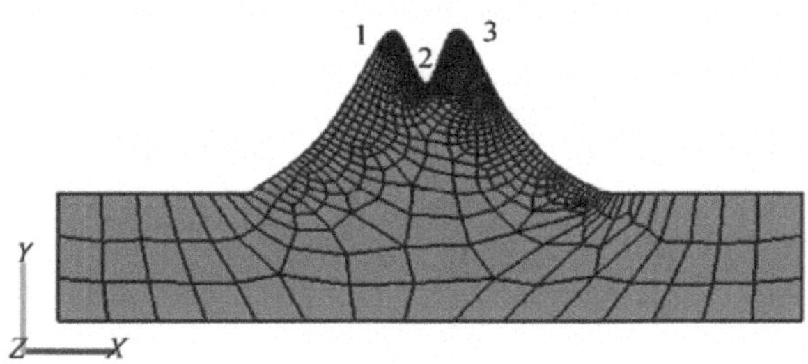

Fig. 3.153 Numerical analysis model of double peak slopes

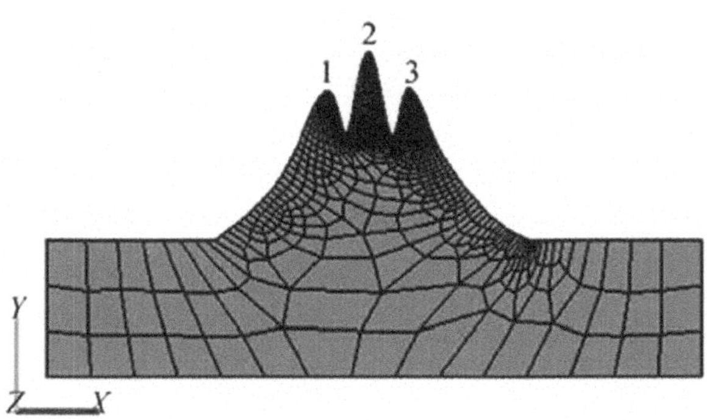

Fig. 3.154 Numerical analysis model of three peak slopes

① Effects of Slope Top Forms on Acceleration Peak Value Amplification Effect along Elevation

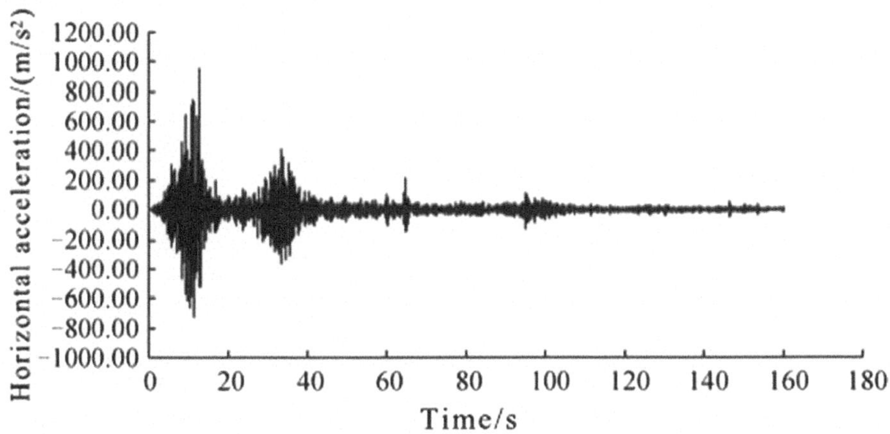

Fig. 3.155 Time-history curve of horizontal acceleration

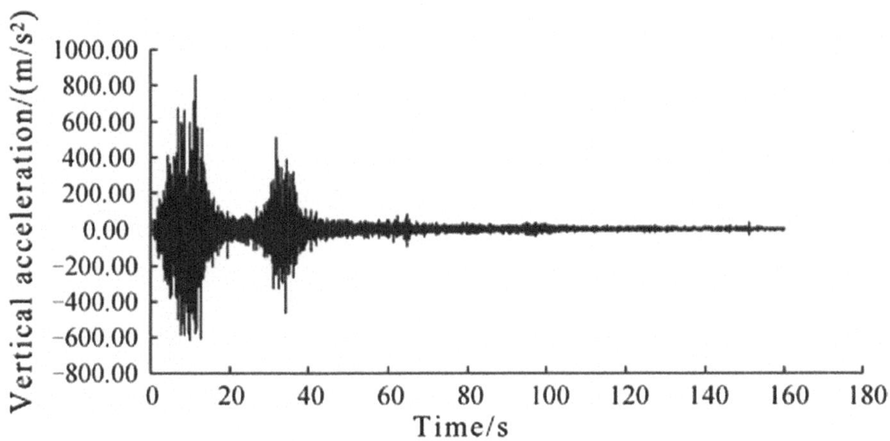

Fig. 3.156 Time-history curve of vertical acceleration

To further study the influence of slope surface forms on acceleration amplification effects along elevation, this chapter picks up the altitude of horizontal displacement, velocity, and acceleration of every monitoring points (see Table 3.16).

Through the comprehensive analysis of Table 3.16, it can be known that the acceleration peak value of every test points is basically all above 2.0 g (except test points on Flattop 2#), in which the peak acceleration of test points on three peak slope top 2# amount to 41.2 m/s^2. Compared with the peak acceleration of 9.57 m/s^2 of input seismic wave, the peak acceleration of different slope forms amplifies to 2.0–4.5 times. The monitoring results of flattop show that the peak acceleration of Test Points 1# and 3# on the two sides is 21.35, 18.65, and 21.35 m/s^2, which indicates that the seismic response near free face direction is more intense. The acceleration

Table 3.16 Altitude response characteristics of slope tops with different forms

Slope top topography	Horizontal altitude	Slope top altitude		
		1	2	3
Flattop	Displacement (m)	0.147	0.146	0.147
	Velocity (m/s)	0.739	0.717	0.739
	Acceleration (m/s^2)	21.35	18.65	21.35
Single peak slope top	Displacement (m)	0.153	0.169	0.153
	Velocity (m/s)	0.818	0.970	0.818
	Acceleration (m/s^2)	20.4	25.1	20.4
Double peak slope top	Displacement (m)	0.171	0.153	0.171
	Velocity (m/s)	1.61	1.42	1.61
	Acceleration (m/s^2)	39.35	27.4	39.35
Three peak slope top	Displacement (m)	0.15	0.21	0.15
	Velocity (m/s)	1.47	1.74	1.43
	Acceleration (m/s^2)	30.2	41.2	30.95

amplitude of three peak slope top, double peak slope top, single peak slope top, and flattop are, respectively, 41.2, 39.35, 25.1, and 21.35 m/s^2, in which the maximum peak acceleration of three peak slope top is about two times of that of flattop, which fully explains that the local topography of slope top has significant influence on the intensity of seismic ground motion response. Through the comprehensive analysis of the maximum peak velocity, acceleration, and displacement comparing results of test points on different slope top forms at the same elevation, it can be known that the seismic ground motion response rules are as follows: three peak slope top > double peak slope top > single top slope top > flattop.

② Influence of Slope Top Forms on Acceleration Fourier Spectrum Amplification Effect along Elevation

To study on the influence of slope top forms on acceleration Fourier spectrum, this chapter will, based on the previous research results about acceleration amplitude, select test point with the maximum acceleration amplitude, i.e. 1# of flattop, 2# of single peak slope top, 1# of double peak top, and 2# of three peak top. See Figs. 3.157, 3.158, 3.159, and 3.160 for detailed calculation results.

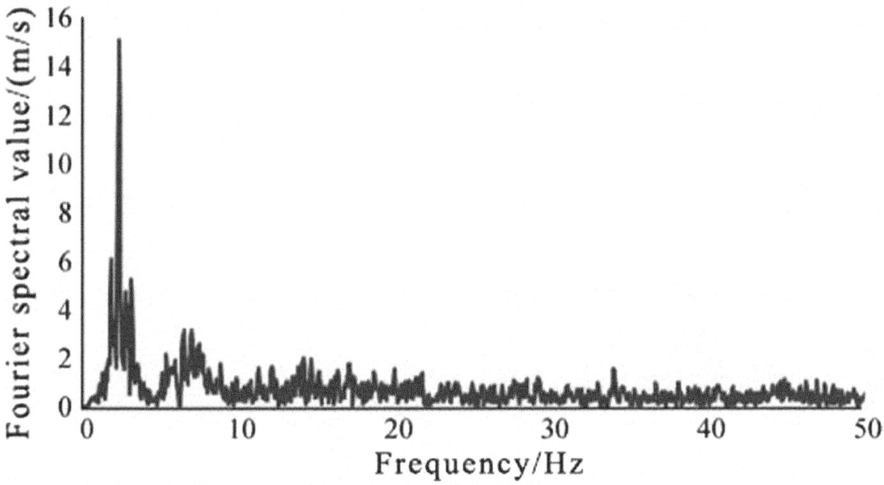

Fig. 3.157 Acceleration Fourier spectrum of Test Point 1# on flattop

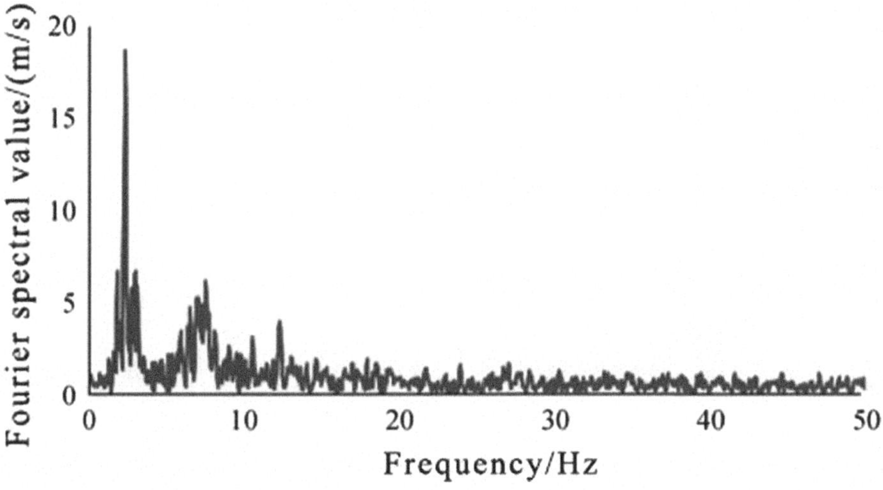

Fig. 3.158 Acceleration Fourier spectrum of Test Point 2# on single peak slope top

It can be known from the comprehensive analysis of Figs. 3.157, 3.158, 3.159, and 3.160 that the slope top forms have major influence on the frequency spectrum components of slope top acceleration Fourier spectrum, which mainly exemplifies as follows: The high-frequency components of acceleration Fourier spectrum on slope top of flattop, single, double, and three peak slopes are gradually strengthened, and the second step frequency gradually becomes obvious. The reason for the above phenomenon may be that the horizontal section area of slopes with flattop changes evenly along elevation, while that of slopes with single, double, or three peak slope

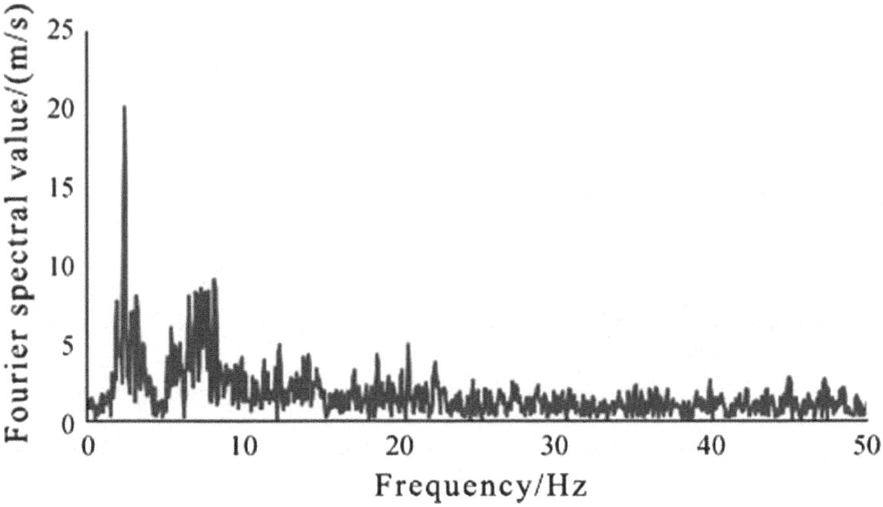

Fig. 3.159 Acceleration Fourier spectrum of Test Point 1# on double peak slope top

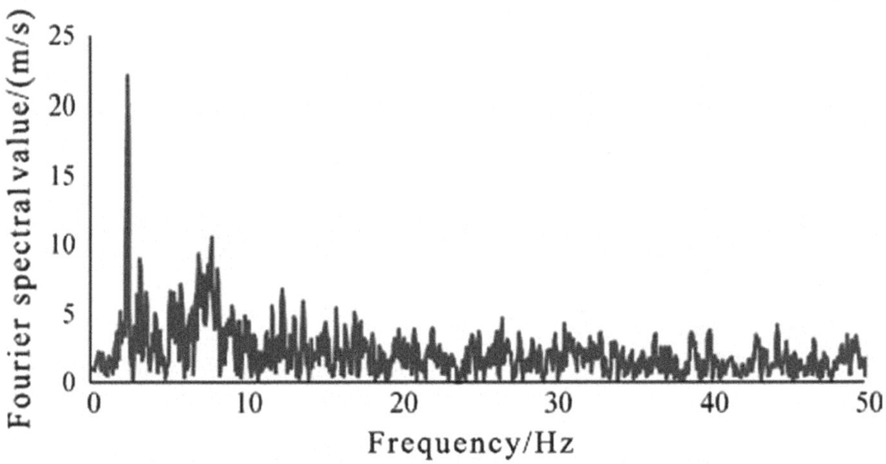

Fig. 3.160 Acceleration Fourier spectrum of Test Point 2# on three peak slope top

tops suddenly decrease at the slope tops, which lead to large phase difference between slope top action and the whole action of the slope mass in the earthquake or even cause whiplash effect, which further causes large peak acceleration, more of high-frequency components and more obvious of the second step frequency.

③ Influence of Slope Top Forms on Acceleration Response Spectrum Amplification Effect along Elevation

To study on the influence of slope top forms on acceleration response spectrum, this chapter will still select Test Point 1# of flattop, 2# of single peak slope top, 1# of

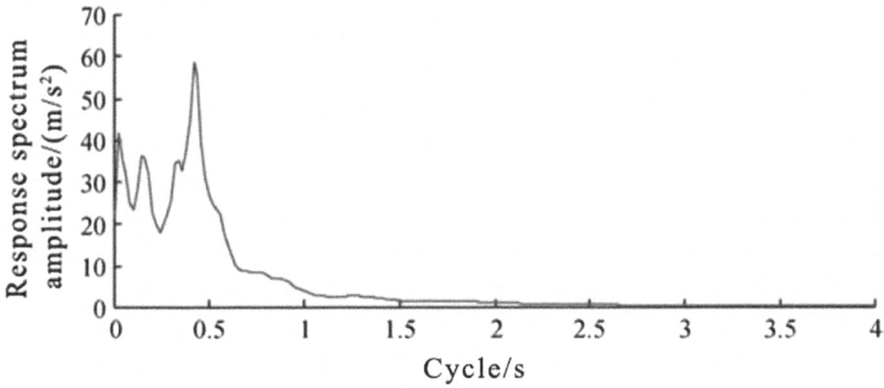

Fig. 3.161 Acceleration response spectrum of Test Point 1# on flattop

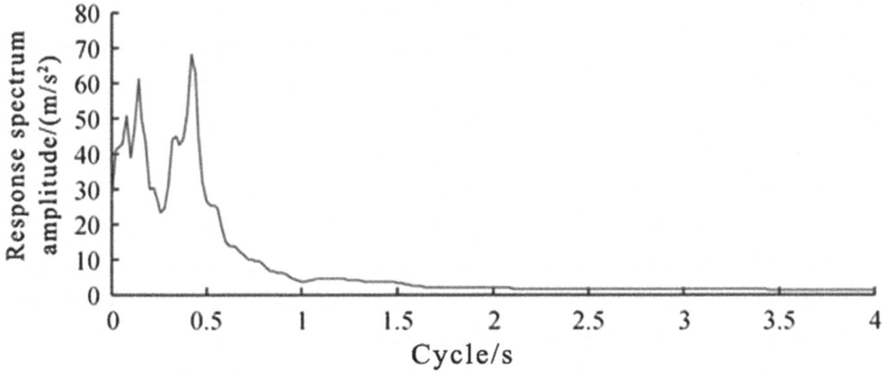

Fig. 3.162 Acceleration response spectrum of Test Point 2# on single peak slope top

double peak top, and 2# of three peak top. See Figs. 3.161, 3.162, 3.163, and 3.164 for detailed calculation results.

It can be known from the comprehensive analysis of Figs. 3.161, 3.162, 3.163, and 3.164 that the slope top forms have major influence on the frequency spectrum components of slope top acceleration response spectrum, which mainly exemplifies as follows: The characteristic periods of acceleration response spectrum of flattop, single, double, and three peak slopes are gradually transferred from long period to short period, and the double peak value of response spectrum gradually becomes obvious, which is basically identical with the calculation results of Fourier spectrum.

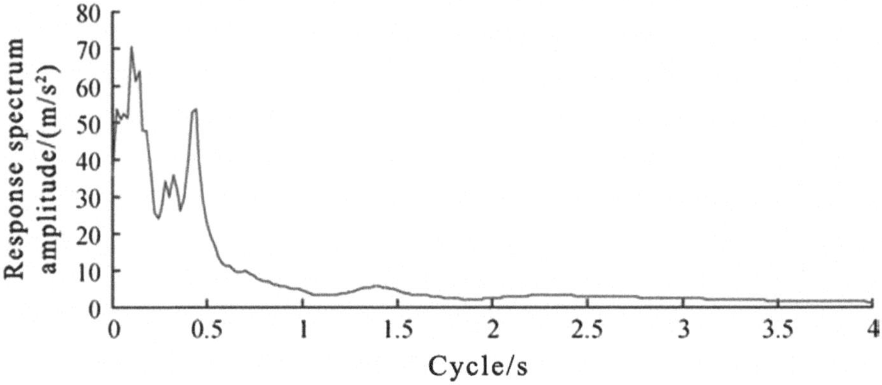

Fig. 3.163 Acceleration response spectrum of Test Point 1# on double peak slope peak

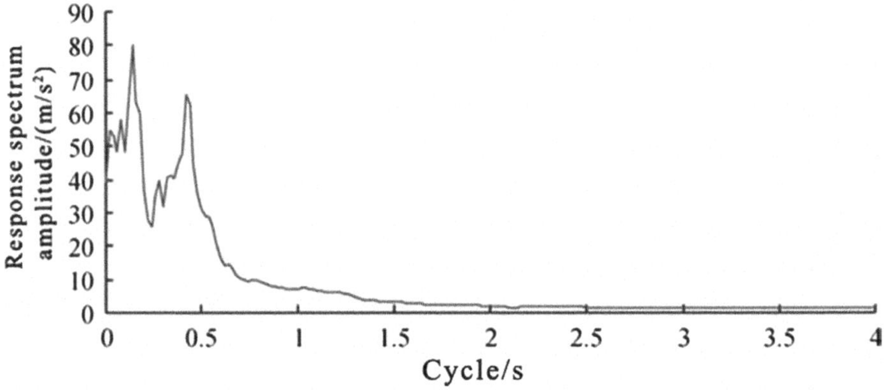

Fig. 3.164 Acceleration response spectrum of Test Point 2# on three peak slope top

3.3.6 Brief Summary

This chapter focused on the two major landslide hazard spots on the left side of G213, built the numerical analysis model based on the shaking table test model, and used the new numerical calculation software—GDEM—to conduct research on the general dynamic response characteristics of high and steep rock slopes with single and double sides under intense seismic effects. At the same time, it studied on the local topography effects of rock slope seismic dynamic response rules on micro-level and came to the following conclusions:

(1) No matter for single side high and steep slope or for dual side high and steep slope, the PGA of different positions along elevation amplifies to a different extent, and it shows as follows: the amplification effects of dual side high and steep slope on acceleration > the amplification effects of single side high and steep slope on acceleration; PGA amplification effects on slope sur-

face > PGA amplification effects inside slope mass; the horizontal acceleration coefficient > the vertical acceleration coefficient; and acceleration coefficient on steep slope topography > acceleration coefficient on gentle slope topography.

(2) The acceleration Fourier spectrum along elevation of single and dual side high and steep slope surfaces shows the changing rules of low-frequency components amplification, and the changing rules are influenced by slope topography and bevel angles. No matter for horizontal acceleration or vertical acceleration, the acceleration response spectrum along elevation of single side slope surfaces shows single peak value, while that of dual side slope surfaces followed the process of converting from single peak value to double peak value.

(3) The local topography of slope top has a significant influence on the intensity of seismic response, and the detailed rules are as follows: three peak slope top > double peak slope top > single peak slope top > flattop; the high-frequency components of all of them gradually strengthen, and the second step frequency becomes gradually obvious. At the same time, the characteristic periods of acceleration response spectrum of flattop, single, double, and three peak slopes are gradually transferred from long period to short period, and the double peak value of response spectrum gradually becomes obvious. The peak acceleration of different slope topographies amplifies to 2.0–4.5 times, and the maximum peak acceleration of three peak slope top is about two times of that of flattop.

Bibliography

Cui Peng, Wei Fangqiang, Chen Xiaoqing, et al. Secondary mountain hazards in Wenchuan earthquake and its countermeasures [J]. Science and technology disaster relief, 2008, 23(4): 317–323.

Huang Runqiu, Tang Chuan, Li Yong, et al. Study on geological hazards in Wenchuan earthquake [M]. Beijing: Science Press, 2009.

Jiang Liangwei. Experimental theory and technology research on large scale shaking table model test of soil slope [D]. Postdoctoral scientific research report from Southwest Jiaotong University, 2009.

Xu Suning, Li Wenpeng. Image study of typical geological hazards in the 5.12 Wenchuan earthquake [M]. Beijing: Geological Press, 2009.

Xu Qiang, Pei Xiangjun, Huang Runqiu. Study on large scale landslide in Wenchuan earthquake [M]. Beijing: Science Press, 2009.

Zhang Gongwei, Wang Sijing, Zhang Daoyuan. Engineering geology in China [M]. Beijing: Science Press, 2000.

Chapter 4
Time–Frequency Analysis Theory of Seismic Stability of High and Steep Rock Slopes

The seismic stability analysis of high and steep rock slopes under seismic effects involves multiple subjects including engineering geology, rock–soil mechanics, geophysics, and seismic engineering, which is of great difficulty. Up to now, there are still many problems that are in urgent need of solutions, and the special geographical and geological conditions and seismic configuration make it especially pertinent in China. Currently, the infrastructure construction such as highway, railway and water conservancy is in rapid developing period, which draws higher demand on instability mechanism and stability judgment of slope engineering, but the existing theories cannot well meet this requirement. For the seismic ground motion characteristics of rock slopes, comprehensive and in-depth research has been conducted through seismic array monitoring at Xishan Park, Zigong, shaking table test, and numerical analysis from the qualitative perspective, and the results show that the seismic ground motion response of dual side rock slope is more intense than that of single side rock slope. In practical engineering, to evaluate something comprehensively and systematically needs qualitative cognition and quantitative analysis. Both of them complement each other, and none is dispensable. Based on the previous study results of the author, it could be known that the existing problems in seismic stability analysis theories of high and steep rock slopes are mainly as follows:

(1) The current studies on seismic dynamic characteristics of rock slopes under seismic effects are still limited in field observation, shaking table test and numerical engineering, while studies on theoretical solutions are not too many. However, the existing theoretical solutions of seismic dynamic characteristics of rock slopes are derived through simple harmonic waves, and frequency and amplitude are stable in the whole time history of seismic ground motion, while that for real seismic waves are changing with time. That causes limitations in the application of the above theoretical solutions.
(2) In the aspects of deformation features and instability mechanism of slopes under seismic effects, the traditional research methods are mainly time-domain analysis method and frequency-domain analysis method. The time-domain analysis

C. Yang et al., *Three Dimensional Space-Time Analysis Theory of Geotechnical Seismic Engineering*, https://doi.org/10.1007/978-981-13-3356-9_4

method could only consider the seismic response in time-domain aspects, such as PGA, velocity, and displacement, while frequency-domain analysis method could only explain the seismic response in frequency-domain aspects, such as Fourier spectrum and energy spectrum. However, seismic wave is a kind of complex non-stationary signal, whose frequency and amplitude change with time, and its time, frequency, and amplitude form a large coupling of three-dimensional space system. Therefore, singular study the deformation features and instability mechanism of slopes under seismic effects in the aspect of frequency-domain information or time-domain information will miss information and lead to one-sidedness of the analysis results.

(3) In the aspect of evaluation method of seismic stability of slopes, because the probability analysis method started late and it is immature, and the numerical simulation method and experimental method are complex in operation and high in cost and consume large amount of time, these three methods are hard to popularize; although the sliding block analysis method is widely used, it only considers the time-domain characteristics without overall consideration of the seismic characteristics, so it is low in accuracy and may cause misjudge in practical engineering; the pseudo-static method is the most common used, but its defect is that it pays no consideration of the frequency and duration characteristics of seismic motion, and in most circumstances, the seismic wave features are described with three major factors like peak value, frequency, and duration, which largely bring down the recognition accuracy of pseudo-static method.

In this manner, this book will set the Wenchuan earthquake as background, focus on the existing relevant scientific problems, and take the dynamic characteristics of slopes under seismic effects as the mainline, and the shaking table test, numerical simulation, and theoretical analysis as measurements, and adopt the time–frequency analysis theory to derive theory of seismic dynamic characteristics of high and steep rock slopes, analyze its disaster mechanism, so as to, on this basis, propose a set of seismic stability analysis method of rock slopes which could consider the influence of all the three major factors (amplitude, frequency, and duration).

4.1 Theoretical Time–Frequency Analysis of Dual Side Rock Slope Acceleration Elevation Amplification Effects

This chapter will, based on the previous qualitative research results of the author, using Hilbert–Huang transform (HHT for short) combining with elastic wave theory and dual side rock slope analysis model, propose the time–frequency analysis solution of rock slope acceleration elevation amplification effects, achieve transition from qualitative analysis to quantitative analysis of acceleration amplification effects so as to provide accurate reference to research and design of slope seismic engineering.

4.1.1 *Derivation of Time–Frequency Analysis Formula for Acceleration Elevation Amplification Effects*

The dual side rock slope acceleration elevation amplification effects time–frequency method is derived, based on elastic wave theory and horizontal slice method, through building force equilibrium differential equation of micro-unit, using Hilbert–Huang transform and boundary conditions, and applying modal analysis method and normal mode theory.

4.1.1.1 The General Thinking and Basic Assumption

The time–frequency analysis method derivation of dual side rock slope acceleration elevation amplification effects mainly follows the following process: Firstly, analyze the model and divide it into micro-units of N horizontal slices; then select anyone of the micro-units, use HHT and elastic wave theory to solve the wave power when seismic wave passing through it, build force equilibrium differential equation and combine boundary conditions; lastly apply modal analysis method and normal mode theory to work out time–frequency analysis solution of dual side rock slope acceleration elevation amplification effects. The derivation of this method will base on the following hypothesis: ① the slope materials are homogeneous and isotropic elastomer; ② the bedrock is of rigid body so we can pay no attention to the bedrock deformation; ③ due to that wave-type conversion will occur when seismic wave arrives at the slope surface with its major energy transferring into Rayleigh surface wave moving along the slope surface, and Rayleigh surface wave moves with elliptic polarization, so there will turn out two equal stress components of opposite direction within the range of one elliptic polarization. This method supposes that the slope material is homogeneous and isotropic elastomer. Therefore, the total stress component produced by Rayleigh wave on the slope surface is near 0, and so the stress of Rayleigh wave will not be taken consideration of in formula calculation of this dissertation.

4.1.1.2 Generalized Analysis Model

Based on the hazard research results of Wenchuan earthquake, we generalize the analysis model of dual side rock slopes (see Fig. 4.1). Figure 4.1 shows that at any elevation z, the shear displacement is v, shear stress is $\gamma = \partial v/\partial z$, the slope ratio of the two sides are 1:$n1$ and 1:$n2$, the total slope ratio is $n = n1 + n2$, the physico-mechanical parameters of slope mass are damping ratio c, density ρ, shear modulus G, and the acceleration time-history of slope bottom input seismic wave is $\ddot{v}_g(t)$. At the same time, divide the slope mass with horizontal slice method into N micro-units of horizontal slices and select micro-unit A as study object. Figure 4.2 is the force analysis model of micro-unit A under SV wave effects. The height of micro-unit

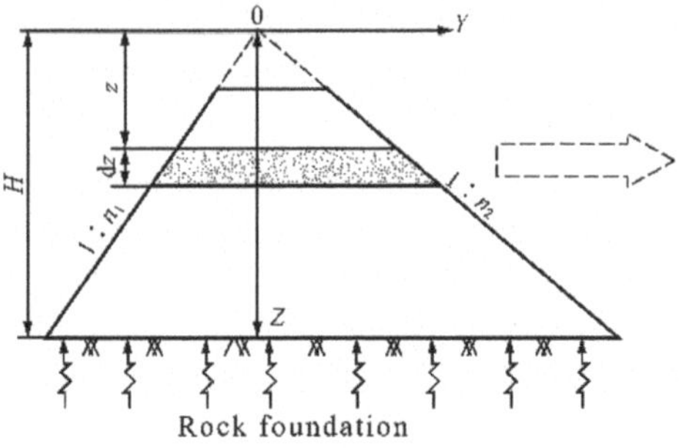

Fig. 4.1 Generalized analysis model of dual side rock slope

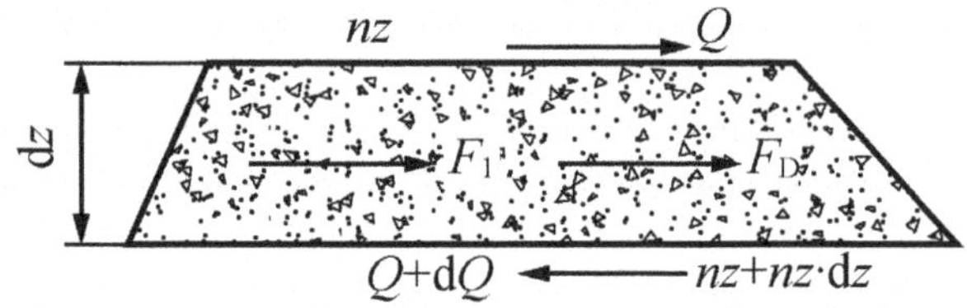

Fig. 4.2 Analysis model of micro-unit *A*

A is dz, the width of the upper surface is nz, and the width of the lower surface is $nz + nz \cdot \mathrm{d}z$.

4.1.1.3 Derivation Process

Figure 4.2 shows that the top shear stress of micro-unit *A* is Q, the bottom shear stress is Q +dQ, the inertia force of earthquake on micro-unit *A* is $F1$, and the damping force is FD. See Formulas (4.1)–(4.4).

$$F_1 = \rho nz \cdot \mathrm{d}z\left(\frac{\partial^2 v}{\partial t^2} + \ddot{v}_g(t)\right) \tag{4.1}$$

$$F_D = cnz \cdot \mathrm{d}z \cdot \frac{\partial v}{\partial t} \tag{4.2}$$

$$Q = \tau nz = G\gamma nz = G\frac{\partial v}{\partial z}nz \tag{4.3}$$

$$\frac{\partial Q}{\partial z}\mathrm{d}z = Gn\left(z\frac{\partial^2 v}{\partial z^2} + \frac{\partial v}{\partial z}\right)\mathrm{d}z \tag{4.4}$$

Build the force equilibrium equation of micro-unit *A* according to the stress conditions in Fig. 4.2 (see Formula 4.5). Then substitute Formulas (5.1)–(5.4) into Formula (5.5), we get Formula (5.6):

$$F_1 + F_D - \frac{\partial Q}{\partial z}\mathrm{d}z = 0 \tag{4.5}$$

$$\frac{\partial^2 v}{\partial t^2}+\frac{c}{\rho}\frac{\partial v}{\partial t}-\frac{G}{\rho}\left(\frac{\partial^2 v}{\partial z^2}+\frac{1}{z}\frac{\partial v}{\partial z}\right)=-\ddot{v}_g(t) \tag{4.6}$$

Using separation of variables method, that is to make $v=\phi(z)Y(t)$, in which $\phi(z)$ is dimensionless vibration mode function, and $Y(t)$ is generalized coordinates or canonical coordinates, i.e. the time function of vibration mode. At the same time, separate Formula (4.6) into two differential equations only with variables t and z (see Formula 4.7):

$$\sum_1^{\infty}\phi\frac{\partial^2 Y}{\partial t^2}+\frac{c}{\rho}\sum_1^{\infty}\phi\frac{\partial Y}{\partial t}-\frac{G}{\rho}\left(\sum_1^{\infty}\frac{\partial^2 \phi}{\partial z^2}Y+\frac{1}{z}\sum_1^{\infty}\frac{\partial \phi}{\partial z}Y\right)=-\ddot{v}_g(t) \tag{4.7}$$

Multiplied by $\rho z\phi$ within $\sum$, and we get Formula (4.8):

$$\sum_1^{\infty}\phi\rho z\phi\frac{\partial^2 Y}{\partial t^2}+c\sum_1^{\infty}\phi z\phi\frac{\partial Y}{\partial t}-\frac{G}{\rho}\left(\sum_1^{\infty}\rho z\phi\frac{\partial^2 \phi}{\partial z^2}Y+\frac{1}{z}\sum_1^{\infty}\rho z\phi\frac{\partial \phi}{\partial z}Y\right)$$
$$=-\dot{v}_g(t)\rho z\phi \tag{4.8}$$

Because of mass $m=\rho nz\cdot \mathrm{d}z$, $c=cnz\cdot \mathrm{d}z$; therefore,m and$c$ are all proportional to z, so ρz can represent m, and cz can represent c. According to orthogonality of vibration mode, we can know: When $m\neq n$, $\sum\varphi mm\varphi n=0$, $\sum\varphi mc\varphi n=0$; on the contrary, when $m=n$, $\sum\varphi mm\varphi n\neq 0$, $\sum\varphi mc\varphi n\neq 0$, so the above formula only suits for some type of vibration mode. In this way, we can separate $-\partial^2 Y/\partial t^2$, $\partial Y/\partial t$ and Y from $\sum$. $\ddot{v}_g(t)$ has no relation to vibration mode, so it should not be in $\sum$ (see Formula 4.9).

$$\left(\sum_0^H\phi^2\rho z\right)\frac{\partial^2 Y}{\partial t^2}+\left(\sum_0^H\phi^2 cz\right)\frac{\partial Y}{\partial t}-\frac{GY}{\rho}\left(\sum_0^H\rho z\phi\frac{\partial^2\phi}{\partial z^2}+\frac{1}{z}\sum_0^H\rho z\phi\frac{\partial\phi}{\partial z}\right)$$
$$=-\left(\sum_0^H\rho z\phi\right)\ddot{v}_g(t) \tag{4.9}$$

Bring $c=2\rho\lambda w$ into Formula (4.9), then use $\sum_0^H\phi^2\rho z$ and $\frac{GY}{\rho}$ to divide each item into Formula (4.9), respectively, so Formula (4.9) can be separated into:

$$\frac{\partial^2 Y}{\partial t^2}+2\lambda w\frac{\partial Y}{\partial t}+R\frac{G}{\rho}Y=-\left(\sum_0^H\phi z/\sum_0^H\phi^2 z\right)\ddot{v}_g(t)=-\eta\ddot{v}_g(t) \tag{4.10}$$

$$\frac{\partial^2\phi}{\partial z^2}+\frac{1}{z}\frac{\partial\phi}{\partial z}+R\phi=0 \tag{4.11}$$

Due to $w^2=\frac{k}{\rho}$, and k of pure shear component is the function of G and component geometric size, so $\frac{RG}{\rho}=w^2$, in which R is the undetermined constant. So Formula (4.10) can be rewritten into Formula (4.12):

$$\frac{\partial^2 Y}{\partial t^2} + 2\lambda w \frac{\partial Y}{\partial t} + w^2 Y = -\left(\sum_0^H \phi z / \sum_0^H \phi^2 z\right) \ddot{v}_g(t) \tag{4.12}$$

① Solution of vibration mode function $\varphi(z)$.

Through comprehensive analysis of Formulas (4.11) and (4.12), it can be known that we need firstly to solve Formula (4.11) to get R before solving Formula (4.12). Now we make $R = (\frac{\xi}{z})^2$, so we will get Formulas (4.13) and (4.14):

$$\frac{\partial \phi}{\partial z} = \frac{\partial \phi}{\partial \xi} \cdot \frac{\partial \xi}{\partial z} = \frac{\xi}{z} \frac{\partial \phi}{\partial \zeta} \tag{4.13}$$

$$\frac{\partial^2 \phi}{\partial z^2} = \frac{\xi^2}{z^2} \frac{\partial^2 \phi}{\partial \xi^2} \tag{4.14}$$

Bring Formulas (4.13) and (4.14) into Formula (4.11), we will get Formula (4.15), while Formula (4.15) is zero-order Bessel function, whose general solution is Formula (4.16).

$$\frac{\partial^2 \phi}{\partial \xi^2} + \frac{1}{\xi} \cdot \frac{\partial \phi}{\partial \xi} + \phi = 0 \tag{4.15}$$

$$\phi = A_1 J_0(\xi) + A_2 N_0(\xi) = A_1 J_0(\sqrt{R}z) + A_2 N_0(\sqrt{R}z) \tag{4.16}$$

In Formula (4.16), J_0 is the first type of zero-order Bessel function, and N_0 is the second type of zero-order Bessel function.

According to boundary condition: $z = 0$, $\partial \phi / \partial z = 0$, take partial derivative of z from Formula (4.16), and we will get:

$$\partial \phi / \partial z = -\sqrt{R}[A_1 J_1(\sqrt{R}z) + A_2 N_1(\sqrt{R}z)] \tag{4.17}$$

In Formula (4.17), J_1 is the first type of zero-order Bessel function, and N_1 is the second type of zero-order Bessel function.

Bring boundary conditions into Formula (4.17), and because when z approaches zero, $N_1(0)$ approaches infinity, and at the same time, when $z = 0$, $J_1(0) = 0$. Therefore, if we want to make the above formula equal zero, then it must have $A_2 = 0$, so Formula (4.16) can be written as:

$$\phi = A_1 J_0(\sqrt{R}z) \tag{4.18}$$

Because vibration mode is the dimensionless relative value, we take $A_1 = 1$.

According to boundary condition: $z = H$, $\phi = 0$, we bring it into Formula (5.18) and get: $J_0(\sqrt{R}z) = 0$. Due to that there are countless null points in the first type of zero-order Bessel function, so this dissertation adopts β_0, n to represent. Corner mark n represents modal order. At the same time, $\sqrt{R}H = \beta_{0,n}$, so $R = \beta_{0,n}^2 / H^2$.

Therefore, there are countless R values, and bring them into Formula (4.18), we will get vibration mode function (see Formula 4.19). It is noteworthy that β_0 and w appearing in the following texts all omit corner mark n.

$$\phi = A_1 J_0\left(\beta_0 \cdot \frac{z}{H}\right) \tag{4.19}$$

② Solution of time function of vibration mode $Y(t)$

Through comprehensive analysis and simplification of Formula (4.12), we will get:

$$\frac{\partial^2 Y}{\partial t^2} + 2\lambda w \frac{\partial Y}{\partial t} + w^2 Y = -\left(\sum_0^H \phi z / \sum_0^H \phi^2 z\right) \ddot{v}_g(t) = -\eta \ddot{v}_g(t) \tag{4.20}$$

In Formula (4.20) $w^2 = \frac{RG}{\rho}$, $R = \frac{\beta_{0,n}^2}{H^2}$ and η are modal participation coefficients. For continuum, there is $\eta = \frac{\int_0^H \phi z \cdot \mathrm{d}z}{\int_0^H \phi^2 z \cdot \mathrm{d}z}$, and by bringing it into Formula (4.18), we can get modal participation factor:

$$\eta = \int_0^1 \frac{z}{H} J_0\left(\beta_0 \frac{z}{H}\right) \mathrm{d}\left(\frac{z}{H}\right) / \int_0^1 \frac{z}{H} J_0^2\left(\beta_0 \frac{z}{H}\right) \mathrm{d}\left(\frac{z}{H}\right) = \frac{2}{\beta_0 J_1(\beta_0)} \tag{4.21}$$

Because Formula (4.20) is a non-homogeneous constant coefficient second-order linear differential equation, its general solution is the general solution of its accompanied homogeneous equation plus the particular solution of an inhomogeneous equation (see Formula 4.22).

$$Y = \frac{-\eta}{w'} \int_0^t \ddot{v}_g(\tau) \mathrm{e}^{-\lambda w (t-\tau)} \sin w'(t-\tau) \mathrm{d}\tau \tag{4.22}$$

$$w' = \frac{\beta_0}{H} v_s \sqrt{1-\lambda^2} \tag{4.23}$$

③ Solution of Slope Mass Displacement, Velocity, and Acceleration Response

Seismic wave is a nonlinear non-stationary signal, and its frequency and amplitude gradually change over time. Therefore, based on the elastic wave theory, the seismic ground motion time-history can be seen as the combination of N (the total number of sampling points) harmonic wave with different amplitude and frequency, so the harmonic equation of any time t (the mth sampling point) can be expressed by elastic displacement in an unified way (see Formula 4.24), and the acceleration amplitude can see Formula (4.25).

$$V_{g,m}(t) = V_{g,m}(0) \mathrm{e}^{i\omega_m t} \tag{4.24}$$

$$\left|\ddot{V}_{g,m}(t)\right| = \left|\ddot{V}_{g,m}(0)\omega_m^2\right| = \ddot{V}_g(\tau) \tag{4.25}$$

In Formulas (4.24) and (4.25), $V_{g,m}(t)$ and $\left|\ddot{V}_{g,m}(t)\right|$, respectively, represent the elastic displacement and acceleration amplitude of slope bottom seismic wave at t (the mth sampling point). $V_{g,m}(0)$, $\left|\ddot{V}_{g,m}(0)\right|$ and ω_m represent the elastic displacement, acceleration amplitude and instantaneous frequency slope bottom seismic wave at t (the mth sampling point).

Through comprehensive analysis of Formulas (4.19)–(4.26), we can get the time–frequency analysis solution of shear displacement response, velocity response, and acceleration response (see Formulas 4.26–4.28).

$$v = \sum_1^n J_0\left(\beta_0 \frac{z}{H}\right) \cdot \frac{-\eta}{w'} \int_0^t \ddot{v}_g(\tau)\omega_\tau^2 \mathrm{e}^{-\lambda w(t-\tau)} \sin w'(t-\tau)\mathrm{d}\tau \tag{4.26}$$

$$\frac{\partial v}{\partial t} = \sum_1^n J_0\left(\beta_0 \frac{z}{H}\right) \cdot \left[-\eta \int_0^t \ddot{v}_g(\tau)\omega_\tau^2 e^{-\lambda w(t-\tau)} \cos w'(t-\tau)\mathrm{d}\tau - \lambda\omega Y\right] \tag{4.27}$$

$$\begin{aligned}\frac{\partial^2 v}{\partial t^2} &= -\ddot{v}_g(t)\omega_\tau^2 + \sum_1^n J_0(\beta_0 \frac{z}{H})\eta\left[\frac{(1-2\lambda^2)w}{\sqrt{1-\lambda^2}} \int_0^t \ddot{v}_g(\tau)\omega_\tau^2 \mathrm{e}^{-\lambda w(t-\tau)}\right.\\ &\quad \left. \cdot \sin w'(t-\tau)\mathrm{d}\tau + 2\lambda w \int_0^t \ddot{v}_g(\tau)\omega_\tau^2 \mathrm{e}^{-\lambda w(t-\tau)} \cdot \cos w'(t-\tau)\mathrm{d}\tau\right]\end{aligned} \tag{4.28}$$

4.1.1.4 Time–Frequency Analysis Process of Seismic Waves

Based on the above derivation, it can be known that the time–frequency effects of seismic waves mainly embody in the elastic displacement amplitude and frequency of input seismic waves. Therefore, this dissertation the following analysis process of seismic wave time–frequency effects: Firstly, break up seismic wave time–frequency into several of IMFs (time history curve) through Empirical Mode Decomposition (EMD for short); secondly, carry out Channel Switch, i.e. switch multi-channel signal composed of multiple IMFs into single-channel signal made up of single IMF; thirdly, solve the instant frequency of each channel and draw the time–frequency curve of each IMF. Lastly, bring each IMF and their instant frequency into Formulas (4.26)–(4.28) to work out the results and superimpose the results, so we can get the seismic response of slopes like displacement, velocity, and acceleration.

4.1.1.5 Practical Application of Seismic Wave Time–Frequency Effects

In order to more deeply and concretely explaining the practical application of seismic wave time–frequency effects, this dissertation will, based on the analysis ideas of

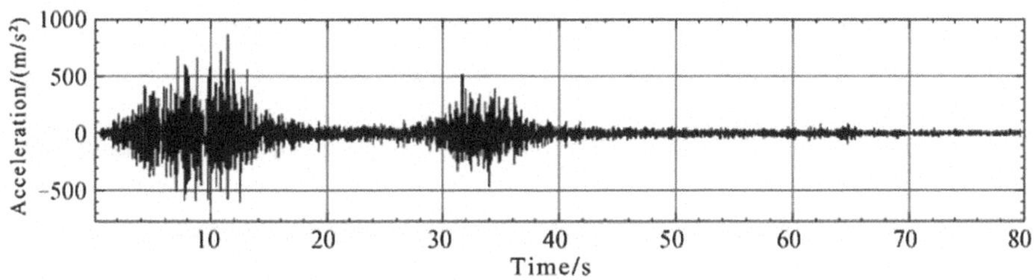

Fig. 4.3 NS direction Wenchuan wolong seismic wave

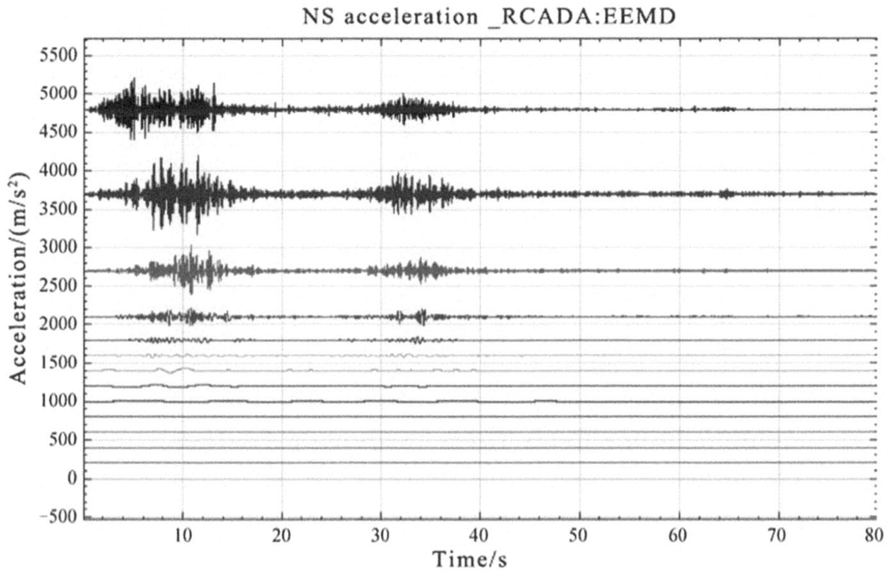

Fig. 4.4 IMF component of NS direction Wenchuan wolong seismic wave

seismic wave time–frequency effects, conduct input wave time–frequency analysis and take the example of NS direction component of Wenchuan wolong seismic wave to explain. The details are as follows: Firstly, input the NS direction Wenchuan wolong seismic wave (see Fig. 4.3); secondly, carry out EEMD decomposition and get every IMF (see Fig. 4.4); lastly, solve the instant frequency of every IMF (see Fig. 4.5). It is noteworthy that, due to that the duration of Wenchuan wolong seismic wave is a bit longer, so Figs. 4.3, 4.4, and 4.5 only select calculation results of $T = 0$–80 s including peak seismic ground motion acceleration and demonstrates it clearly. At the same time, the Y-coordinate in Figs. 4.4 and 4.5 only represents physical characters, but not size.

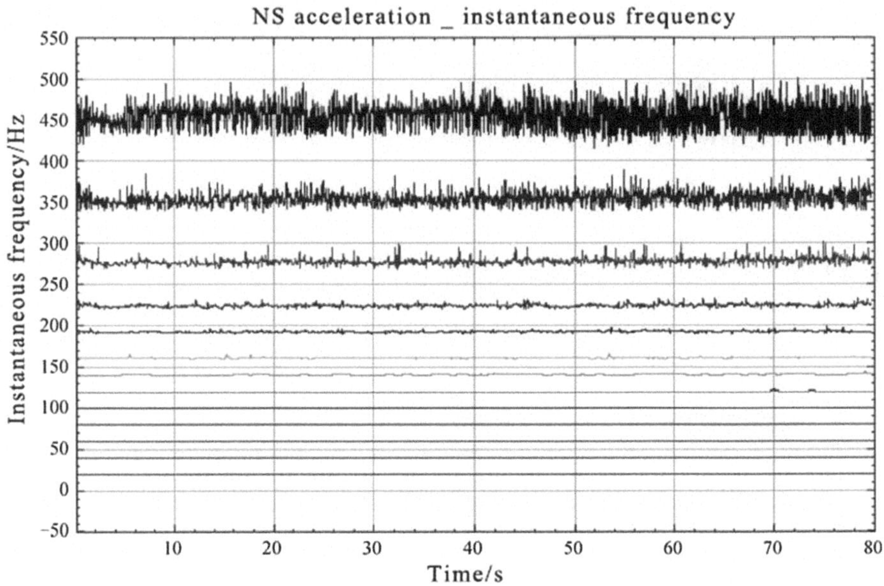

Fig. 4.5 IMF component instant frequency of NS direction Wenchuan wolong seismic wave

4.1.2 Solving Approach of Acceleration Elevation Amplification Effects Time–Frequency Analysis Formula

To sum up, this chapter will generalize the solving approach of dual side rock slope acceleration elevation amplification effects time–frequency analysis method, and the details are as follows:

(1) According to the study object, we sum up appropriate geological analysis model and decide the basic physical parameters such as geometric size, mechanical parameters of rock and soil mass;
(2) Decide the design response spectrum curve of this region according to *Code for Seismic Design of Buildings* and synthesize artificial ground motion time history based on seismic ground motion time history fitting method proposed by Hu Yuxian, et al., so as to decide the seismic ground motion acceleration time history of this region;
(3) Based on seismic ground motion acceleration time history curve, work out IMF of bottom input seismic wave and the correspondent frequency time history curve.

Lastly, bring the obtained results into Formulas (4.26)–(4.28) to calculate and superimpose the results, so we can get the seismic response of slopes like displacement, velocity, and acceleration.

Table 4.1 Physico-mechanical parameters of bedrock

Model name	Constitutive model	Unit weight (kN/m^3)	Cohesive force (kPa)	Internal friction angle (°)	Elasticity modulus (MPa)	Poisson's ratio
Bedrock	Mohr–Coulomb	22	93.6	37.9	450	0.25

4.1.3 Verification of Dual Side Rock Slope Acceleration Elevation Amplification Effects Time–Frequency Analysis Method

Before verifying the correctness of dual side rock slope acceleration amplification effects along elevation time–frequency analysis results, this chapter needs to explicitly explain the verification approach, and the details are as follows: First, build numerical analysis model by geometric similarity ratio of 1:1 to the shaking table test model of dual side rock slope and compare the numerical calculation results with the experimental results so as to verify the correctness of numerical analysis method. This part of work has been finished in Sect. 3.3.3. Second, generalize the numerical analysis model based on the prototype of dual side rock slope shaking table test, adopt the same numerical calculation method, and set up acceleration monitoring points at slope top and mid-slope so as to verify the correctness of acceleration amplification effects along elevation time–frequency analysis method.

4.1.3.1 Overview of Numerical Analysis

In order to accurately verify the correctness of acceleration amplification effects along elevation time–frequency analysis method, this dissertation will, based on the prototype of shaking table test model, restore the shaking table test model into prototype slope according to the similarity system. It will pay no attention to the influence of slope volume, adjust the bevel angles on two sides into 50°, adopt the restored rock and soil mass parameters according to the similarity ratio, and input only the horizontal Wenchuan wolong seismic wave so as to produce only the vertical propagating shear wave. Set up horizontal acceleration and shear displacement monitoring points at the top and medium of prototype slope numerical model to observe the displacement and acceleration time history changes at top and medium positions under seismic effects.

See Fig. 4.6 for numerical analysis model, Table 4.1 for physico-mechanical parameters of slope mass, and Figs. 4.7 and 4.8 for time history curve of Wenchuan wolong seismic elementary wave.

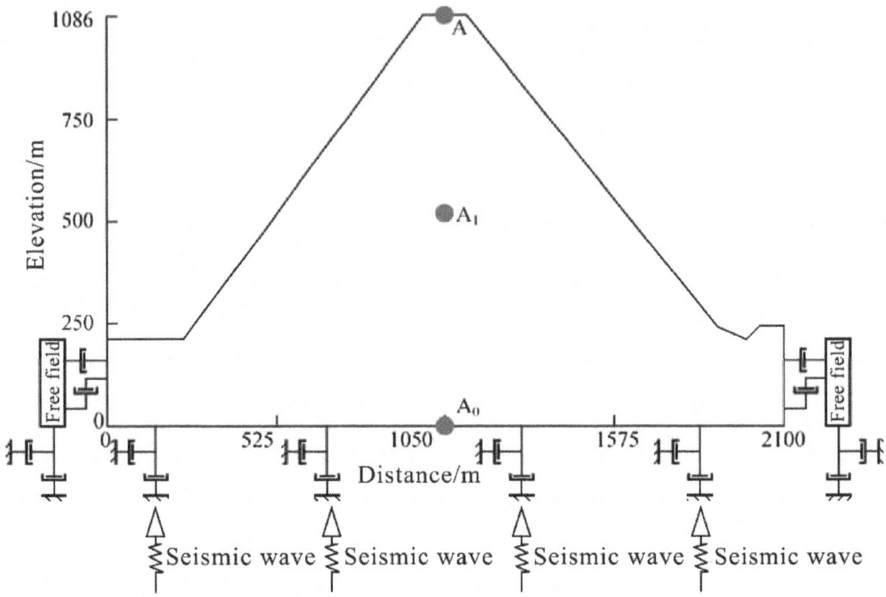

Fig. 4.6 Prototype slope model

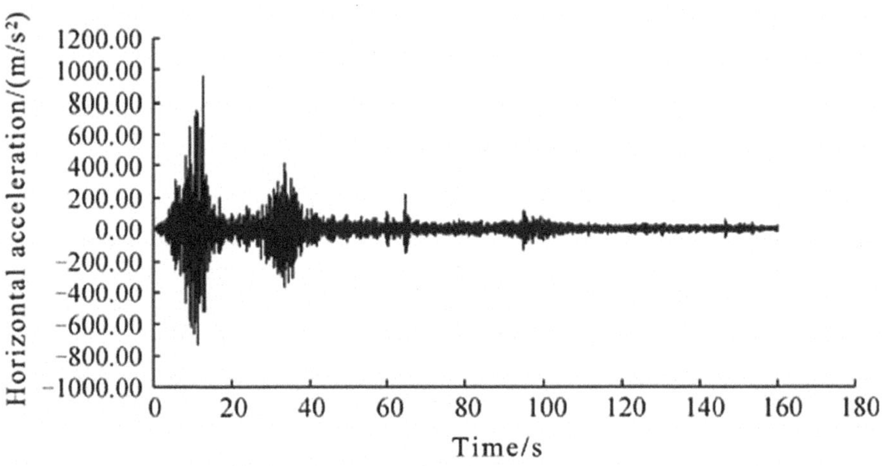

Fig. 4.7 Time-history curve of horizontal acceleration

4.1.3.2 Correctness Verification of Time–Frequency Analysis Method of Dual Side Rock Slope Acceleration Elevation Amplification Effects

Based on the time–frequency analysis method solving approach of dual side rock slope acceleration amplification effects along elevation (hereinafter referred to as

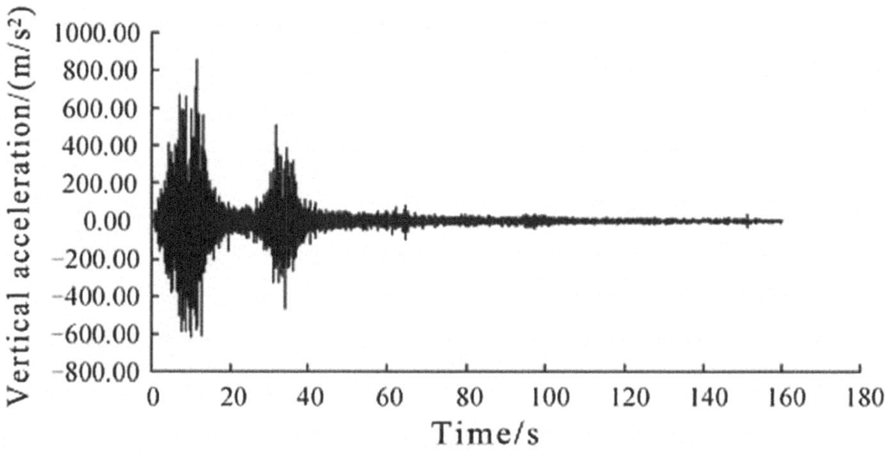

Fig. 4.8 Time-history curve of vertical acceleration

Table 4.2 Comparison of time–frequency analysis method and numerical calculation of shear displacement and horizontal acceleration amplification coefficient at slope top

PGA (g)	Shear displacement (m)			Horizontal acceleration amplification coefficient		
	Time–frequency analysis method	Numerical calculation	Error (%)	Time–frequency analysis method	Numerical calculation	Error (%)
0.2	0.51	0.55	7.3	2.12	2.073	−2.2
0.4	0.57	0.62	8.1	1.92	1.85	−3.7
0.6	0.64	0.706	9.3	1.73	1.62	−6.8

time–frequency analysis method), bring the relevant parameters and the above calculation results into Formulas (4.26)–(4.28) to work out the shear displacement and horizontal acceleration of each point. See Tables 4.2 and 4.3 for detailed calculation results.

It can be known from Tables 4.2 and 4.3 that, under the effects of Wenchuan wolong seismic wave with PGA = 0.2, 0.4, and 0.6, the numerical calculation results of shear displacement at slope top and mid-slope are all greater than calculation results worked out by time–frequency analysis method, and the error of these two distributed, respectively, in the range of 7.3–9.3% and 4.8–8.6%. The numerical calculation results of horizontal acceleration at slope top and mid-slope are all less than that of calculation results at slope top and mid-slope, and the error of these two distributed, respectively, in the range of 2.2–6.8% and 2.0–4.1%. Therefore, with the increase in height, the calculation error of time–frequency analysis method also increases; with the increase of PGA, the calculation error of time–frequency analysis method also increases, too. The reason for the above phenomenon may be that the time–frequency analysis method of dual side high and steep rock slope acceleration

Table 4.3 Comparison of time–frequency analysis method and numerical calculation of shear displacement and horizontal acceleration amplification coefficient at mid-slope

PGA (g)	Shear displacement (m)			Horizontal acceleration amplification coefficient		
	Time–frequency analysis method	Numerical calculation	Error (%)	Time–frequency analysis method	Numerical calculation	Error (%)
0.2	0.20	0.22	4.8	1.51	1.48	−2.0
0.4	0.26	0.28	7.1	1.35	1.32	−2.3
0.6	0.32	0.35	8.6	1.26	1.21	−4.1

Note Horizontal acceleration amplification coefficient in Tables 4.2 and 4.3 = acceleration of test point/horizontal acceleration of test point at model bottom

elevation effects are derived based on elastic wave theory without consideration of nonlinear characteristics of rock and soil mass. With the increase of seismic intensity, the nonlinear characteristics of rock and soil mass become more obvious, which weaken the acceleration amplification effects and strengthen shock insulation and shock absorption effects, which cause major difference in calculation results of these two.

In general, no matter the shear displacement or horizontal acceleration amplification coefficient, the calculation results of time–frequency analysis method and the numerical analysis results of dual side high and steep rock slope acceleration elevation effects are basically the same, with error controlled within 10%, which fully verify the correctness of the calculation results and feasibility of calculation method in this dissertation.

4.1.4 Parameter Studies on Time–Frequency Analysis Method of Acceleration Elevation Amplification Effects

The previous research results based on field seismic array monitoring, shaking model test, and numerical analysis show that the amplitude and frequency of input wave and bevel angles have significant influence on rock slope acceleration amplification effects. Therefore, this chapter will adopt time–frequency analysis method used by dual side rock slope elevation amplification effects and carry out research of influence of bevel angles, peak value of input wave and frequency on shear displacement and acceleration from the perspective of theoretical analysis.

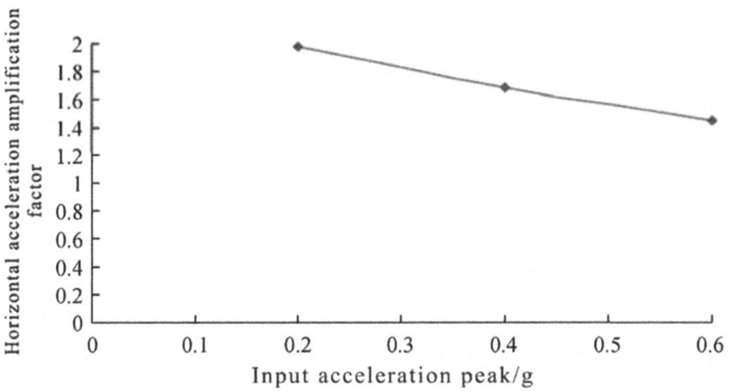

Fig. 4.9 Acceleration amplification coefficient at slope top VS input PGA

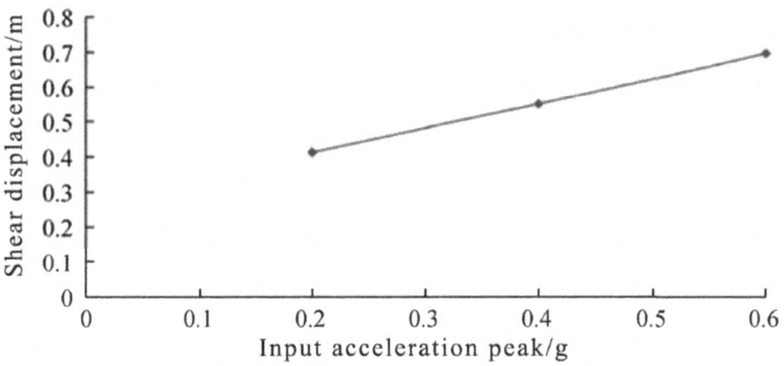

Fig. 4.10 Maximum shear displacement at slope top VS input PGA

4.1.4.1 Research of Influence of Input Wave Peak Value on Rock Slope Acceleration Elevation Amplification Effects

In order to study on the influence of input wave peak value on rock slope acceleration elevation amplification effects, now we take the following example to explain: The slope height is 1000 m, the bevel angles of the slope sides are all 60°, the damping ratio is 5%, the elasticity modulus is 20,000 Mpa, and Poisson's ratio is 0.21. We input sine wave with PGA = 0.1, 0.2, 0.4, and 0.6, and its frequency is 0.8 Hz. See Figs. 4.9 and 4.10 for detailed calculation results. It is noteworthy that in Fig. 4.10, the acceleration amplification coefficient at slope top = slope top acceleration/slope bottom acceleration.

Through comprehensive analysis of Figs. 4.9 and 4.10, it can be known that, with the effects of sine wave with PGA = 0.1, 0.2, 0.4, and 0.6, the acceleration amplification coefficient at slope top gradually decreases with the increase of input GPA, and it is within the range of 1.4–2.0. The maximum shear displacement gradually increases at the slope top, and it is within the range of 0.4–0.7 m.

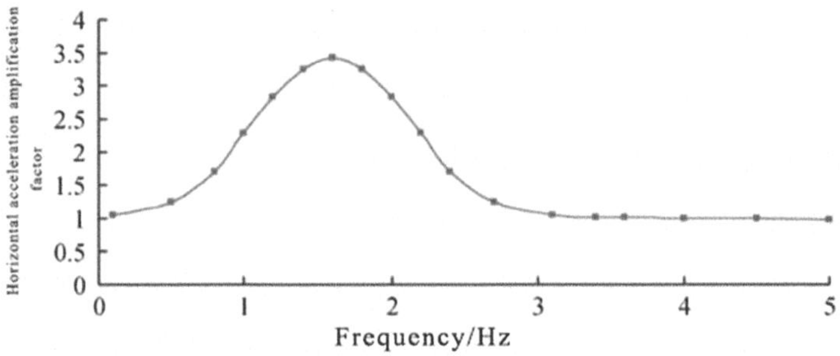

Fig. 4.11 Acceleration amplification coefficient at slope top VS input wave frequency

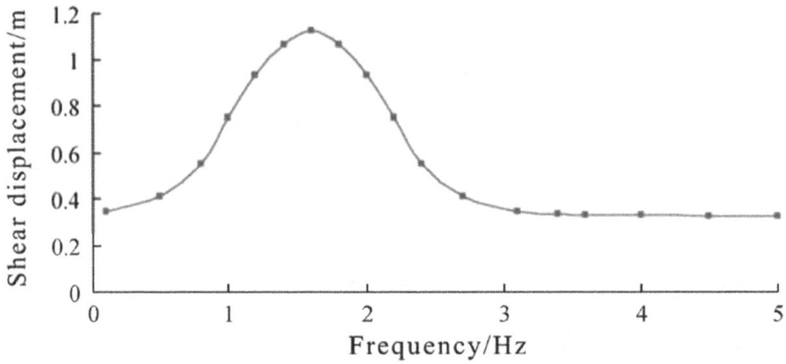

Fig. 4.12 Maximum shear displacement at slope top VS input wave frequency

4.1.4.2 Research of Influence of Input Wave Peak Value on Rock Slope Acceleration Elevation Amplification Effects

In order to study on the influence of input wave frequency on rock slope acceleration elevation amplification effects, now we adopt example of the same geometric size with the above example, and input sine wave with PGA = 0.4 g, whose frequency changes are as follows: 0.1, 0.5, 0.8, 1.0, 1.2, 1.4, 1.6, 1.8, 2.0, 2.2, 2.4, 2.7, 3.1, 3.4, 3.6, 4.0, 4.5, and 5.0 Hz. See Figs. 4.11 and 4.12 for detailed calculation results.

The results show that, with the increase of the input wave frequency, the maximum shear displacement and acceleration amplification coefficient at the slope top all distribute in saddle shape and reach the maximum value with $f = 1.6$ Hz. This may be because when the input wave frequency is 1.6 Hz, it is close to the natural frequency ($f = 1.56$ Hz) of the slope mass, so resonance occurs in the slope mass and intensifies the seismic effects of the slope mass, which lead the maximum shear displacement and acceleration at the slope top reaching the maximum value at this input wave frequency.

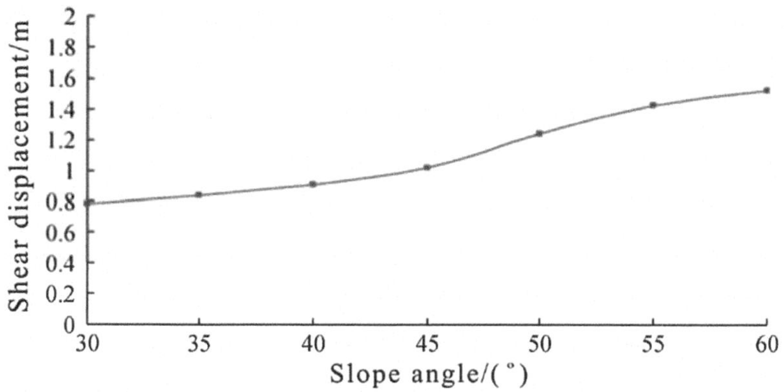

Fig. 4.13 Acceleration amplification coefficient at slope top VS input wave frequency

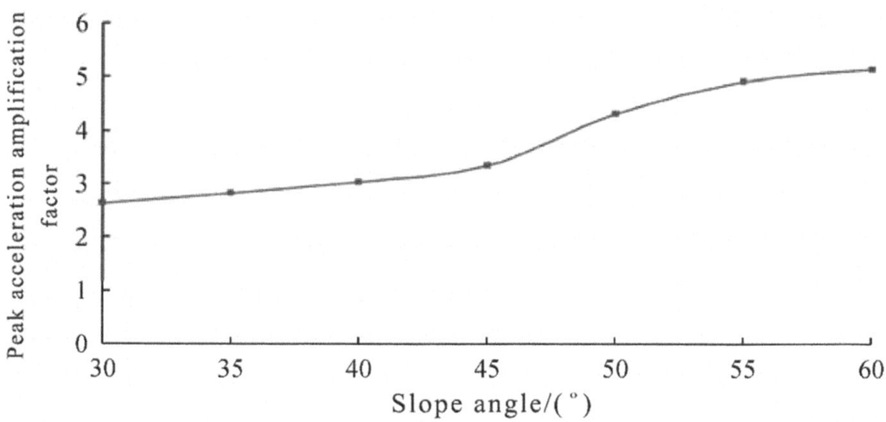

Fig. 4.14 Maximum shear displacement at slope top VS input wave frequency

4.1.4.3 Research of Influence of Bevel Angles on Rock Slope Acceleration Elevation Amplification Effects

In order to study on the influence of bevel angles on rock slope acceleration elevation amplification effects, now we adopt example of the same geometric size with the above example, and input sine wave with PGA = 1.0 g and frequency = 1.5 Hz, and study on bevel angles of 30°, 35°, 40°, 45°, 50°, 55°, and 60°. See Figs. 4.13 and 4.14 for detailed calculation rcsults.

Through comprehensive analysis of Figs. 4.13 and 4.14, we can get the following rules: With the increase of bevel angles, the maximum shear displacement and acceleration amplification coefficient at the slope top all gradually increase and distribute in "step form"; i.e. in the ranges of 30°–45° and 50°–60°, the amplification coefficient increases slightly; while in the range of 45°–50°, it increases suddenly,

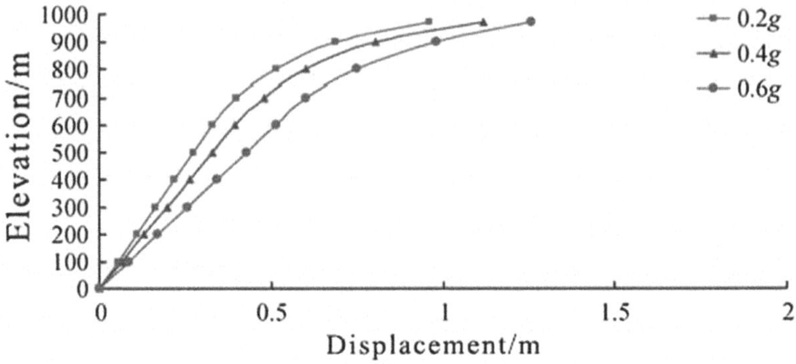

Fig. 4.15 Acceleration amplification coefficient at slope top VS input wave frequency

which could fully explain that most landslides in Wenchuan earthquake happen in mountains with bevel angles over 40°.

4.1.4.4 Research of "Whiplash Effect" of High and Steep Rock Slope Under Seismic Effects

Under seismic effects, because the mass and stiffness of the protruding parts of the buildings are small, there will come across faster velocity and larger displacement at moment of each seismic load turning back and forth, which is called "whiplash effect." While the top of high and steep rock slope would often come across the above phenomenon under strong seismic effects and cause geometric hazards like sliding of the rock mass at the slope top, which has been verified in the field survey of geological disasters in Wenchuan earthquake. This chapter still selects the above example to study and exerts sine wave with PGA = 0.2, 0.4, and 0.6 g and with frequency = 1.6 Hz. See Fig. 4.15 for detailed calculation results.

Through comprehensive analysis of Fig. 4.15, it can be known that "whiplash effect" occurs at the dual side high and steep rock slope top when the input wave frequency $f = 1.6$ Hz, and the shear displacement and acceleration coefficient at the slope top all suddenly increase in multiple times. Therefore, the time–frequency analysis method of dual side high and steep rock slope acceleration elevation amplification effects can explain the "whiplash effect" of slope mass from the perspective of theoretical calculation.

4.1.5 Comparative Analysis of Acceleration Elevation Amplification Effects Time–Frequency Analysis Method with Standard Calculation Method

The acceleration elevation amplification effects are the core part of slope engineering seismic design in the seismic code in China. This chapter, with full knowledge of rock slope seismic ground motion characteristics, focuses on the acceleration elevation amplification effects calculation method of some seismic design code in China and comparatively analyzes the differences between time–frequency analysis method of rock slope acceleration elevation amplification effects proposed in this chapter and the calculation methods of acceleration elevation amplification coefficient proposed in other codes, so as to put forward some modification suggestions on the calculation methods of acceleration elevation amplification effects in the seismic code in China.

4.1.5.1 Brief Introduction of Some Seismic Design Code

This chapter will mainly study on the specifications and explanations about seismic amplification effects codes in China, which mainly include *Code for Seismic Design of Buildings* GB50011-2010 (*Seismic Code* for short), *Specifications for Seismic Design of Hydraulic Structures* DL/5074.2000 (*Hydraulic Seismic Specifications* for short), and Design Specification for Slope of *Hydropower and Water Conservancy Project* SL/386-2007 (*Slope Design Specifications* for short). It clearly states out in *Seismic Code* stressing the influence of topography on seismic amplification coefficient, while it puts forward in *Hydraulic Seismic Specifications* the amplification coefficient calculation of earth and rock dam. For the amplification coefficient of the first or second hydraulic slopes, the regulations of *Slope Design Specifications* can refer to *Hydraulic Seismic Specifications*.

① *Code for Seismic Design of Buildings* GB50011-2010

According to the regulation of *Seismic Code*, it should consider the amplification effects of earthquake on topographies like slopes when build buildings, and the maximum acceleration amplification coefficient should multiply by certain amplifying coefficient to meet the design needs. The amplification coefficient value could multiply by amplifying coefficient in the range of 1.1–1.6 according to the concrete height and bevel angles of slopes. While the clause explanation makes it clear that the amplifying coefficient in the regulations is obtained through summary of calculation results of seismic response analysis of large amount of rock and soil mass. The general situations can be reflected through seismic response analysis and large amount of slope seismic damage cases as follows:

1. The response is more intense as the height of high and protruding topography from the base level is higher.
2. The response is slighter as the distance from the slope and slope top boundary is larger.

3. From the perspective of rock and soil composition, the dynamic response of soil mass is stronger than that of the rock mass in the same terrain.
4. The response far away from the central part of the boundary becomes slighter as the top surface of high and protruding topography become broader.
5. The steeper the slope is, the amplification effects at its top are stronger.

Based on the above qualitative changing trend, with slope elevation difference = H, and tangent values of bevel angles = H/L, we put forward the calculation formula of slope acceleration amplification coefficient:

$$\lambda = 1 + \alpha \tag{4.29}$$

In the formula, λ is the acceleration amplification coefficient of the slope top;

α is the increase gradient of acceleration amplification coefficient of the slope top, and it is adopted by Table 4.4.

② *Specifications for Seismic Design of Hydraulic Structures* DL/5073-2000

Clause 5.1.3 in *Hydraulic Seismic Specifications specifically requires that the seismic stability calculation of earth and rock dam should adopt pseudo-static method. The earth and rock dam are similar to slopes, so its dynamic amplification coefficient can refer to the seismic design of slopes.* The clause regulates: When $H \leq 40$ m, the dynamic amplification coefficient distributes in trapezoid, with the maximum value occurring at dam top and we take 1 for the dam bottom amplification coefficient, and 2.0–3.0 for the dam top coefficient; when $H > 40$ m, the dynamic amplification coefficient $\beta(T) = 1 + [\beta(T)\text{max} - 1]/3$ at 0–0.6H and distributes in trapezoid at 0.6H–1.0H, with the maximum value 2.0–3.0 occurring at dam top.

③ *Design Specification for Slope of Hydropower and Water Conservancy Project* SL/386-2007

Table 4.4 Seismic influence coefficient increase gradient of local protruding topography

Slope elevation difference $H(m)$	Non-rock topography	$H < 5$	$5 \leq H < 15$	$15 \leq H < 25$	$H \geq 25$
	Rock topography	$H < 20$	$20 \leq H < 40$	$40 \leq H < 60$	$H \geq 60$
Tangent values of bevel angles (H/L)	$H/L < 0.3$	0	0.1	0.2	0.3
	$0.3 \leq H/L < 0.6$	0.1	0.2	0.3	0.4
	$0.6 \leq H/L < 1$	0.2	0.3	0.4	0.5
	$H/L \geq 1$	0.3	0.4	0.5	0.6

Note The maximum amplifying amplitude 0.6 regulated by the clause is given according to analysis results and comprehensive judgment, and this clause can be applied to all kinds of topographies including mounds, mountain ridges, cliffs, and steep slopes

Table 4.5 Comparison table of calculation results of acceleration elevation amplification coefficient

Input PGA	Position	Calculation method		
		method of this dissertation	*Seismic code*	*Hydraulic seismic specifications*
0.2 g	Mid-slope	1.51	1.6	1.33–1.67 Average value :1.5
0.4 g		1.35	1.6	
0.2 g	Slope top	2.12	1.6	2.0–3.0 Average value :2.5
0.4 g		1.92	1.6	

Clause D.2.4 in *Slope Design Specifications* regulates that, referring to the provisions in *Hydraulic Seismic Specifications* the dynamic amplification coefficient of particles to slopes of first and second grade could amplify from slope bottom to top after argument.

4.1.5.2 Comparative Analysis of High and Steep Rock Slope Time–Frequency Analysis Method with Standard Calculation Method

Now we take the example of numerical analysis model, rock and soil mass parameters and external input ground motion in 3.3.3.2 *Dual Side Rock Slope Acceleration Elevation Amplification Effects Time–frequency Analysis Method*, use the regulated acceleration elevation amplification effects analysis method in *Code for Seismic Design of Buildings* GB50011-2010, *Specifications for Seismic Design of Hydraulic Structures* DL/5073-2000, and *Design Specification for Slope of Hydropower and Water Conservancy Project* SL/386-2007 to calculate the acceleration elevation amplification coefficient of this model. See Table 4.5 for concrete results. Because the calculation results of *Slope Design Specifications* are identical with that of *Hydraulic Seismic Specifications*, this chapter will take comparison of these two.

It can be known from Table 4.5 that, under the effect of Wenchuan wolong seismic wave with PGA = 0.2 g, the acceleration elevation amplification coefficient at rock slope mid-slope is 1.51, which is slightly less than the calculation results of 1.6 in *Seismic Code*, and it is within the range of 1.33–2.0 in *Hydraulic Seismic Specifications*, which is close to average value of 1.5; the acceleration elevation amplification coefficient at slope top is 2.12, which is more than the calculation results of 1.6 in *Seismic Code*, and it is within the range of 1.33–2.0 in *Hydraulic Seismic Specifications* and less than its average value of 2.5. Under the effect of Wenchuan wolong seismic wave with PGA = 0.4 g, the acceleration elevation amplification coefficient at rock slope mid-slope is 1.35, which is slightly less than the calculation results of 1.6 in *Seismic Code*, and it is within the range of 1.33–2.0 in *Hydraulic Seismic Specifications*, and less than its average value of 1.5; the acceleration elevation amplification coefficient at slope top is 1.92, which is more than the calculation results of 1.6 in *Seismic Code*, and it is within the range of 1.33–2.0 in *Hydraulic Seismic Specifications* and less than its average value of 2.5. It can be known from this that, under effects of seismic wave with PGA = 0.2 g, the calculation results of calculation method in this dissertation are basically identical with that in *Seismic Code*, and in *Hydraulic Seismic Specifications*, while under effects of seismic wave with PGA = 0.4 g, the calculation results of calculation method for parts lower than mid-slope in this dissertation are less than that in *Seismic Code* and are identical with that in *Hydraulic Seismic Specifications;* and for parts higher than mid-slope, that in this dissertation is more than that in *Seismic Code* and less than that in *Hydraulic Seismic Specifications*. The reason for this can be that the methods in the specifications do not consider the changes of acceleration amplification with seismic intensity, and the importance of the industry served by each specification varies from each other. Therefore, it is suggested that the calculation of rock slope acceleration elevation amplification coefficient in *Seismic Code, Hydraulic Seismic Specifications,* and *Slope Design Specifications should take consideration of the influence of seismic intensity, so as to improve the rationality of structural seismic design.*

4.2 Slope Deformation Characteristics and Formation Mechanism of High and Steep Rock Slopes Under Seismic Effects

The seismic stability of slopes has been the hotspot issues of geological disaster study, and especially after the 5.12 Wenchuan earthquake, it has become the study focus of specialists and scholars. While seismic instability mechanism is the key for slope seismic stability study, the seismic ground motion characteristics are the precondition for seismic instability mechanism study. Therefore, based on the research results of seismic ground motion characteristics of slopes, it is quite necessary to conduct research on slope deformation characteristics and instability mechanism under seismic effects, and it has significant importance for post-disaster reconstruction.

After 5.12 Wenchuan earthquake, members in our research group immediately carried out large amount of research on collapses and sliding disasters in the seismic area and conducted statistical analysis to the total distribution of collapses and sliding disasters and distribution rules of classic regional collapses and sliding in the seismic area. A large amount of field survey results show that slope structure is still the main factor to control slope deformation and damage. Because of Wenchuan earthquake region located in deep canyon area with precipitous terrain, with effects of long period of gravity and rainfall, the middle and upper slopes always develop thin cover layer and strong-weathered and relief relaxed rock mass, which is often located in the middle and upper slopes, and easily come across instability damage

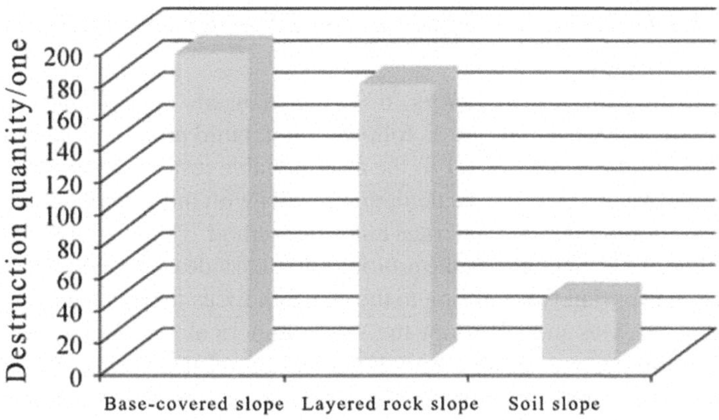

Fig. 4.16 Quantity distribution map of rock slope instability structure

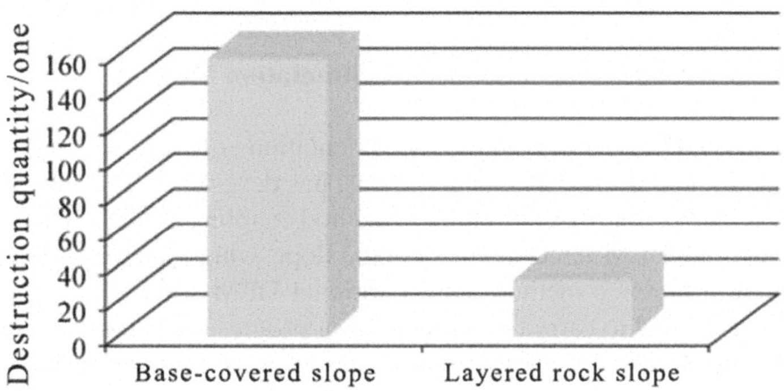

Fig. 4.17 Quantity distribution map of bedrock and overburden slope instability structure

under seismic effects. That makes the bedrock and overburden slopes made of soil layer and intensive-weathered layer the most common slope structure with maximum instability quantity in seismic areas (see Figs. 4.16 and 4.17). Therefore, this section will, based on the previous research results of rock slope seismic ground motion characteristics, focus on bedrock and overburden slopes, select classic working condition, adopt approach of shaking table test combining with numerical analysis, and conduct in-depth study of bedrock and overburden slope deformation characteristics and formation mechanism under seismic effects from the perspective of time-history and time–frequency.

4.2.1 Brief Introduction of Numerical Simulation Method

Before conducting the research work, this chapter needs to explicitly explain the study approach, and the details are as follows: First, build numerical analysis model by geometric similarity ratio of 1:1 to the shaking table test model of dual side rock slope and compare the numerical calculation results with the experimental results so as to verify the correctness of numerical analysis method. This part of work has been finished in Sect. 3.3.3. Second, for the prototype of dual side rock slope shaking model test, adjust it appropriately according to the research focus in this chapter, generalize the numerical analysis model, adopt the same numerical calculation method with that of the prototype, and set up evenly large amount of test points of acceleration, displacement, stress and strain on sliding mass, sliding beds and the sliding mass surface, so as to conduct in-depth research on the deformation characteristics and formation mechanism of bedrock and overburden slope.

4.2.1.1 Brief Introduction of Numerical Simulation Method

In recent years, GDEM, a new numerical calculation software which can be well applied to geological disaster damage analysis, has developed. The software makes finite element method and discrete element method coupling together, i.e. using FEM inside the slope and DEM at the boundary of the slope, which can well solve the limits of FEM, finite difference method, and traditional DEM software. Therefore, this chapter adopts GDEM to carry out simulation of progressive failure and movement process after instability of slope jointed rock mass under seismic effects.

4.2.1.2 Numerical Analysis Model Construction

In order to deeply study on the seismic landslide mechanism of bedrock and overburden slope, this chapter builds numerical analysis model according to the prototype slope of shaking table test model, whose geometric characteristics are as follows: The height of single side bedrock and overburden slope is 1312 m, its width is 2240 m, bevel angles are 50°–60°, the width of river valley at the slope foot is 103.8 m, the riverbed is slightly inclined, the bevel angle of the river valley right bank slope is in the range of 30°–40°, and the slope top is partly covered with seriously weathered layer; the height of dual side bedrock and overburden slope is 1086 m, its width is 2100 m, which include steep slope topography and gentle slope topography with bevel angles, respectively, in the range of 50°–60° and 42°–50°. At the same time, the field survey results of slope sliding show that deposit body at mid-slope is the most serious, so we simplify the soil layer and weathered layer on the slope surface into deposit body at mid-slope according to the field survey results to better study the sliding mechanism of the slope. In addition, large amounts of monitoring points were set up inside and on the surface of the numerical model to monitor the dynamic

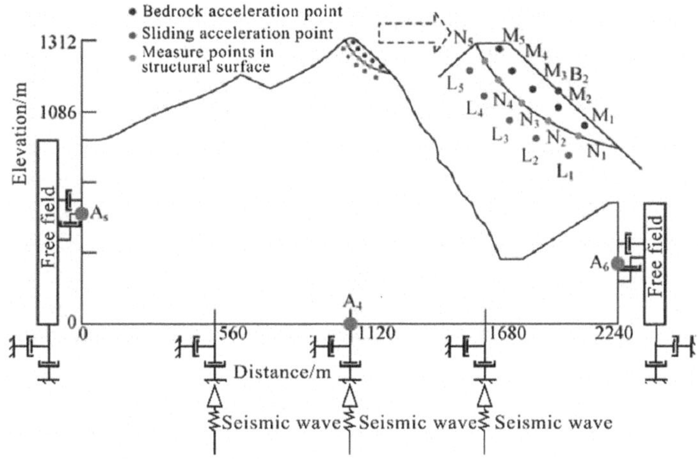

Fig. 4.18 Numerical analysis model of single side bedrock and overburden slope

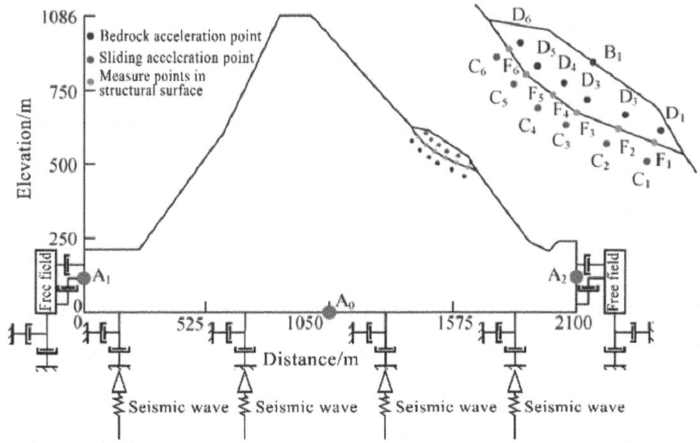

Fig. 4.19 Numerical analysis model of dual side bedrock and overburden slope

response time history changes of slope acceleration, stress and strain of sliding body and sliding mass under seismic effects. See Figs. 4.18 and 4.19 for details.

4.2.1.3 The Determination of Numerical Constitutive Model and Material Parameters

In the calculation of DEM, the rock and soil mass of sliding bed and sliding mass mainly come across plastic deform before instability, so elastoplastic model and Mohr–Coulomb failure principle (M-C failure principle) are adopted by both of them.

After instability occurs, this chapter adopts rigid constitutive model for soil particles inside the sliding mass mainly because: First, after instability occurs, this chapter mainly focuses on the motion features of sliding mass such as the trajectory of sliding mass and deposition; second, because the displacement of sliding mass may amount to several hundred meters, and deformation of the sliding mass differs many orders of magnitude to that, the sliding mass displacement is much more than deformation of soil particles inside the sliding mass; third, if we still adopt Mohr–Coulomb failure principle after sliding occurs, it will cause slower convergence rate of slope system, which may lead to energy dissipation and non-convergence of system calculation. For sliding bed, because of its large geometric size and plastic deformation under seismic effects, this chapter, therefore, adopts elastoplastic model and M-C failure principle.

Because failure phenomenon like slips and cracks may occur on sliding mass structural surface and among soil particles inside sliding mass under seismic effects, this chapter will adopt brittle fracture constitutive model and M-C failure principle for all the mentioned positions, adopting the stochastic structural surface model in GDEM to achieve order, complexity, and randomness of structural surface inside the sliding mass. Among blocks and blocks, it sets up joint points mixed units to achieve independent calculation of particle motion of both structural surface spring and particle DEM, which can effectively solve the calculation of DEM initial states. The numerical simulation of this chapter adopts the exactly same material with the prototype slope, and the related parameters of sliding mass structural surface can be solved by relevant formulas in GDEM. See Table 4.6 for concrete physico-mechanical parameters and constitutive model of materials.

4.2.1.4 Selection of Boundary Conditions and Decision of Seismic Ground Motion Input Conditions

The selection of boundary conditions in this part uses the same method with Sect. 3.3.2.3. In addition, because the Wenchuan wolong seismic station nears the prototype slope, this chapter selects the observed seismic wave by Wenchuan wolong seismic station with horizontal and vertical PGA of 957 and 857 cm/s/s, and the duration of 163 s. See Figs. 3.142 and 3.143 for detailed horizontal and vertical acceleration time history curve.

4.2.2 Deformation Characteristics and Disaster Mechanism of Bedrock and Overburden Slopes

In order to study on the seismic sliding mechanism of single and dual side bedrock and overburden slopes, this chapter sets up acceleration Test Points L1–L5 and C1–C6 inside the slope mass of single and dual side prototype numerical analysis model,

Table 4.6 Physico-mechanical parameters of sliding mass, bedrock, and structural surface

Numerical analysis method	Constitutive model	Model name	Unit weight (kN/m^3)	Cohesive force (kPa)	Internal friction angle (°)	Elasticity modulus (MPa)	Poisson's ratio
Single and double-sided bedrock and overburden slopes	M-C principle	Bedrock	22	93.6	37.9	450	0.25
		Slope deposit body	21	49.8	28.8	60	0.35
	Constitutive model	Structural surface location	Normal stiffness (MPa)	Shear stiffness (MPa)	Internal friction angle (°)	Tensile strength (kPa)	Cohesive force (kPa)
Single side bedrock and overburden slope	Brittle fracture constitutive model	Slope deposit body inside	3600	1800	32	25.98	48.8
		Structural surface	27,000	13,800	32	6.30	13.8
Dual side bedrock and overburden slope		Slope deposit body inside	600	300	32	4.33	8.1
		Structural surface (slope foot)	4500	2300	32	1.21	2.3
		Structural surface (mid-slope)	4500	2300	32	1.05	2.3
		Structural surface (slope top)	6200	6200	36	3.86	7.2

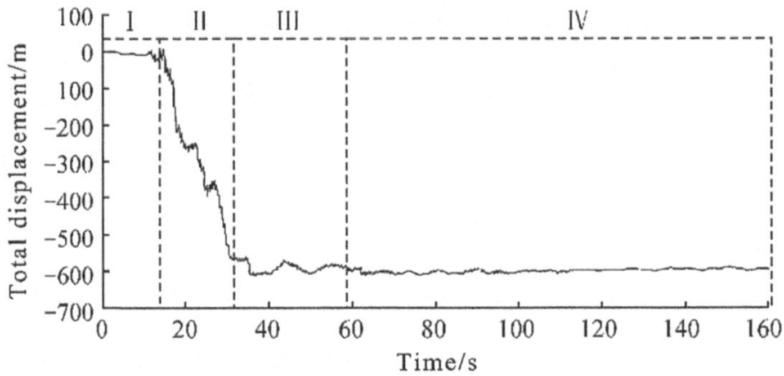

Fig. 4.20 Total displacement time history curve of B1

acceleration of Test Points M1–M5 and D1–D6 inside the slope mass, stress and strain Test Points N1–N5 and F1–F6 on slope mass structural surface and displacement Test points B1 and B2, so as to conduct targeted study on dynamic characteristics and sliding mechanism of slope deposit body on single and dual side prototype slope under seismic effects.

4.2.2.1 The Whole Landslide Process

In order to study on the whole landslide process of bedrock and overburden slopes in earthquake, this chapter only sets up displacement monitoring points B1 and B2 on deposit body surface of single and dual side bedrock and overburden slopes. Due to that the sliding process revealed by the monitoring results of B1 and B2 are basically the same; thus, this section selects the total displacement curve of B1 to analyze, and the results are shown in Fig. 4.20. Figure 4.20 shows that the total displacement of B1 stays at around 0 when T = 0–15.75 s, while it suddenly increases at T = 15.75–33.12 s, and its volatility gradually decreases at T = 33.12–58.18 s, ending at about 600 m. According to the above failure phenomenon, this chapter divides the sliding process into four phases: starting phase, high-speed sliding phase, deposit phase, and self-stabilizing phase. Therefore, this chapter will select seismic response of starting phase at T = 0–15.75 s to further study the sliding mechanism.

4.2.2.2 Landslide Hazard Phenomenon

When GDEM carries out numerical solutions, it could update calculation mode in real time according to unit deformation and motion. Finally, shearing strain and tensile strain distribution of every structural surface inside the single and dual side bedrock and overburden slope models of different times in the earthquake were obtained through calculation (see Fig. 4.21). It is noteworthy that blue in the figure represents

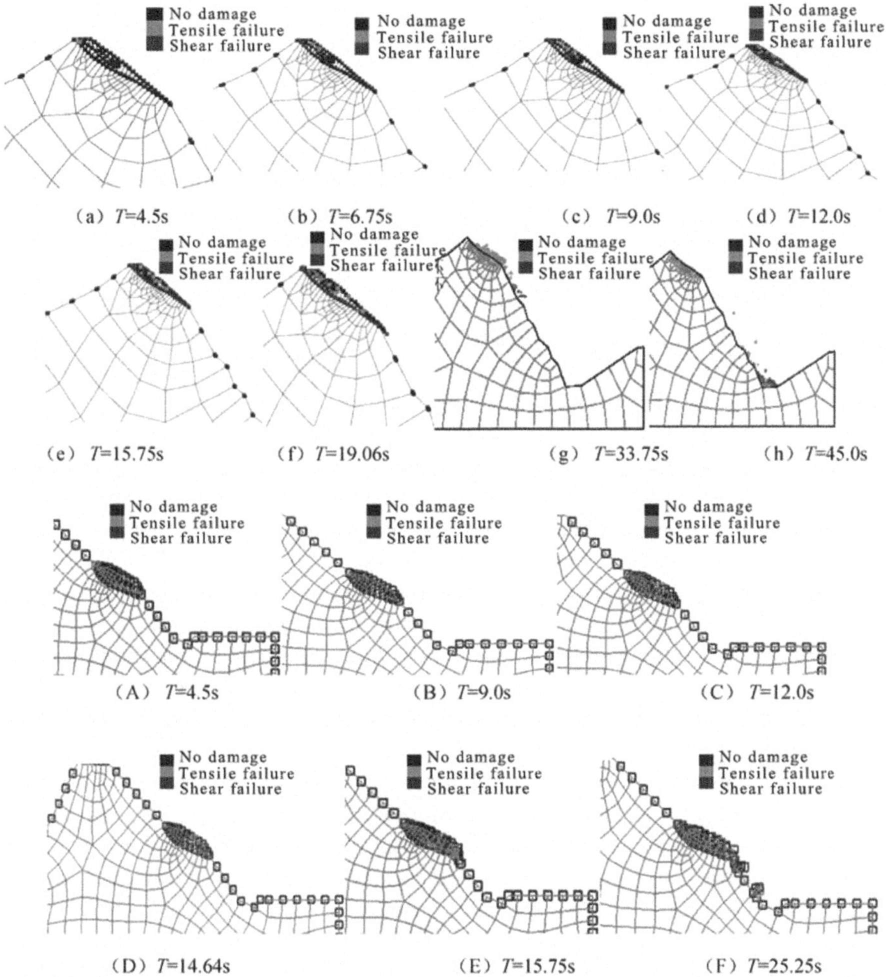

Fig. 4.21 Shear and tensile failure points distribution map of single and dual side bedrock and overburden slopes

non-damage of contact elements, red represents shear failure, and green represents tensile failure.

Through comprehensive analysis of Fig. 4.21, it can be known that the whole process of deposit body sliding from deformation to failure of single and dual side bedrock and overburden slopes is basically the same, with slight difference. Firstly, the difference of the two: At $T = 15.75$ s, the slope top sliding structural surface of single side bedrock and overburden slope is basically in complete connection, while the sliding mass at mid-slope of dual side bedrock and overburden slope has already come across landslides, which fully demonstrates that, with the same external conditions such as input seismic wave and sliding structural surface parameters, the seismic response of dual side bedrock and overburden slopes is stronger than that of the single side bedrock and overburden slopes, and the seismic damage is also more serious. The above phenomenon is not only identical with the research results of rock slope seismic ground motion characteristics, but also with the survey results of Wenchuan earthquake damage. Secondly, the similarity of the two: First, at the beginning of seismic ground motion, a few tensile and shear damage failure points occur at the sliding mass top; with the increase of input ground motion, the shear failure points at the sliding mass structural surface gradually develop downwards and form large amount of discontinuous tensile and shear failure points; with the continuous increase of input ground motion, the tensile and shear failure points on the sliding mass structural surface gradually connect and form continuous shear failure slip crack surface and develop to lock-fixed section of sliding mass leading edge; lastly, with input seismic wave acceleration reaching the peak value, the shear failure points on the sliding mass structural surface rapidly extend and form continuous shear crack surface connecting the whole sliding mass structural surface. The above failure phenomenon is not only identical with the shaking table test phenomenon, but also with the research results of most sliding mass occurrence time in 5.12 Wenchuan earthquake. The reasons for the above phenomenon may be as follows: (1) At the beginning of gravity and seismic power, tensile stress concentration occurs at the sliding mass top, causing tensile deformation of sliding mass along sliding mass structural surface trailing edge, which causes tensile failure at this position; (2) under seismic effects, shear strength inside sliding bed and on the sliding mass surface will gradually decrease with increase of seismic ground motion acceleration (see Section ③ Dynamic Characteristics of Soil Particles inside Sliding Mass for details); (3) with the increase of seismic ground motion acceleration, the inconsistency of motion between sliding bed and sliding mass gradually increases which intensifies the development of shear and tensile failure points on sliding mass structural surface (see Section ④ Research on Inconsistency of Motion between Sliding Bed and Sliding Mass); (4) with the increase of seismic ground motion acceleration, the difference of seismic energy distribution and dissipation gradually enlarges between sliding bed and sliding mass, which intensifies the damage of sliding mass structural surface and causes slope sliding (see Section ⑤ Research on Difference of Seismic Energy Distribution and Dissipation between Sliding Bed and Sliding Mass).

To sum up, the bedrock and overburden slopes coming across landslides are a balanced and progressive process. Under effects of gravity and seismic power, tensile stress concentration firstly occurs at sliding mass top, causing deformation of sliding mass along sliding mass structural surface trailing edge, which causes tensile and shear failure points at this position. Then, with continuous seismic power, the shear failure points on sliding mass structural surface gradually develop to lock-fixed section of sliding mass leading edge with increasing amount of tensile failure points on sliding mass surface, which finally causes progressive damage of lock-fixed section, crack surface connecting to sliding zone and sliding mass slipping from shear crack into landslide.

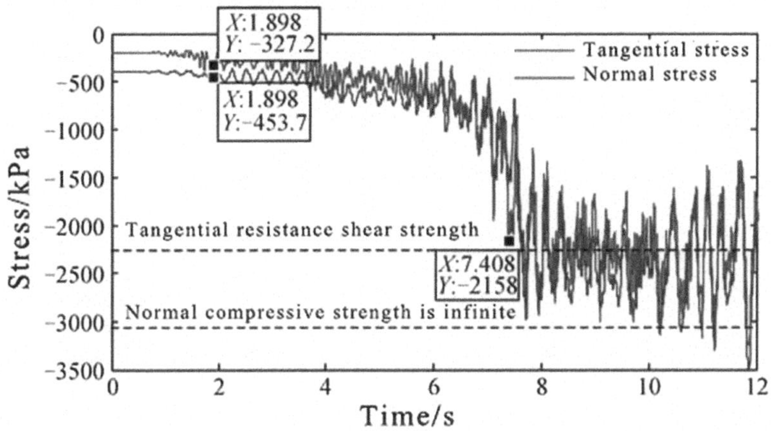

Fig. 4.22 Normal and tangent stress time history curve

4.2.2.3 Landslides Mechanism Analysis

This chapter will, based on the previous numerical calculation results, analyze from the micro-perspective of stress–strain development on sliding mass structural surface in earthquakes, including time-domain and joint time–frequency-domain problems like static and dynamic triaxial test, acceleration time history, and Hilbert spectrum of rock and soil mass, which could study on the occurrence mechanism of earthquake inducing landslides more deeply and comprehensively.

① Stress–Strain Relations of Sliding Mass Structural Surface

In order to carry out more in-depth study on landslides mechanism of bedrock and overburden slopes under seismic effects, this chapter, respectively, sets up stress Test Points N1–N5 and strain Test Points F1–F6 on sliding mass structural surface of single an dual side bedrock and overburden slopes, so as to analyze the failure mechanism of sliding mass structural surface under seismic effects from the perspective of stress–strain. Because the sliding process of single and dual side bedrock and overburden slopes are basically the same, this chapter selects representative Test Point F3 of stress and strain on dual side sliding mass structural surface as analyzing object. See Figs. 4.22 and 4.23 for detailed results. It is noteworthy that, Fig. 4.23 shows Test Point F3 has come across shear failure at $T = 12.0$, so this chapter only selects seismic response before 12.0 s to analyze.

Because all the mechanic indexes of soil mass decrease under seismic effects, tangent shear strength of sliding mass also decreases. Therefore, this chapter takes the strength index of soil mass under its own weight as basis to partially consider the safety. In the process of working out the strength, the following hypothesis was made: The soil mass is the ideal elastoplastic model and completely obeys M-C failure principle. The lower bedrock has incompressibility; i.e. the normal compressive strength of structural surface is infinite. The tangent shear strength calculation for-

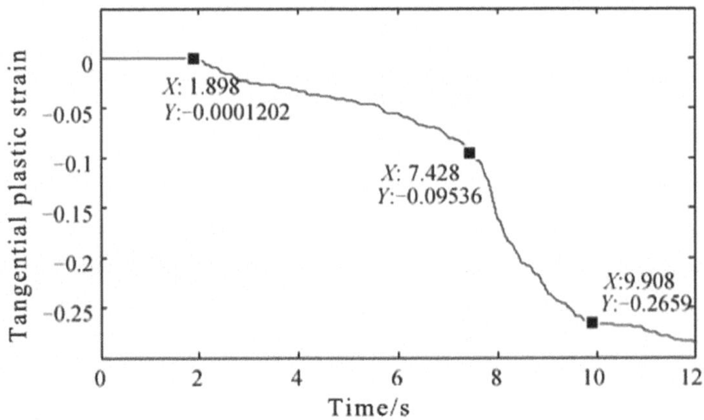

Fig. 4.23 Tangent plastic strain time history curve

mula of sliding mass surface is obtained through calculation (see Formula 4.30). At the same time, due to that GDEM could only adopt stress and strain data in X- and Y-directions when carrying out numerical calculation, so this chapter will change coordinates according to Formulas (4.31) and (4.32).

$$P_\tau = C + \rho g h \cdot \cos\theta \cdot \tan\varphi \tag{4.30}$$

$$P_n' = T_x \cdot \sin\theta + T_y \cos\theta \tag{4.31}$$

$$P_\tau' = T_y \cdot \sin\theta - T_x \cos\theta \tag{4.32}$$

In the above formulas, P_τ is the tangent shear strength of sliding mass surface, C stands for cohesive force of structural surface, φ stands for its internal friction angle, h represents the vertical thickness of upper covering layer of test points, θ is the bevel angle of slope surface and the horizontal surface, ρ is the density of upper covering layer, P_n' and P_τ', respectively, represent normal and shear stress and strain of sliding mass structural surface, and T_τ and T_y, respectively, stands for stress and strain in X- and Y-directions on sliding mass structural surface.

Bring the relevant parameters into the above formulas, it will get that P_τ equals 2322 kPa, the shear and normal stress time history curve of sliding mass structural surface see Fig. 4.22. Through comprehensive analysis of Fig. 4.22, it can be known that under seismic ground motion effects with duration $T = 12.0$ s, the normal stress of test points has always been negative, which demonstrates that the structural surface has always been in compressed state, and could not come across tensile failure. Therefore, this chapter will not consider the normal plastic strain time history curve of test points. See Fig. 4.23 for calculation results of plastic strain.

Through comprehensive analysis of Figs. 4.22 and 4.23, it can be known that before input seismic power $T = 1.898$ s, the tangent stress of test points is small, and the correspondent tangent plastic strain is almost 0, which shows that test points are

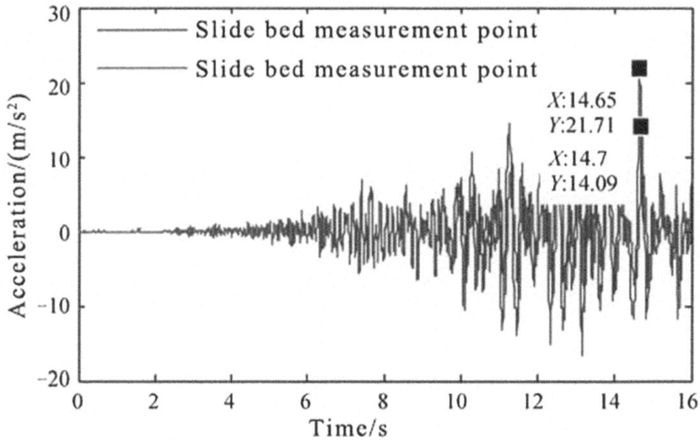

Fig. 4.24 Time-history curve of horizontal acceleration

still in elastic stage when T =0–1.898 s; when T = 1.898–7.408 s, the tangent stress of test points gradually increases, so is tangent plastic stress, and the growth rate is on the rise; when T =7.408 s, the tangent stress of test points is over its tangent shear strength for the first time, the correspondent tangent plastic strain comes across the turning point at T =7.428 s and suddenly increase afterward. At T =7.408–12.0 s, the tangent stress of test points fluctuate around tangent shear strength line, and the tangent plastic stress time history curve comes across turning point at T =9.908 s, with tangent plastic strain staying at 0.26 afterward. At the same time, it can be known from the characteristics of brittle fracture constitutive model and M-C failure principle that the test point has already been damaged, which is basically identical with the failure model of this test point in Fig. 4.21.

② Vibration Acceleration Difference of Two Sides of Sliding Mass Structural Surface

In order to study on the seismic wave propagating difference on sliding bed and inside the sliding mass, this chapter, respectively, sets up acceleration Test Points L1–L5 and C1–C6 inside the slope mass of single and dual side bedrock and overburden slopes, and acceleration Test Points M1–M5 and D1–D6 inside the sliding mass to observe the propagating characteristics of seismic wave on the two sides of structural surface. Because the sliding process of single and dual side bedrock and overburden slopes are basically the same, this chapter selects representative Test Points (C3, D3) as analyzing objects. See Figs. 4.24 and 4.25 for detailed results.

Through comprehensive analysis of Figs. 4.24 and 4.25, it can be known that there exists big difference of horizontal acceleration of the two sides of sliding mass structural surface. Firstly, when seismic ground motion acceleration is small, the horizontal acceleration inside the sliding mass is over that inside the sliding bed; with the increase of seismic ground motion acceleration, the horizontal acceleration

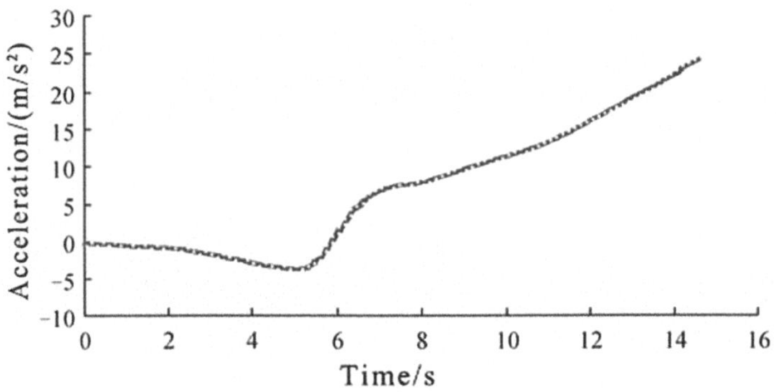

Fig. 4.25 Time-history envelop of horizontal acceleration difference

inside the sliding bed gradually surpasses that inside the sliding mass; secondly, there is difference in the time of the two sides of sliding mass structural surface reaching the horizontal PGA with sliding mass slightly lagging behind. The reason for the above phenomenon may be that at the beginning of earthquake, the seismic ground motion acceleration is small, few discontinuous shear failure points occur on the structural surface, and the soil mass on the two sides of sliding mass structural surface did not come across relative slip. When seismic wave reaching the sliding mass structural surface, the PGA amplification effects occur inside the sliding mass. Based on the propagating theory of seismic wave in semi-infinite space, this chapter believes that reason for PGA amplification effects occurring inside the sliding mass may be as follows: Firstly, when seismic wave propagating from bedrock (sliding bed) to upper cover layer (sliding mass), the reflected wave comes across wave field separating phenomenon, for example, *P*-wave reflects into *P*-wave and SV wave and SV wave reflects into SV wave and *P*-wave, and reflected wave of all wave types overlap together to form complex seismic wave field, which enlarges the motion of upper cover layer; secondly, the energy of seismic wave is mainly provided by *P*-wave, SH wave, and SV wave, and they can all enlarge the motion of upper layer when spreading from bedrock to upper cover layer. With the increase of seismic ground motion intensity, the acceleration amplification effects become more obvious, and acceleration difference of the two sides of structural surface also increases, while that will cause more shear failure points on the sliding mass structural surface, and they gradually connect together, forming shear failure surface. Because the soil mass on the two sides of structural surface come across relative slip, the above theory loses efficacy, and at the same time, this phenomenon causes seismic wave energy dissipation at this position. With the increase of slippage, there will be more energy dissipation which causes sharp decreases of acceleration inside the sliding mass and leads to the acceleration inside the sliding bed exceeding that inside the sliding mass.

③ Dynamic Characteristics of Soil Particles inside Sliding Mass

Fig. 4.26 Dynamic triaxial test of saturated sample

Table 4.7 Table of dynamic shear strength test results of sliding mass materials

Vibration times N		12	20	30
Corresponding level		7	7.5	8
		dynamic shear stress ratio τ_d/σ_{Sc}		
Consolidation confining pressure	100	0.407	0.382	0.362
	200	0.395	0.369	0.35
	400	0.373	0.335	0.322

In order to study on the dynamic characteristics of soil particles inside the sliding mass under the effects of dynamic loading, this chapter carries out dynamic triaxial test focusing on the sliding mass materials on the high and steep prototype slope and obtains the shear strength of soil particles inside the sliding mass under different seismic intensity. See Figs. 4.26 and 4.7 for detailed results. It is noteworthy that because the dynamic response rules of sliding mass materials of single and dual side bedrock and overburden slopes are basically the same under effects of dynamic circulative loading.

Through comprehensive analysis of Table 4.7, it can be known that, in the same vibration cycle, the growth amplitude of dynamic shear strength τd gradually increases with rising confining pressure; under the same confining pressure, the dynamic shear strength τd gradually decreases with increase of vibration cycle N, i.e. increase of correspondent seismic grade. Therefore, the shear strength inside the sliding bed and of sliding mass structural surface will gradually decrease with the growth of seismic ground motion acceleration.

④ Inconsistency of Motion between Sliding Bed and Sliding Mass

In order to study on the inconsistency of motion between sliding bed and sliding mass, this chapter, respectively, sets up acceleration Test Points L1–L5 and C1–C6 inside the slope mass of single and dual side high and steep bedrock and overburden slopes, and acceleration Test Points M1–M5 and D1–D6 inside the sliding mass.

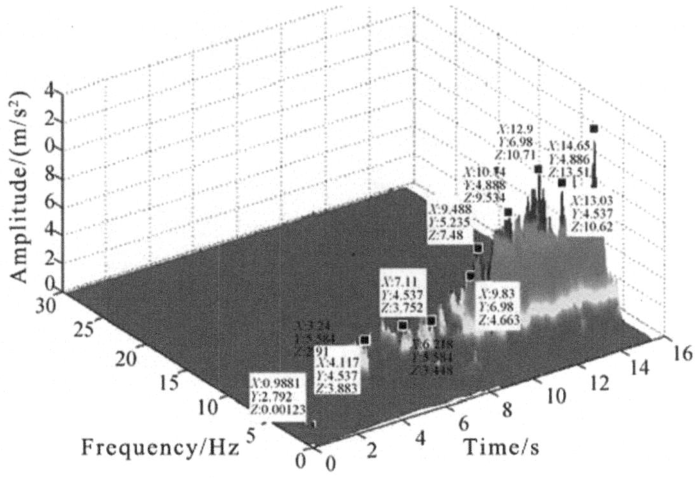

Fig. 4.27 Hilbert spectrum of horizontal acceleration inside the sliding bed

Because the sliding process of single and dual side bedrock and overburden slopes are basically the same, this chapter selects representative Test Points (C3, D3) on the two sides of sliding mass structural surface of dual side bedrock and overburden slope as analyzing objects. The inconsistency of motion between sliding bed and sliding mass mainly involves frequency, amplitude and time of them, and the current time-domain analysis and frequency-domain analysis could only conduct time–amplitude or frequency–amplitude research, but could not organically combine amplitude, frequency and time together, so this chapter will adopt Hilbert–Huang transform, and select horizontal acceleration time history curve of C3 and D3 to study sliding bed and sliding mass motion inconsistency from the perspective of joint time–frequency-domain (see Figs. 4.27 and 4.28).

First, explain the definition of instant frequency adopted by the dissertation according to Hu Yuxian, et al.: The instant frequency $\omega(t)$ at any time is frequency ω to make instant Hilbert spectrum value $H(t, \omega)$ of this moment reaching its maximum and that is the dominant frequency of the instant Hilbert spectrum value $H(t, \omega)$ of this moment. The changing process of this instant dominant frequency is the instant frequency process of initial acceleration record.

It can be known from Figs. 4.27 and 4.28 that in the horizontal acceleration Hilbert spectrum of the two sides of sliding mass structural surface exists certain similarities and differences. **Similarities**: The horizontal acceleration of the two sides of sliding mass structural surface mainly concentrates in 0–10 Hz, and although some acceleration exists in 10–30 Hz, but they are with small amplitude, which is identical with the calculation results of Fourier spectrum. **Differences**: First, at the beginning of earthquake, the amplitude of horizontal acceleration inside the sliding mass is over that inside the sliding bed; with the increase of seismic intensity, the amplitude of horizontal acceleration inside the sliding bed surpassing that inside the

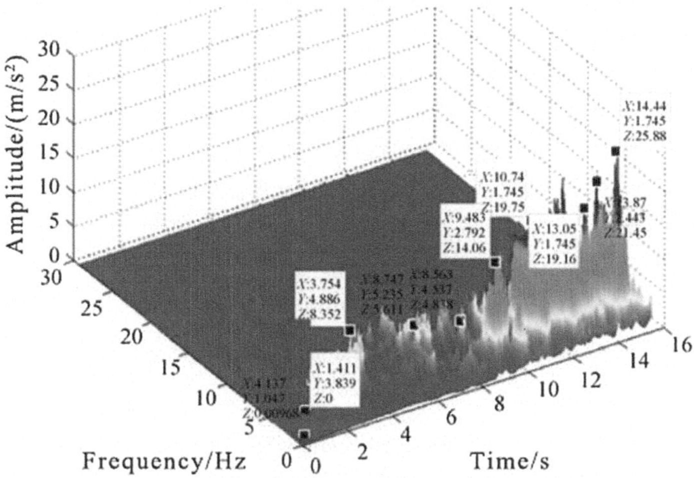

Fig. 4.28 Hilbert spectrum of horizontal acceleration inside the sliding mass

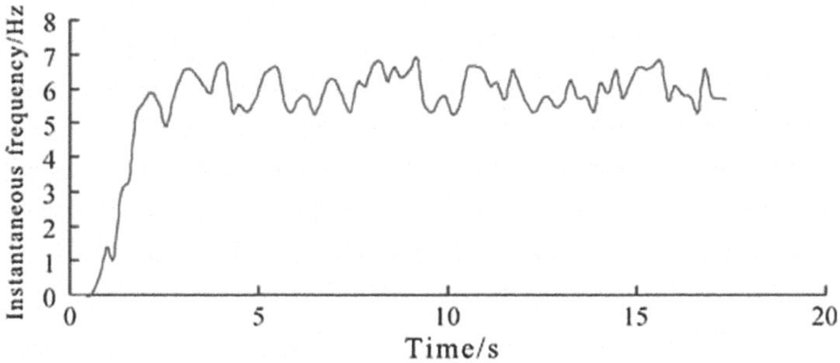

Fig. 4.29 Instant frequency of horizontal acceleration inside the sliding bed

sliding mass; second, based on the previous definition of instant frequency, select the acceleration amplitude in Hilbert spectrum at any time and find out its correspondent frequency to draw the time–frequency time history curve inside the sliding bed and sliding mass (see Figs. 4.29 and 4.30). It can be known from Figs. 4.29 and 4.30 that, when seismic ground motion is small, the instant frequency of horizontal acceleration inside the sliding bed and sliding mass stabilizes in certain range, with the former slightly higher than the latter; with the increase of seismic ground motion, the instant frequency of horizontal acceleration inside the sliding bed stays unchanged, while that inside the sliding mass gradually decreases and finally becomes stable.

The above phenomenon may lie in the following reasons: At the beginning of earthquake, the seismic ground motion acceleration is small, and the sliding mass surface has enough shear strength. Therefore, few discontinuous shear failure points appear at the sliding mass structural surface, and the soil mass on the two sides of sliding mass structural surface does not come across relative slip, and the motion of

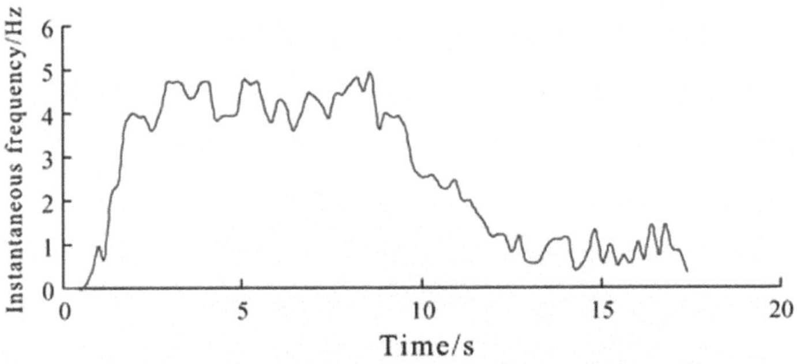

Fig. 4.30 Instant frequency of horizontal acceleration inside the sliding mass

sliding bed and sliding mass is basically the same which causes the instant frequency of horizontal acceleration inside the sliding bed and sliding mass stabilize in a certain range with the former slightly higher than the latter. When seismic wave reaching the sliding mass structural surface, the wave impedance of deposit is smaller than that of the sliding bed, which causes horizontal acceleration coming across amplification effects inside the sliding mass. With the increase of seismic ground motion intensity, the acceleration amplification effects become more obvious, and acceleration difference of the two sides of structural surface also increases, while that will cause more shear failure points on the sliding mass structural surface, and they gradually connect together, forming shear failure surface. The soil mass of the two sides of structural surface comes across relative slip, the instant frequency of horizontal acceleration inside the sliding mass gradually decreases, and the seismic wave dissipates at this position. With the increase of slippage, there will be more energy dissipation which causes sharp decreases of acceleration inside the sliding mass and leads to the acceleration inside the sliding bed exceeding that inside the sliding mass. The instant frequency of horizontal acceleration inside the sliding mass also stabilizes in certain range.

It can be known from the above general analysis that, at the beginning of earthquake, the seismic ground motion acceleration is small, and the motion between sliding bed and sliding mass has strong consistency. With the increase of seismic ground motion strength, the difference of vibration frequency and amplitude between sliding bed and sliding mass gradually enlarges, and the motion inconsistency gradually strengthens. However, the strengthening of motion inconsistency will intensify the development of tensile and shear failure points on structural surface and lead to landslides. Therefore, the inconsistency of motion between sliding bed and sliding mass is a main controlling factor leading to landslides of bedrock and overburden slopes.

⑤ Difference of Seismic Energy Distribution and Dissipation between Sliding Bed and Sliding Mass

In order to study on the difference of seismic energy distribution and dissipation between sliding bed and sliding mass, this chapter, respectively, sets up acceleration

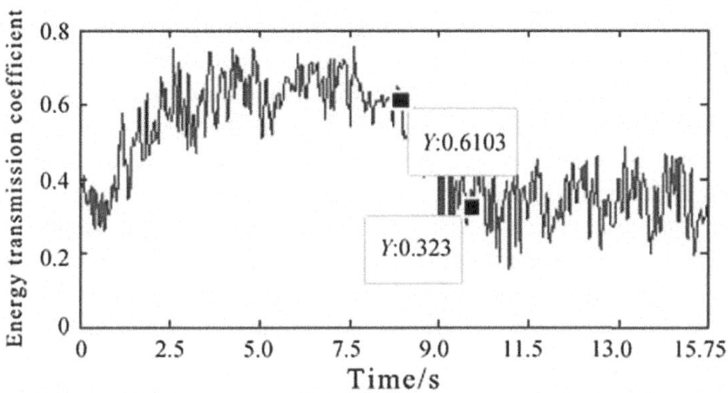

Fig. 4.31 Time–frequency curve of transmission coefficient of the sliding mass

Test Points L1–L5 and C1–C6 inside the slope mass of single and dual side high and steep bedrock and overburden slopes, and acceleration Test Points M1–M5 and D1–D6 inside the sliding mass. Because the sliding process of single and dual side bedrock and overburden slopes are basically the same, this chapter selects representative Test Points (C1–C6, D1–D6) on the two sides of sliding mass structural surface of dual side bedrock and overburden slope as analyzing objects. At the same time, the seismic energy inside the sliding mass mainly comes from transmission energy of seismic wave inside sliding bed passing through the sliding structural surface. Therefore, the transmission coefficient of sliding mass structural surface could well reflect the difference of seismic ground motion energy between sliding bed and sliding mass.

Therefore, this chapter will use Hilbert–Huang transform and work out the energy distribution of all Test Points (C1–C6, D1–D6) from the joint time–frequency-domain so as to obtain the time–frequency curve of transmission coefficient of the whole sliding mass structural surface (see Fig. 4.31). It is noteworthy that the spectrum values in the Fourier spectrum only stand for the amount of energy and have no physical meaning. This chapter defines transmission coefficient as follows: Energy transmission coefficient = energy inside the sliding mass/ energy inside the sliding bed.

It can be known from Fig. 4.31 that there is big difference of energy distribution between sliding bed and sliding mass. First, when the seismic motion is small, the energy transmission coefficient of seismic motion inside the sliding mass stabilizes at 0.55–0.75; with the increase of seismic intensity, it gradually decreases and stabilizes at 0.20–0.40. According to Figs. 4.29 and 4.30, it can be known that the seismic motion energy inside the sliding bed mainly distributes in the dominant frequency range of 5–7 Hz, while that inside the sliding mass mainly distributes in the dominant frequency ranges of 0.5–1.5 Hz and 4–5 Hz, with the lower dominant frequency range controlling the major power of seismic wave. Therefore, with the increase of seismic intensity, the energy transmission coefficient decreases, and the controlling frequency range of seismic motion energy inside the sliding mass transferring from high-frequency range to low-frequency range. The reason for the above phenomenon may be that when seismic ground motion is small, there is no relative slip between sliding bed and sliding mass, the instant frequency stabilizes at 5–7 Hz, and the energy transmission coefficient is high, stabilizing in the range of 0.55–0.75. At the same time, using elastic wave theory can get the calculation result 0.665 of energy transmission coefficient. These two are basically of the same situation. Therefore, at the beginning of earthquake, the energy of seismic ground motion inside the sliding bed mainly concentrates in the range of 5–7 Hz, and that of the sliding mass mainly concentrates in the range of 4–5 Hz. With the increase of seismic intensity, the energy of input seismic wave gradually rising and the shear and tensile failure points on the sliding mass structural surface also become more, which gradually form the connecting failure surface. Relative slip occurs between sliding bed and sliding mass, which causes the increase of energy dissipation of seismic wave on the structural surface, leading to decrease of energy transmission coefficient, which finally stabilizes in the range of 0.20–0.40. At the same time, with the appearance of inconsistency between sliding bed and sliding mass, the instant frequency of seismic acceleration inside the sliding mass gradually decreases and stabilizes in the range of 0.5–1.5 Hz. Thus, after relative slip between sliding bed and sliding mass occurs, the controlling frequency range of seismic motion energy inside the sliding mass gradually stabilizes in the range of 0.5–1.5 Hz. Therefore, with the increase of seismic intensity, the damage of sliding mass structural surface enlarges, the relative slippage between the sliding bed and sliding mass is also on the rise, the energy dissipation on the structural surface becomes more, energy transmission coefficient decreases, and the controlling frequency range of seismic motion energy inside the sliding mass transferring from high-frequency range to low-frequency range.

Through the comprehensive analysis above, it can be known that, at the beginning of seismic ground motion, energy transmission coefficient of sliding mass structural surface is high, large amount of seismic motion energy is stored inside the cover layer, while the thickness of the cover layer is thin, which causes the energy intensity of storage and dissipation inside the cover layer is much more than that in the lower sliding bed, leading to tensile and shear failure of sliding mass structural surface. With the increase of seismic intensity, the difference of energy between sliding bed and sliding mass gradually enlarges, which intensifies the damage of sliding mass structural surface and leads to landslides. Thus, difference of seismic energy distribution and dissipation between sliding bed and sliding mass is a key factor to induce landslides, which can fully explain the seismic damage phenomenon of "peeling type" landslides widely spreading in the soft rock distribution region and hard rock superficial weathered zone in Wenchuan earthquake.

4.3 Time–Frequency Analysis Theory of Seismic Stability of High and Steep Rock Slopes

The seismic stability of slopes has been the hotspot issues of geological disaster study, and especially after the 5.12 Wenchuan earthquake, it has become the study focus of specialists and scholars. For slope seismic stability analysis, the current common used methods include pseudo-static method, sliding block analysis method, and numerical simulation, in which pseudo-static method only takes consideration of the influence of PGA on slope stability, but not the influence of frequency and duration of seismic waves; sliding block analysis method can solve neither the instant stability coefficient of slopes under seismic loading effects, nor can it accurately evaluate the seismic stability in real engineering, so it is not convenient to use; although numerical simulation can carry out time–frequency analysis of rock slope stability under seismic effects, the operation is complex, especially for high and steep slopes with large difference in rock layer distribution. However, the results of 4.2 Slope Deformation Characteristics and Formation Mechanism of High and Steep Rock Slopes under Seismic Effects show that bedrock and overburden slopes coming across landslides are induced by duration, frequency, and amplitude of seismic ground motion, and these three are an organic whole, so singly considering one of them will lead to partial recognition of slope seismic stability. Therefore, it is quite necessary to study on the influence of time, frequency, and amplitude on slope seismic stability time–frequency analysis.

According to the field research results and formation mechanism analysis results of large-scale landslides disasters in Wenchuan earthquake, it can be known that SV wave (elastic wave whose vibration direction and propagation direction is vertical to the slope cross section) is the most disastrous and causes the most serious damage. In addition, the major energy of seismic wave is controlled by SV wave, while the duration of P-wave is short and its energy is small. Therefore, we can approximately take it that the slope stability under earthquake is the slope stability under SV wave. But SV wave is a complex non-stationary signal, whose frequency feature changes with time. At the same time, bedrock and overburden slope are the most common slope structure with maximum instability quantity in seismic areas. Therefore, this chapter will focus on bedrock and overburden slope, base on elastic wave theory, adopt Hilbert–Huang transform and general analysis model of slopes, and propose time–frequency analysis method of bedrock and overburden slope seismic stability which could take consideration of three factors of seismic waves so as to provide new ideas to bedrock and overburden slope stability recognition.

4.3.1 *Time–Frequency Analysis Formula Derivation of Seismic Stability of Bedrock and Overburden Slopes*

Based on elastic wave theory and general geological analysis model, we adopt Hilbert–Huang transform, propose time–frequency analysis method of bedrock and overburden slope seismic stability, and verify its rationality through shaking table test and numerical analysis results.

4.3.1.1 The General Thinking and Basic Assumption

The time–frequency analysis method derivation of bedrock and overburden slopes mainly follows the following process: Generalize the analysis model, make sliding mass structural surface meshing, adopt HHT and elastic wave theory to solve the instant stress components due to reflection and transmission when seismic wave passing through this point; take consideration of gravitational stress field, use M-C failure principle, discuss its damage, and work out the stress component of other meshing points in turn; draw the instant stress distribution of each point on the structural surface and finally work out the instant stability coefficient of bedrock and overburden slopes under earthquake. The derivation of this method will base on the following hypothesis: ① the closely connected medium on the two sides of sliding mass structural surface is homogeneous and isotropic elastomer; ② the sliding mass structural surface conforms to M-C failure principle; ③ due to that the seismic waves continuously coming across refraction and transmission when transmitting from seismic source upwards, which causes they close to the vertical direction when reaching the earth surface, so this chapter selects seismic waves transmitting vertically upward from bedrock to slope mass as input wave; ④ because wave-type conversion will occur when seismic wave arrive at the cover layer surface with its major energy transferring into Rayleigh surface wave moving along the cover layer surface, and the material damping inside the cover layer will largely consume the seismic wave energy, which will finally lead to reflective seismic wave on the cover layer surface bringing very weak seismic energy to the sliding mass structural surface, this chapter will pay no attention to the influence of reflective seismic wave on cover layer surface to sliding mass structural surface.

4.3.1.2 Generalized Analysis Model

It can be known from the research results of Wenchuan earthquake hazard that bedrock and overburden slope are the most common slope structure with maximum instability quantity in seismic areas, and a common feature of slopes of this kind is: Obvious structural surface exists inside the slope mass, the material property on its two sides has certain difference, and structural surface is the major controlling factor of slope landslides. Therefore, this chapter builds the generalized analysis model

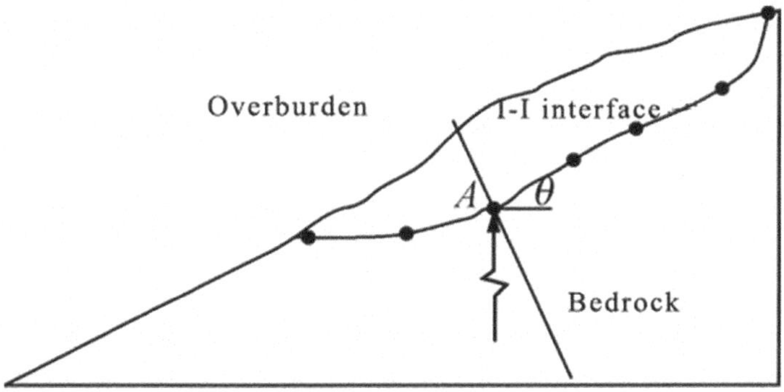

Fig. 4.32 Generalized model of bedrock and overburden slopes

based on the above thinking (see Fig. 4.32). Figure 4.32 shows the upper surface of cover layer, I-I interface is of any shape, the tangential obliquity of I-I interface at Micro-unit A is θ, and evenly mesh the sliding mass surface into 7 points (take this as example), and select A as study object. Figure 3.35 shows the reflective and transmission model of SV wave at Point A, and the physico-mechanical parameters of middle–upper medium of this model (bedrock) are, respectively, ρ_1, C_{p1}, C_{s1}, and G1, G1 (density, P-wave speed, S-wave speed, and shear modulus) and the physico-mechanical parameters of lower medium are ρ_2, C_{p2}, C_{s2} and G2 (density, P-wave speed, S-wave speed, and shear modulus). In there, I-I interface is the sliding mass structural surface; when SV wave S^1 transmitted onto the I-I interface, it will generate reflective SV wave S^2, reflective P-wave S^3, transmission SV wave S^4, and transmission P-wave S^5, while α_1, α_1', β_1 and β_1', respectively, represent the incident angle of SV wave, reflection angle of P-wave and transmission angle of P-wave. Note: The incident angle and reflection angle of SV wave are equal.

4.3.1.3 The Reflective and Transmission Coefficient of Sliding Mass Structural Surface

In the XOZ plane in Fig. 4.33, the potential function of all the stress waves can be represented by elastic displacement amplitude, and the potential function of incident S-wave, reflected S-wave, reflected P-wave, transmitted S-wave, transmitted P-wave could be, respectively, written into Formulas (4.33)–(4.36).

$$S^i = S_0^i e^{j(k_x^{i1} X + k_z^{i1} Z - \omega t)}, i = 1, 2, 3, 4, 5 \tag{4.33}$$

In the formulas, $i = 1, 2, 3, 4, 5$, respectively, stand for incident S-wave S^1, reflected S-wave S^2, reflected P-wave S^3, transmitted S-wave S^4, and transmitted P-wave S^5; and k_x^{i1} and k_z^{i1}, respectively, represent wave vector of incident wave,

Fig. 4.33 Reflective and transmission analysis model of Point *A*

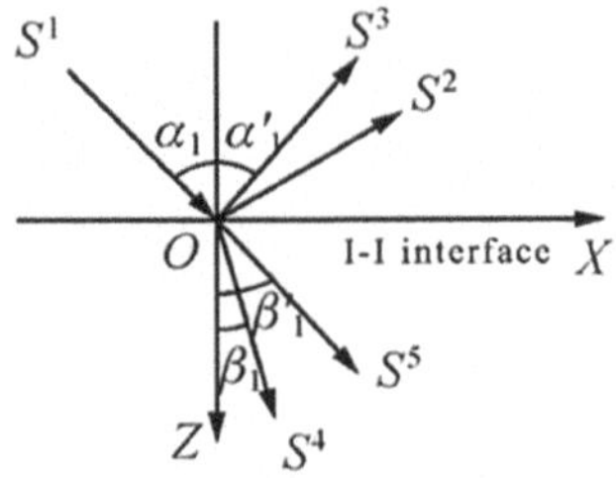

reflected wave, and transmitted wave along *X*- and *Z*-direction, according to Snell Theorem $k_x^{11} = k_x^{21} = k_x^{31} = k_x^{41} = k_x^{51}$.

$$\begin{aligned} k_x^{11} &= \frac{\omega}{V_{s1}} \sin\alpha_1;\ \ k_x^{21} = \frac{\omega}{V_{s1}} \sin\alpha_1;\ \ k_x^{31} = \frac{\omega}{V_{p1}} \sin\alpha_1'; \\ k_x^{41} &= \frac{\omega}{V_{s2}} \sin\beta_1;\ \ k_x^{51} = \frac{\omega}{V_{p2}} \sin\beta_1' \end{aligned} \tag{4.34}$$

$$\begin{aligned} k_z^{11} &= \frac{\omega}{V_{s1}} \cos\alpha_1;\ \ k_z^{21} = -\frac{\omega}{V_{s1}} \cos\alpha_1;\ \ k_z^{31} = -\frac{\omega}{V_{p2}} \cos\alpha_1'; \\ k_z^{41} &= \frac{\omega}{V_{s2}} \cos\beta_1;\ \ k_z^{51} = \frac{\omega}{V_{p2}} \cos\beta_1' \end{aligned} \tag{4.35}$$

$$\frac{\text{Sin}\alpha_1}{V_{s1}} = \frac{\sin\alpha_1'}{V_{p1}} = \frac{\sin\beta_1}{V_{s2}} = \frac{\sin\beta_1'}{V_{p2}} \tag{4.36}$$

In the formulas, V_{p1}, V_{s1}, respectively, stand for the speed of *S*-wave and *P*-wave in upper medium; V_{p2}, V_{s2}, respectively, stand for the speed of *S*-wave and *P*-wave in lower medium.

When incident wave is on the XOY plane, i.e. when taking effects at $Z = 0$, it should meet the following conditions:

$$\begin{cases} u_{1z}(x,z,t) = u_{2z}(x,z,t); u_{1x}(x,z,t) = u_{2x}(x,z,t) \\ \sigma_{1z}(x,z,t) = \sigma_{2z}(x,z,t); \tau_{1z}(x,z,t) = \tau_{2z}(x,z,t) \end{cases} \tag{4.37}$$

It can be known from Fig. 4.33 that, in the upper medium, the horizontal and vertical displacement components are:

$$\begin{cases} u_1 = S^1\cos\alpha_1 - S^2\cos\alpha_1 + S^3\sin\alpha_1' \\ \omega_1 = -S^1\sin\alpha_1 - S^2\sin\alpha_1 - S^3\cos\alpha_1' \end{cases} \tag{4.38}$$

In the lower medium, the horizontal and vertical displacement components are:

$$\begin{cases} u_2 = S^4\cos\beta_1 + S^5\sin\beta_1' \\ \omega_2 = -S^4\sin\beta_1 + S^5\cos\beta_1' \end{cases} \tag{4.39}$$

Solving simultaneously with Formulas (4.33)–(4.39) can get displacement boundary condition:

$$\begin{cases} S^1\cos\alpha_1 - S^2\cos\alpha_1 + S^3\sin\alpha_1' = S^4\cos\beta_1 + S^5\sin\beta_1' \\ -S^1\sin\alpha_1 - S^2\sin\alpha_1 - S^3\cos\alpha_1' = -S^4\sin\beta_1 + S^5\cos\beta_1' \end{cases} \tag{4.40}$$

Substituting the relevant parameters into Formula (4.40) and (4.41), after simplification, it can get:

$$\begin{cases} S_0^1\cos\alpha_1 - S_0^2\cos\alpha_1 + S_0^3\sin\alpha_1' = S_0^4\cos\beta_1 + S_0^5\sin\beta_1' \\ -S_0^1\sin\alpha_1 - S_0^2\sin\alpha_1 - S_0^3\cos\alpha_1' = -S_0^4\sin\beta_1 + S_0^5\cos\beta_1' \end{cases} \tag{4.41}$$

Calculating the stress on I-I interface with isotropic Hooke's law:

$$\begin{cases} \sigma_z = \lambda\left(\frac{\partial u_x}{\partial x} + \frac{\partial u_z}{\partial z}\right) + 2G\frac{\partial u_z}{\partial z} \\ \tau_{zx} = G\left(\frac{\partial u_x}{\partial z} + \frac{\partial u_z}{\partial x}\right) \end{cases} \tag{4.42}$$

In the formula, λ stands for lame constant and G stands for shear modulus.

Solving simultaneously with Formulas (4.37)–(4.40), and Formula (4.42) can get stress boundary condition:

$$\begin{cases} \lambda_1[S_0^1\cdot k_x^{11}\cdot\cos\alpha_1 - S_0^2\cdot k_x^{21}\cdot\cos\alpha_1 + S_0^3\cdot k_x^{31}\cdot\sin\alpha_1'] \\ +(\lambda_1+2\mu_1)[-S_0^1\cdot k_z^{11}\cdot\sin\alpha_1 - S_0^2\cdot k_z^{21}\cdot\sin\alpha_1 - S_0^3\cdot k_z^{31}\cdot\cos\alpha_1'] \\ = \lambda_2[S_0^4\cdot k_x^{41}\cdot\cos\beta_1 + S_0^5\cdot k_x^{51}\cdot\sin\beta_1'] + (\lambda_2+2\mu_2)[-S_0^4\cdot k_z^{41}\cdot\sin\beta_1 \\ +S_0^5\cdot k_z^{51}\cdot\cos\beta_1']\mu_1[S_0^1\cdot k_z^{11}\cdot\cos\alpha_1 - S_0^2\cdot k_z^{21}\cdot\cos\alpha_1 + S_0^3\cdot k_z^{31}\cdot\sin\alpha_1' \\ -S_0^1\cdot k_x^{11}\cdot\sin\alpha_1 - S_0^2\cdot k_x^{21}\cdot\sin\alpha_1 - S_0^3\cdot k_x^{31}\cdot\cos\alpha_1'] = \mu_2[S_0^4\cdot k_z^{41}\cdot\cos\beta_1 \\ +S_0^5\cdot k_z^{51}\cdot\sin\beta_1' + S_0^5\cdot k_x^{51}\cdot\cos\beta_1' - S_0^4\cdot k_x^{41}\cdot\sin\beta_1] \end{cases} \tag{4.43}$$

Through comprehensive analysis of Formulas (4.40), (4.41), (4.42), and (4.43), it can get:

$$
\begin{cases}
S_0^1 \cos\alpha_1 - S_0^2 \cos\alpha_1 + S_0^3 \sin\alpha_1' - S_0^4 \cos\beta_1 - S_0^5 \sin\beta_1' = 0 \\
-S_0^1 \sin\alpha_1 - S_0^2 \sin\alpha_1 - S_0^3 \cos\alpha_1' + S_0^4 \sin\beta_1 - S_0^5 \cos\beta_1' = 0 \\
[\lambda_1 \cdot k_x^{11} \cdot \cos\alpha_1 - (\lambda_1 + 2\mu_1) \cdot k_z^{11} \cdot \sin\alpha_1] S_0^1 - [\lambda_1 \cdot k_x^{21} \cdot \cos\alpha_1 \\
+(\lambda_1 + 2\mu_1) \cdot k_z^{21} \cdot \sin\alpha_1] S_0^2 + [\lambda_1 \cdot k_x^{31} \cdot \sin\alpha_1' - (\lambda_1 + 2\mu_1) \cdot k_z^{31} \\
\cdot \cos\alpha_1'] S_0^3 + [(\lambda_2 + 2\mu_2) \cdot k_z^{41} \cdot \sin\beta_1 - \lambda_2 \cdot k_x^{41} \cdot \cos\beta_1] S_0^4 - [\lambda_2 \cdot k_x^{51} \\
\cdot \sin\beta_1' + (\lambda_2 + 2\mu_2) \cdot k_z^{51} \cdot \cos\beta_1'] S_0^5 = 0 \mu_1 [k_z^{11} \cdot \cos\alpha_1 - k_x^{11} \cdot \sin\alpha_1] S_0^1 \\
-\mu_1 [k_z^{21} \cdot \cos\alpha_1 + k_x^{21} \cdot \sin\alpha_1] S_0^2 + \mu_1 [k_z^{31} \cdot \sin\alpha_1' - k_x^{31} \cdot \cos\alpha_1'] S_0^3 \\
-\mu_2 [k_z^{41} \cdot \cos\beta_1 - k_x^{41} \cdot \sin\beta_1] S_0^4 - \mu_2 [k_z^{51} \cdot \sin\beta_1' + k_x^{51} \cdot \cos\beta_1'] S_0^5 = 0
\end{cases}
\tag{4.44}
$$

Dividing each item into Formula (4.44) with S_0^1 and making $A' = S_0^2/S_0^1$, $B' = S_0^3/S_0^1$, $C' = S_0^4/S_0^1$, $D' = S_0^5/S_0^1$, it can get transmitted and reflected coefficient equation set after rewriting:

$$
\begin{cases}
\cos\alpha_1 - A' \cos\alpha_1 + B' \sin\alpha_1' - C' \cos\beta_1 - D' \sin\beta_1' = 0 \\
-\sin\alpha_1 - A' \sin\alpha_1 - B' \cos\alpha_1' + C' \sin\beta_1 - D' \cos\beta_1' = 0 \\
[\lambda_1 \cdot k_x^{11} \cdot \cos\alpha_1 - (\lambda_1 + 2\mu_1) \cdot k_z^{11} \cdot \sin\alpha_1] - [\lambda_1 \cdot k_x^{21} \cdot \cos\alpha_1 \\
+(\lambda_1 + 2\mu_1) \cdot k_z^{21} \cdot \sin\alpha_1] A' + [\lambda_1 \cdot k_x^{31} \cdot \sin\alpha_1' - (\lambda_1 + 2\mu_1) \cdot k_z^{31} \\
\cdot \cos\alpha_1'] B' + [(\lambda_2 + 2\mu_2) \cdot k_z^{41} \cdot \sin\beta_1 - \lambda_2 \cdot k_x^{41} \cdot \cos\beta_1] C' - [\lambda_2 \cdot k_x^{51} \\
\cdot \sin\beta_1' + (\lambda_2 + 2\mu_2) \cdot k_z^{51} \cdot \cos\beta_1'] D' = 0 \mu_1 [k_z^{11} \cdot \cos\alpha_1 - k_x^{11} \cdot \sin\alpha_1] \\
-\mu_1 [k_z^{21} \cdot \cos\alpha_1 + k_x^{21} \cdot \sin\alpha_1] A' + \mu_1 [k_z^{31} \cdot \sin\alpha_1' - k_x^{31} \cdot \cos\alpha_1'] B' \\
-\mu_2 [k_z^{41} \cdot \cos\beta_1 - k_x^{41} \cdot \sin\beta_1] C' - \mu_2 [k_z^{51} \cdot \sin\beta_1' + k_x^{51} \cdot \cos\beta_1'] D' = 0
\end{cases}
\tag{4.45}
$$

Simplifying Formula (4.45) into matrix equation:

$$
BX = c \Leftrightarrow \begin{bmatrix} B_1 & B_2 \end{bmatrix} \cdot \begin{bmatrix} A' \\ B' \\ C' \\ D' \end{bmatrix} = \begin{bmatrix} -\cos\alpha_1 \\ \sin\alpha_1 \\ -(\lambda_1 \cdot k_x^{11} \cdot \cos\alpha_1 - (\lambda_1 + 2\mu_1) \cdot k_z^{11} \cdot \sin\alpha_1) \\ -\mu_1 (k_z^{11} \cdot \cos\alpha_1 - k_x^{11} \cdot \sin\alpha_1) \end{bmatrix}
\tag{4.46}
$$

In Formula (4.46), the concrete forms of coefficient B1 and B2 are as follows:

$$B_1 = \begin{bmatrix} -\cos\alpha_1 & \sin\alpha_1' \\ -\sin\alpha_1 & -\cos\alpha_1' \\ -\lambda_1 \cdot k_x^{21}\cos\alpha_1 - (\lambda_1 + 2\mu_1)k_z^{21}\sin\alpha_1 & \lambda_1 \cdot k_x^{31}\sin\alpha_1' - (\lambda_1 + 2\mu_1)\cdot k_z^{31}\cdot\cos\alpha_1' \\ -\mu_1(k_z^{21}\cos\alpha_1 + k_x^{21}\sin\alpha_1) & \mu_1(k_z^{31}\sin\alpha_1' - k_x^{31}\cos\alpha_1') \end{bmatrix} \tag{4.47}$$

$$B_2 = \begin{bmatrix} -\cos\beta_1 & -\sin\beta_1' \\ \sin\beta_1 & -\cos\beta_1' \\ -\lambda_2 \cdot k_x^{41}\cos\beta_1 + (\lambda_2 + 2\mu_2)k_z^{41}\sin\beta_1 & -\lambda_2 \cdot k_x^{51}\sin\beta_1' - (\lambda_2 + 2\mu_2)k_z^{51}\cos\beta_1' \\ -\mu_2[k_z^{41}\cos\beta_1 - k_x^{41}\sin\beta_1] & -\mu_2[k_z^{51}\sin\beta_1' + k_x^{51}\cos\beta_1'] \end{bmatrix} \tag{4.48}$$

In order to verify the correctness of formula derivation, we now work out the reflected and transmitted coefficient of sliding mass structural surface when SV wave vertically transmitted onto the surface.

$$A' = \frac{\rho_1 V_{s1} - \rho_2 V_{s2}}{\rho_1 V_{s1} + \rho_2 V_{s2}};\ B' = 0 \tag{4.49}$$

$$C' = \frac{2\rho_1 V_{s1}}{\rho_1 V_{s1} + \rho_2 V_{s2}};\ D' = 0 \tag{4.50}$$

The calculation results above are exactly the same with the acknowledged calculation results, which demonstrates the derivation results of this chapter are correct.

4.3.1.4 Instant Seismic Stability Coefficient of Bedrock and Overburden Slopes

Because of the sliding mass structural surface adopting the sliding friction model which conforms to the M-C principle, it should meet the following conditions when coming across shear damage at Point 0: $\tau_s \geq \sigma_n \tan\varphi + C$, in which, φ and C are, respectively, the internal friction angle and cohesive force; if tensile failure occurring at Point 0, it should meet the following conditions: $\sigma_n \geq \sigma_{tension}$.

According to Hooke's law and solving the reflected and transmitted coefficient of seismic wave simultaneously with Formulas (4.46)–(4.48), bring the calculation results and relevant parameters into Formula (4.53), which can get the downslide stress, cohesion stress, and tensile stress of SV wave at Point A. The details are as follows:

$$\begin{aligned} \tau_s = \tau_0 + \mu_1[&S_0^1 \cdot k_z^{11} \cdot \cos\alpha_1 - S_0^2 \cdot k_z^{21} \cdot \cos\alpha_1 + S_0^3 \cdot k_z^{31} \cdot \sin\alpha_1' \\ &- S_0^1 \cdot k_x^{11} \cdot \sin\alpha_1 - S_0^2 \cdot k_x^{21} \cdot \sin\alpha_1 - S_0^3 \cdot k_x^{31} \cdot \cos\alpha_1'] \end{aligned} \tag{4.51}$$

$$\begin{aligned}\sigma_0 \tan\varphi + C = \{&\lambda_1[S_0^1 \cdot k_x^{11} \cdot \cos\alpha_1 - S_0^2 \cdot k_x^{21} \cdot \cos\alpha_1 + S_0^3 \cdot k_x^{31} \cdot \sin\alpha_1'] \\ &+ (\lambda_1 + 2\mu_1)[-S_0^1 \cdot k_z^{11} \cdot \sin\alpha_1 \\ &- S_0^2 \cdot k_z^{21} \cdot \sin\alpha_1 - S_0^3 \cdot k_z^{31} \cdot \cos\alpha_1'] + \sigma_0\} \tan\varphi + C\end{aligned} \quad (4.52)$$

$$\begin{aligned}\sigma_n = &\lambda_1[S_0^1 \cdot k_x^{11} \cdot \cos\alpha_1 - S_0^2 \cdot k_x^{21} \cdot \cos\alpha_1 + S_0^3 \cdot k_x^{31} \cdot \sin\alpha_1'] \\ &+ (\lambda_1 + 2\mu_1)[-S_0^1 \cdot k_z^{11} \cdot \sin\alpha_1 - S_0^2 \cdot k_z^{21} \cdot \sin\alpha_1 - S_0^3 \cdot k_z^{31} \cdot \cos\alpha_1'] + \sigma_0\end{aligned} \quad (4.53)$$

Because of the sliding mass structural surface conforms to the M-C principle, it should meet the following conditions when coming across shear damage at Point A: $\tau_s \geq \sigma_n \tan\varphi + C$, in which, $\sigma_{tension}$ and C are, respectively, the internal friction angle and cohesive force; if tensile failure occurring at Point A, it should meet the following conditions: In Formulas (4.51)–(4.53), σ_0 and τ_0 are, respectively, the initial normal and tangential gravitational stress at Point A, so it can solve with slices method. By analogy, the above method could solve the stress of other points on the sliding mass structural surface. Then, collecting the calculation results of each point to solve the total downslide stress SF and the total tensile stress TF on the sliding mass structural surface (see Formula 4.54). Finally, using the safety coefficient KO required by the design to judge the slope safety (see Formula 4.55).

$$F_{滑} = \sum_{i=1}^{n} \tau_{si} \cdot dA_i \,;\, F_{抗} = \sum_{i=1}^{n} (\sigma_{0i} \cdot \tan\varphi_i + C_i) \cdot dA_i \quad (4.54)$$

$$\begin{aligned}&\text{If } K = \text{TF/SF} < K0, \text{ the sliding mass is dangerous;} \\ &\text{if } K = \text{TF/SF} > K0, \text{ the sliding mass is safe.}\end{aligned} \quad (4.55)$$

In Formulas (4.54) and (4.55), τ_{si} stands for the downslide shear stress of the ith meshing point; $A_i = 0.5\ (Li - 1 + Li + 1) \times 1$, which represents the area of the ith meshing point, in which, L_{i-1} and L_{i+1} are, respectively, the side length of the $i-1$th and $i+1$th meshing points; σ_{0i} represents the normal stress of the ith meshing point; φ_i and C_i, respectively, stand for the internal friction angle and cohesive force of the ith meshing point.

Through comprehensive analysis of Formulas (4.51)–(4.55), the seismic stability time–frequency analysis method derived by this chapter could rationally consider (1) gravitational stress field, (2) the internal friction angle, cohesive force and reflected and transmitted coefficient of controlling structural surface, (3) physico-mechanical parameters of bedrock and cover layer, (4) influencing factors like instant frequency, incident angle, acceleration, velocity, and displacement time history curve.

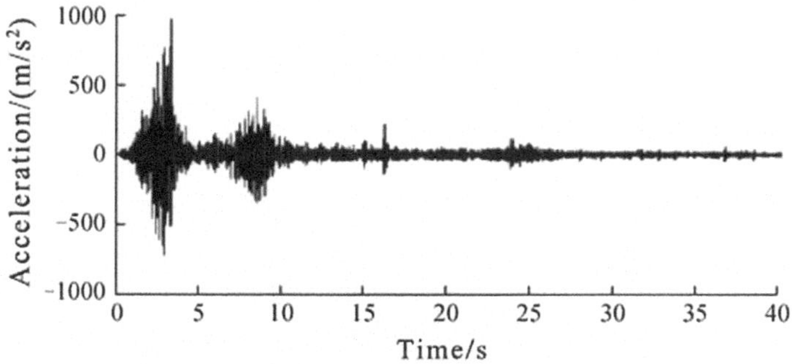

Fig. 4.34 Time history curve of input seismic wave

4.3.1.5 Time–Frequency Analysis Process of Seismic Waves

Through comprehensive analysis of the seismic stability time–frequency analysis method of bedrock and overburden slopes, it can be known that the instant frequency effects of seismic weaves mainly embody in the elastic displacement amplitude s_0^i of incident wave, reflected wave and transmitted wave, and the frequency of wave vector ω, i.e. k_x^{i1} and k_z^{i1}, for example $k_x^{i1} = \omega \cdot \sin \alpha_1 / V_{p2}$. Therefore, this book proposes the analysis process of seismic wave time–frequency effects: Firstly, break up seismic wave time–frequency into several of IMFs (time history curve) through Empirical Mode Decomposition (EMD for short); secondly, carry out Channel Switch, i.e. switch multi-channel signal composed of multiple IMF into single-channel signal made up of single IMF; thirdly, solve the instant frequency of each channel and draw the time–frequency curve of each IMF. Lastly, bring each IMF and their instant frequency into Formulas (4.51)–(4.55) to work out the results and superimpose the results, and we can get the stress component of this point.

4.3.1.6 Practical Application of Seismic Wave Time–Frequency Effects

In order to more deeply and concretely explaining the practical application of seismic wave time–frequency effects, this dissertation will, based on the above analysis ideas, conduct input wave time–frequency analysis and the detail is as follows: Because the follow-up work need combining the shaking table test and numerical analysis model to verify the correctness of this time–frequency method, this chapter will select the original horizontal seismic waves of the shaking table test, i.e. horizontal Wenhcuan wolong seismic wave (see Fig. 4.34), in which, the Hilbert Spectrum see Fig. 4.35; secondly, carry out EEMD decomposition and get every IMF (see Fig. 4.36); lastly, solve the instant frequency of every IMF (see Fig. 4.37). It is noteworthy that, due to that the duration of Wenchuan wolong seismic wave is a bit longer, so Figs. 4.36 and 4.37 only select calculation results of T =0–5 s including peak seismic ground motion acceleration, and demonstrates it clearly.

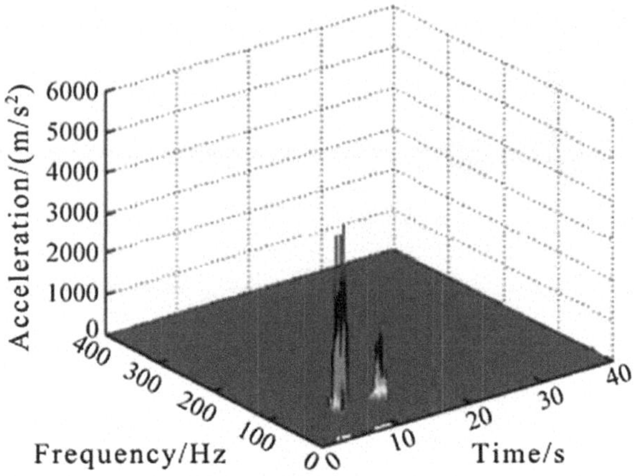

Fig. 4.35 Hilbert Spectrum of horizontal acceleration

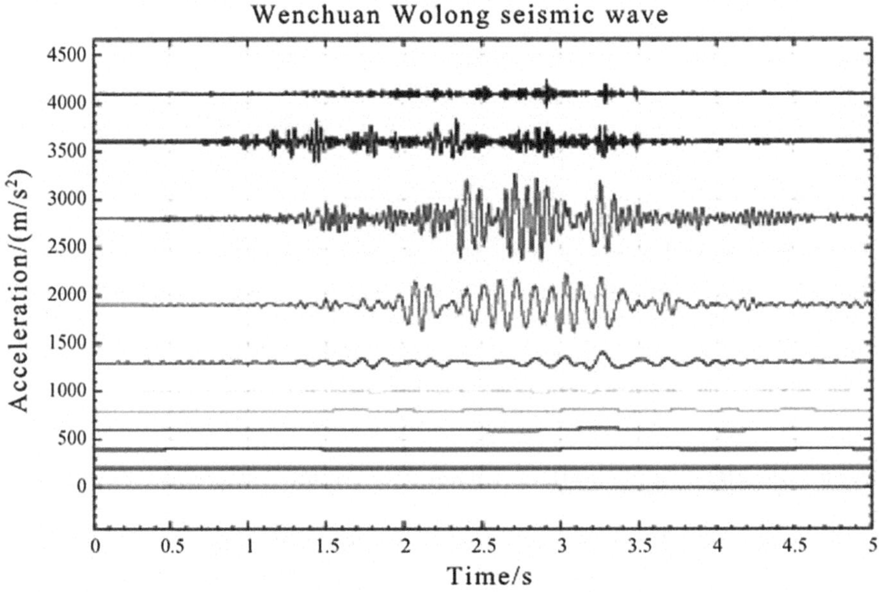

Fig. 4.36 Acceleration time history curve of IMF

Note, the Y-coordinate in Figs. 4.36 and 4.37 only represents physical characters, but not size.

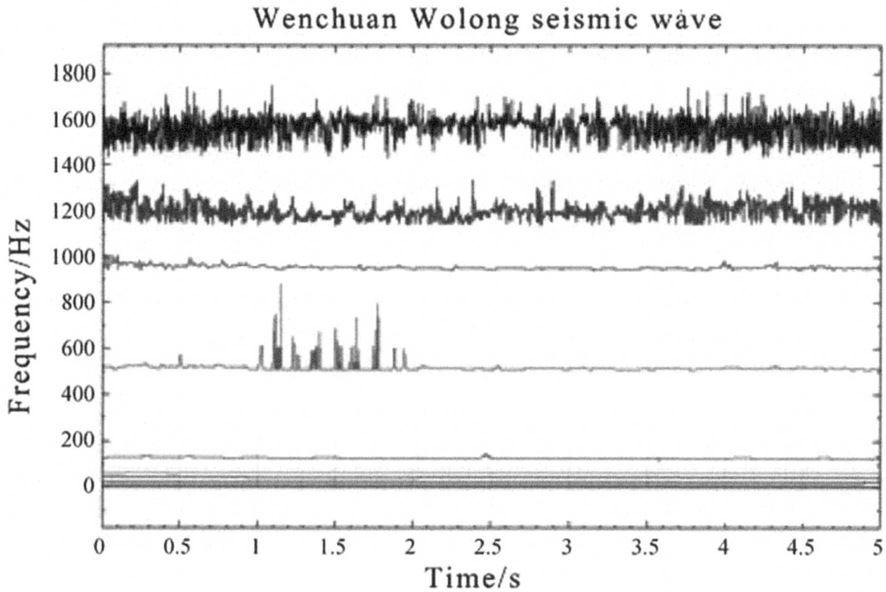

Fig. 4.37 Instant frequency map of IMF

4.3.2 *Time–Frequency Analysis Formula Derivation of Bedrock and Overburden Slopes Seismic Stability*

According to the above analysis, this chapter summarizes the time–frequency analysis formula derivation of bedrock and overburden slopes seismic stability, and it is as follows:

(1) According to the study object, we sum up appropriate geological analysis model and decide the basic physical parameters such as geometric size, mechanical parameters of rock and soil mass and make sliding mass structural surface meshing.
(2) If there exists real seismic wave, then we adopt it as input wave; if not, we adopt artificial seismic wave. The detail is as follows: Decide the design response spectrum curve of this region according to *Code for Seismic Design of Buildings* and synthesize artificial ground motion time history based on seismic ground motion time history fitting method, so as to decide the seismic ground motion acceleration time history of this region.
(3) Decide the displacement time history curve according to the acceleration time history curve, adopt any points in the mesh as study point, and use Formula (3.51) to solve the displacement time history curve of reflected wave and transmitted wave of this point.
(4) Use Hilbert–Huang transform solving IMF of incident wave, reflected wave and transmitted wave as well as the correspondent frequency–time history curve.

(5) Bring the obtained results into Formulas (4.51)–(4.54) to calculate and superimpose the results, so we can get the stress component of this point.
(6) Adopt above method to work out the stress of each point on the sliding mass structural surface, draw the stress field of sliding mass structural surface, use Formula (3.59) to solve SF and TF of sliding mass structural surface and judge the seismic stability of bedrock and overburden slopes under SV wave with Formula (4.55) combining relevant specifications.

4.3.3 Verification of Seismic Stability Time–Frequency Analysis Method of Bedrock and Overburden Slope

Due to that the monitoring points of shaking table test could not directly be set up on the sliding mass structural surface, which cannot accurately judge the damage development of the sliding mass structural surface, so it can only reflect the whole damage of the slope and indirectly reflect the damage of the sliding mass structural surface. Therefore, this section will adopt both methods of shaking table test and numerical simulation to comprehensively verify the correctness of the above time–frequency analysis method from two aspects of sliding mass and structural surface. Before verifying the correctness of seismic stability time–frequency analysis method of bedrock and overburden slopes, this chapter needs to explicitly explain the verification approach, and the details are as follows: First, build numerical analysis model by geometric similarity ratio of 1:1 to the shaking table test model of dual side bedrock and overburden slopes, set up stress and strain test points at the correspondent position, and compare the numerical calculation results with the experimental results so as to verify the correctness of numerical analysis method. This part of work has been finished in Sect. 3.1.3. Secondly, carry out explicit and complete verification of the calculation results of time–frequency analysis method using the shaking table test in 3.1 and the above numerical analysis results.

4.3.3.1 Overview of Numerical Analysis

In order to accurately verify the correctness of seismic stability time–frequency analysis method of bedrock and overburden slopes, this chapter will build the numerical analysis method according to the shaking table test and adopt the same rock and soil parameters and input seismic wave. In addition, the bedrock and cover layer adopt the elastoplastic model and the sliding mass structural surface adopts the brittle fracture constitutive model, and both models conform to M-C failure principle.

Set up stress and strain Test Point A on the sliding mass structural surface, and at the same time, the seismic stability time–frequency analysis method of bedrock and overburden slopes also select Test Point A and B as study objects, so as to accurately and completely verify the correctness of calculation results of single point on the

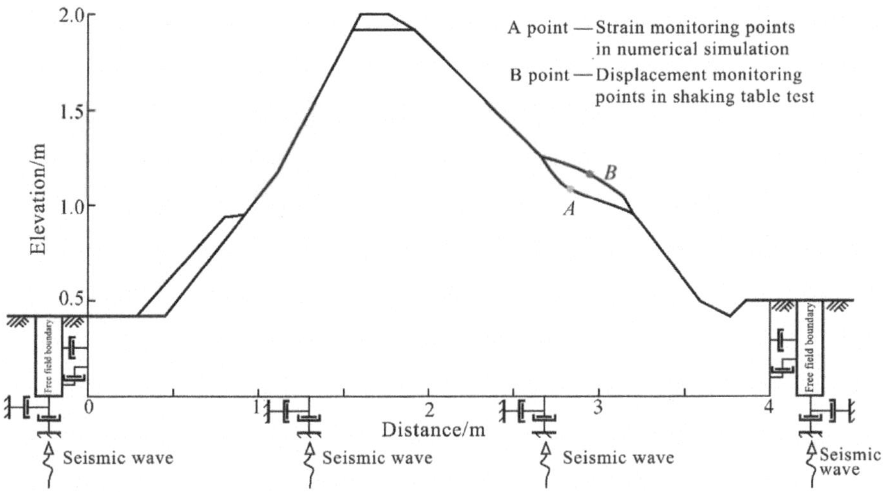

Fig. 4.38 Numerical analysis model

structural surface and the sliding mass. See Fig. 4.38 for test points distribution. It is noteworthy that because the Wenchuan wolong seismic wave mainly includes *P*-wave and *S*-wave, while *S*-wave includes SV wave and SH wave, in which, the seismic wave energy carried by *P*-wave is small, and that carried by *A* wave is big, which plays a key role in the slope seismic stability. At the same time, *P*-wave and SV wave mainly cause damage inside the cross section, while SH wave mainly causes damage outside the cross section. However, the slope problem is often simplified into plane stress problem, which pays no attention to the slope damage outside the cross section, so it does not need to consider the effects of SH wave. In general situation, most researches of slope seismic stability only consider the effects of SV wave, and approximately take the real seismic wave as SV wave to partially consider the safety. Therefore, this shaking model test approximately simulates the effects of SV wave through input of Wenchuan wolong seismic wave.

4.3.3.2 Gravitational Stress Field

This book regulates that the normal stress takes stress as positive and tension as negative; shear stress takes the upward of sliding mass structural surface as positive and upward as negative.

Test Point *A* is 20 cm from the surface of cover layer, the tangential obliquity of sliding mass structural surface at micro-unit *A* is 32°, and this chapter adopts slices method to calculate the gravitational stress field (see Figs. 4.39, 4.40, and 4.41). Lastly, we get the initial normal stress σ and initial shear stress τ of this point are, respectively , 3.29 and 1.90 kPa, so, according to M-C failure principle, it can be

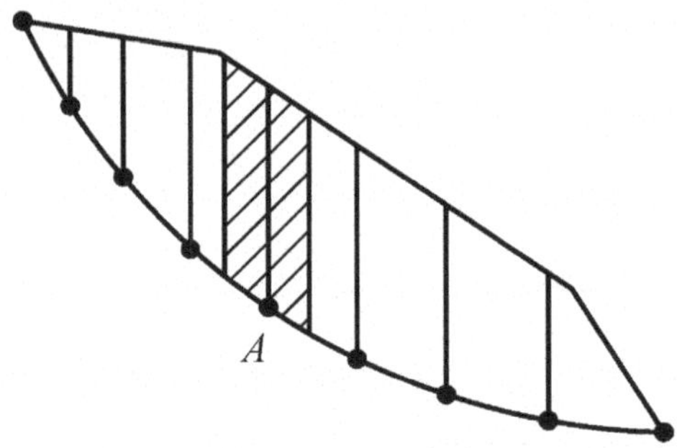

Fig. 4.39 Analysis model of deposit body

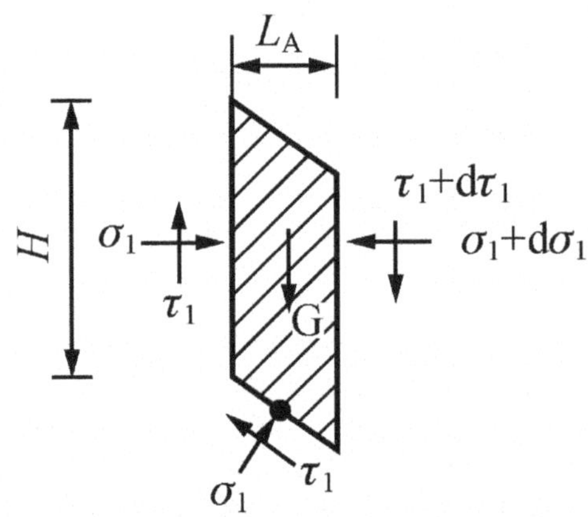

Fig. 4.40 Mechanical model figure of sliding block A

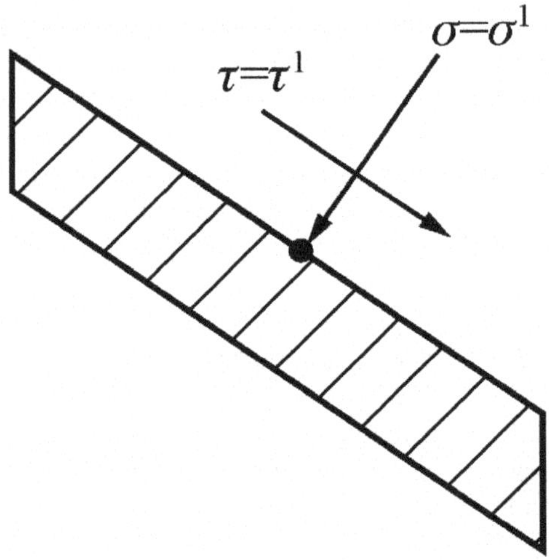

Fig. 4.41 Mechanical model figure of Test Point A

known that the rest sliding shear stress P is:

$$P = \sigma \tan\theta + C - \tau = 3.29\ \text{kPa} \times \tan 32^\circ + 2.3{-}1.9\ \text{kPa} = 2.455\ \text{kPa}$$

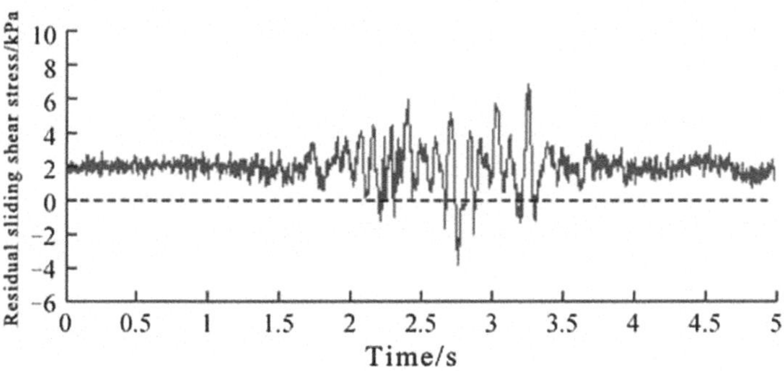

Fig. 4.42 Rest shear stress time history curve of Test Point A

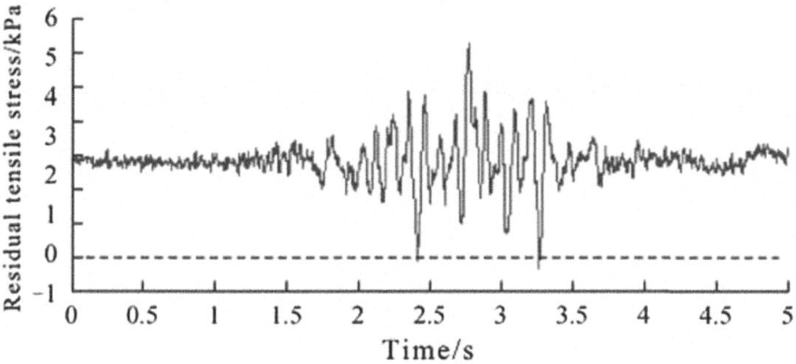

Fig. 4.43 Rest tensile stress time history curve of Test Point A

4.3.3.3 Comparison of Calculation Results

Based on the derivation process of seismic stability time–frequency analysis method of bedrock and overburden slopes, we bring the relevant parameters and the above calculation results into Formulas (4.51)–(4.55) to solve the stress of Test Point A, so as to work out the rest sliding stress of it (see Figs. 4.42 and 4.43 for detailed results). At the same time, the calculation results of plastic stress of Test Point *A* see Fig. 4.44 and work out the rest sliding stress of left points by analogy. Finally, we superimpose the results and get the rest sliding stress time history curve of the whole sliding mass structural surface, and the detailed calculation results see Figs. 4.45 and 4.46. At the same time, the displacement calculation results of test point see Fig. 4.47. Due to that the duration of Wenchuan wolong seismic wave is a bit longer and Test Point *A* is damaged before $T = 5$ s, so Figs. 4.42, 4.43, 4.44, 4.45, 4.46, and 4.47 only select calculation results of $T = 0$–5 s including peak seismic ground motion acceleration.

It can be known from Figs. 4.42 and 4.43 that, Test Point *A* comes across shear failure at $T = 2.25$ s tensile failure at $T = 2.45$ s. Therefore, Test Point *A* will come across shear failure at $T = 2.25$ s **in the earthquake; the numerical analysis results**

Fig. 4.44 Rest tangential plastic stress time history curve of Test Point A

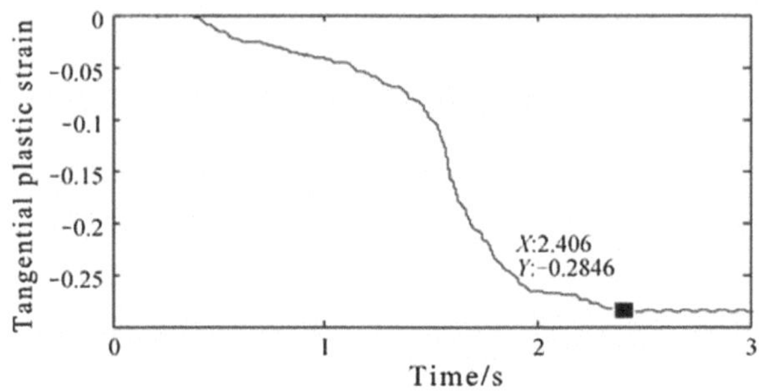

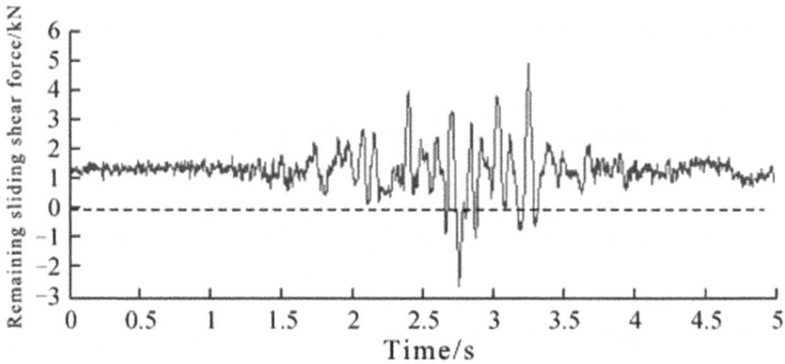

Fig. 4.45 Rest shear stress time history curve of the sliding mass

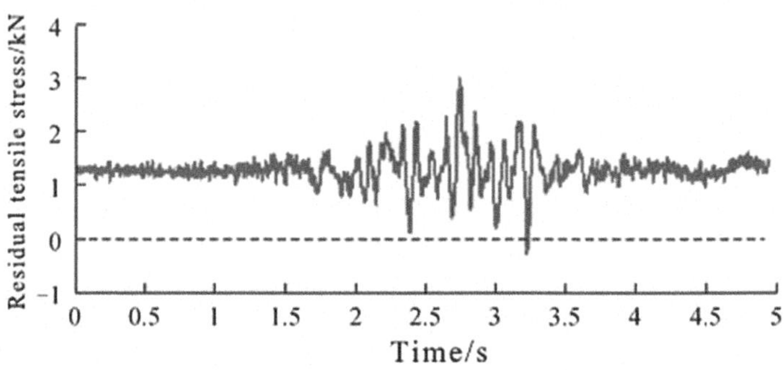

Fig. 4.46 Rest tensile stress time history curve of the sliding mass

in Fig. 4.44 **show that when $T=1$–2 s, the tangential plastic stress of *A* suddenly changes and becomes stable after $T=2.406$ s. According to the features of the brittle fracture constitutive model and M-C failure principle, we can know that the test point has already been damaged**. It can be known from Figs. 4.45 and 4.46 that the sliding mass coming across shear failure at $T=2.75$ s and tensile failure at $T=3.25$ s; thus in the earthquake, the sliding mass will come across shear failure at $T=2.75$ s; the results of the shaking table test in Fig. 4.47 show that the displacement of the test point suddenly changes, which can demonstrate

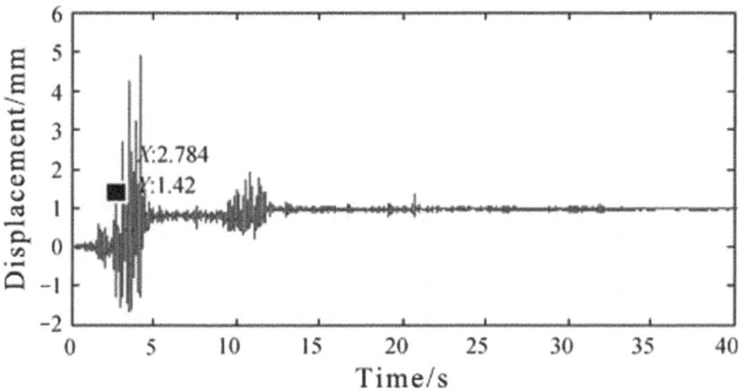

Fig. 4.47 Shaking table test results

that the sliding mass has already been damaged. The difference of the calculation results with the numerical analysis results and the shaking table test results may be because: Firstly, the time–frequency analysis method of rock slope seismic stability was derived based on the elastic wave theory, which paid no attention to the nonlinear characteristics of rock and soil mass as well as the sliding mass structural surface; secondly, the time–frequency analysis method of rock slope seismic stability did not consider the plastic strain and accumulative effect in the earthquake. But in general, no matter the single test point or the whole sliding mass, the calculation results of time–frequency analysis method, the numerical analysis results, and the shaking table test results of rock slope seismic stability are basically the same, which fully verify the correctness of the calculation results and feasibility of calculation method in this dissertation.

Through comprehensive analysis of the above calculation results, it can be known that, under SV wave effects, the sliding mass structural surface mainly comes across shear failure, which is basically the same with the results of field research of Wenchuan earthquake; the time of sliding mass structural surface coming across failure is slightly ahead of the PGA arriving time of $T = 3.25$ s in Wenchuan earthquake, which fully explains that it is a little bit conservative for the current supporting structure seismic design adopting the method of PGA, so this chapter suggest using time–frequency design so as to save and reduce cost.

4.3.4 Study on Parameters of Seismic Stability Time–Frequency Analysis of Bedrock and Overburden Slopes

Through comprehensive analysis of the above derivation, it can be known that the transmission and reflection coefficients of structural surface include factors like the

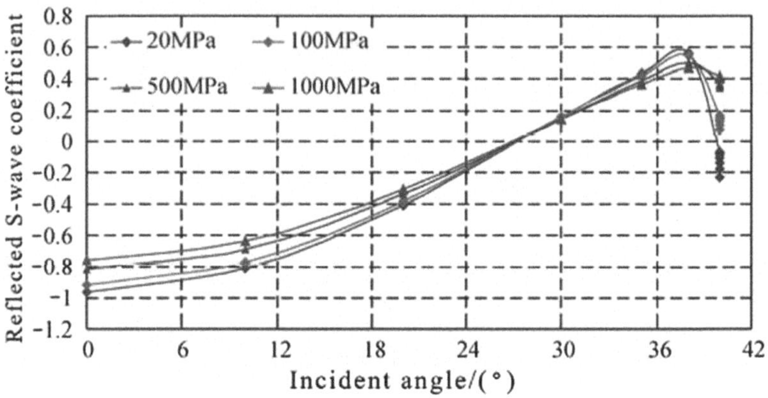

Fig. 4.48 Reflected coefficients of reflected S-wave with different incident angles

elastic displacement amplitude of incident wave, reflected wave, and transmitted wave. Therefore, it is quite necessary to conduct research on the influence factors like transmitted and reflected coefficients of the structural surface in this chapter.

4.3.4.1 The Influence of Incident Angle on Transmitted and Reflected Coefficients of the Structural Surface

In order to study on the influence of incident angle on transmitted and reflected coefficients of the structural surface, this chapter now takes the following example to explain: Take sine wave with amplitude = 1 and frequency = 5 Hz as the incident S-wave, and the incident angle are 10°, 20°, 30°, 40°, 50°, 60°, 70°, 80°, and 90°. When the elasticity modulus of the cover layer, respectively, is 20, 100, 500, and 1000 MPa, the elasticity modulus of the bedrock is 30000 MPa. The detailed results of the reflected and transmitted coefficients of different structural surface parameters which can be Seen from Figs. 4.48, 4.49, 4.50, and 4.51.

Through comprehensive analysis of Figs. 4.48 and 4.49, we can get the following rules: The reflected coefficient of reflected S-wave reaches the maximum when the incident angle is 38°, and the phase position changes when the incident angle is 25°; the reflected coefficient of reflected P-wave slightly grows with the increase of the incident angle when the incident angle is in the range of 0°–38°, while it suddenly increases when the incident angle is 38°, and when the incident angle is in the range of 0°–40°, the phase position of the reflected coefficient stays unchanged; the transmitted coefficient of transmitted S-wave slightly decreases with the increase of the incident angle when the incident angle is in the range of 0°–38° and reaches the maximum when the incident angle is 38°, while it suddenly increases when the incident angle is in the range of 38°–40°, and when the incident angle is in the range of 0°–40°, the phase position of the transmitted coefficient stays unchanged;

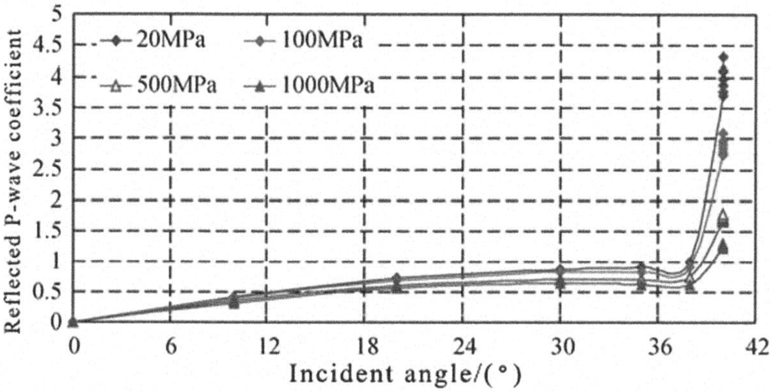

Fig. 4.49 Reflected coefficients of reflected *P*-wave with different incident angles

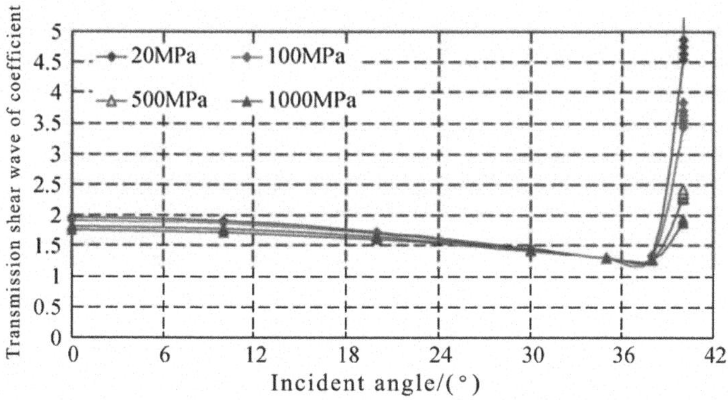

Fig. 4.50 Transmitted coefficients of transmitted *S*-wave with different incident angles

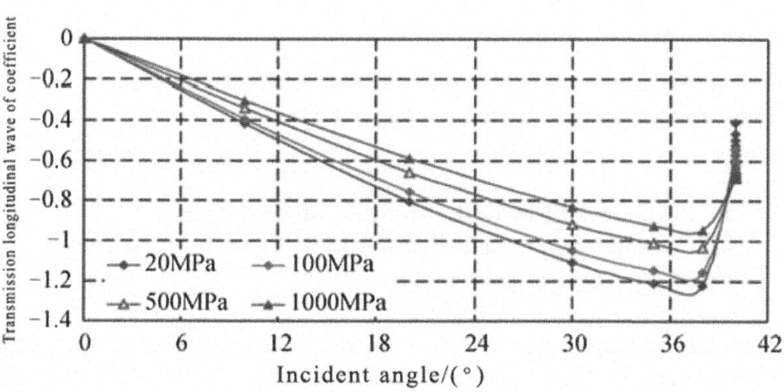

Fig. 4.51 Transmitted coefficients of transmitted *P*-wave with different incident angles

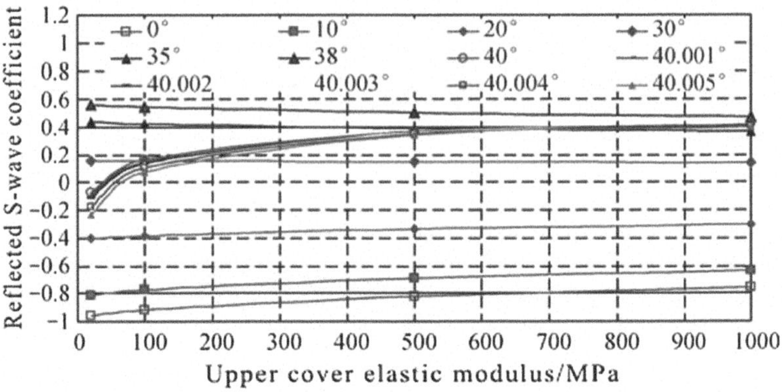

Fig. 4.52 Reflected coefficients of reflected S-wave

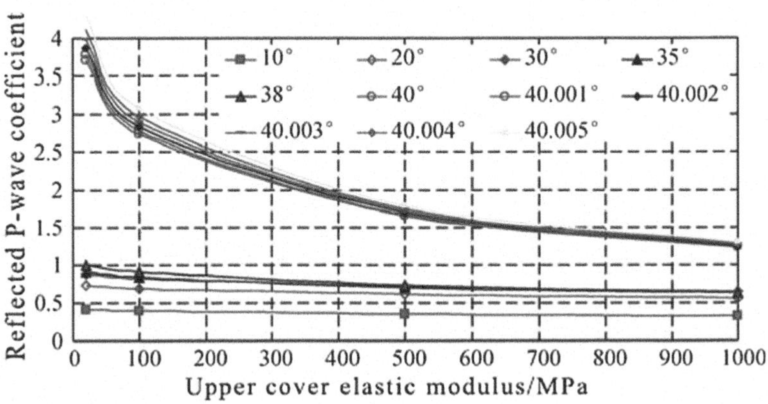

Fig. 4.53 Reflected coefficients of reflected P-wave

the transmitted coefficient of transmitted P-wave reaches its maximum when the incident angle is 38° and suddenly increases afterward.

4.3.4.2 The Influence of Elasticity Modulus of Cover Layer on Transmitted and Reflected Coefficients of the Structural Surface

In order to study on the influence of elasticity modulus of cover layer on transmitted and reflected coefficients of the structural surface, this chapter now still takes the above example to explain, the elasticity modulus of the cover layer, respectively, are 20, 100, 500, and 1000 MPa, and the elasticity modulus of the bedrock is 30,000 MPa. See Figs. 4.52, 4.53, 4.54, and 4.55 for detailed results.

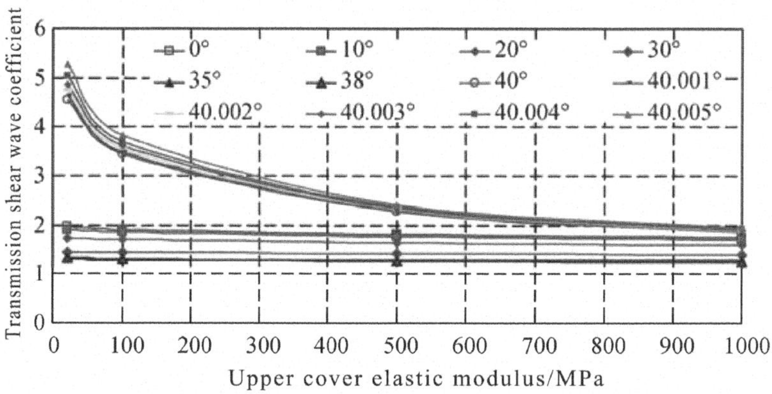

Fig. 4.54 Transmitted coefficients of transmitted *S*-wave

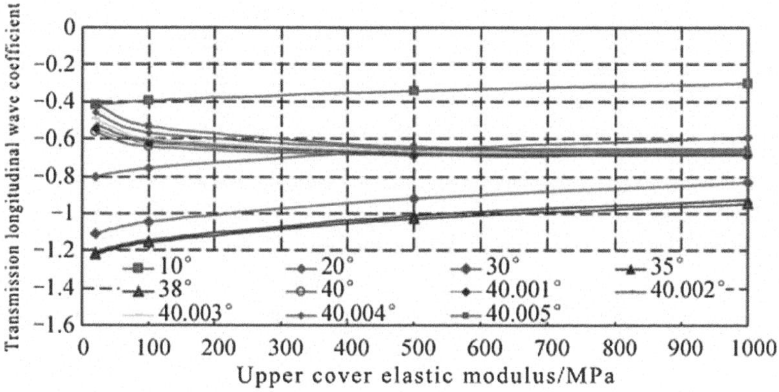

Fig. 4.55 Transmitted coefficients of transmitted *P*-wave

Through comprehensive analysis of Figs. 4.52, 4.53, 4.54, and 4.55, we can come to the following rules: With the gradual increase of the elasticity modulus of the cover layer, the reflected coefficients of reflected *S*-wave, the reflected coefficients of reflected *P*-wave, the transmitted coefficients of transmitted *S*-wave, and the transmitted coefficients of transmitted *P*-wave all come to certain balance, which are approximately in hyperbolic curve or inverted hyperbolic curve.

4.3.4.3 The Influence of Incident Angle on Transmitted and Reflected Coefficients of the Structural Surface

In order to study on the influence of incident angle on transmitted and reflected coefficients of the structural surface, this chapter now takes the above example to

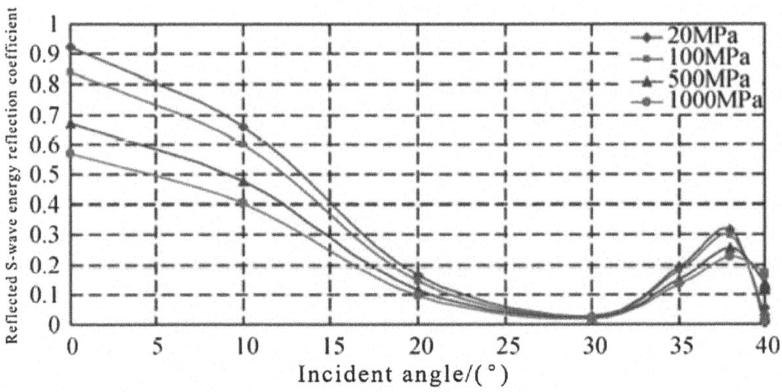

Fig. 4.56 Energy reflected coefficient of reflected S-wave

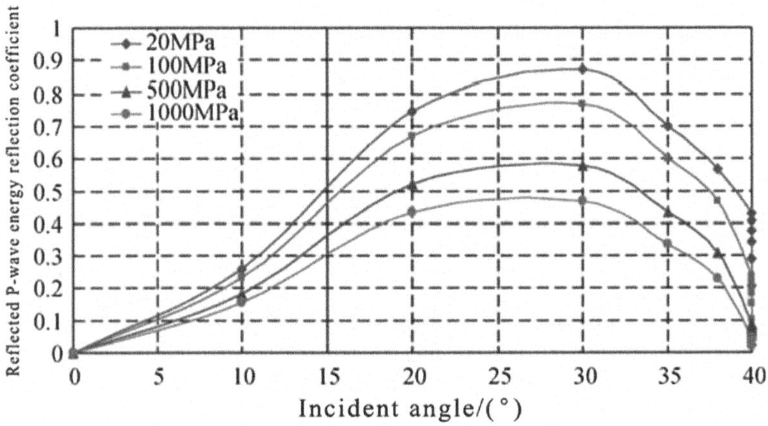

Fig. 4.57 Energy reflected coefficient of reflected P-wave

explain and the incident angle are 10°, 20°, 30°, 40°, 50°, 60°, 70°, 80°, and 90°. See Figs. 4.56, 4.57, 4.58, and 4.59 for detailed results.

Through comprehensive analysis of Figs. 4.56, 4.57, 4.58, and 4.59, we can come to the following rules: The energy reflected coefficient of reflected S-wave reaches its maximum when the incident angle is 0° and arrives at a small peak when the incident angle is 38°, but it is close to 0 when the incident angle is 30°. The energy reflected coefficient of reflected P-wave reaches its maximum when the incident angle is 30°. The energy transmitted coefficient of transmitted S-wave reaches its maximum when the incident angle is 0° and lightly decreases when the incident angle is in the range of 0°–38°, but when the incident angle is in the range of 38°–40°, it suddenly increases. The energy transmitted coefficient of transmitted P-wave reaches its maximum when the incident angle is 38°, it increases with the increase of incident angle when the

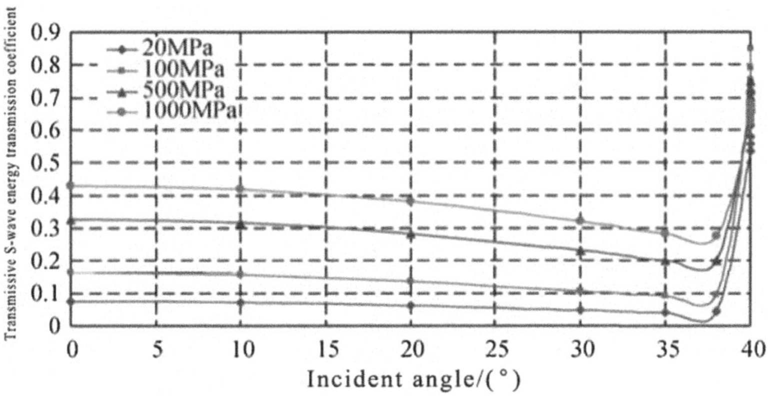

Fig. 4.58 Energy transmitted coefficient of transmitted *S*-wave

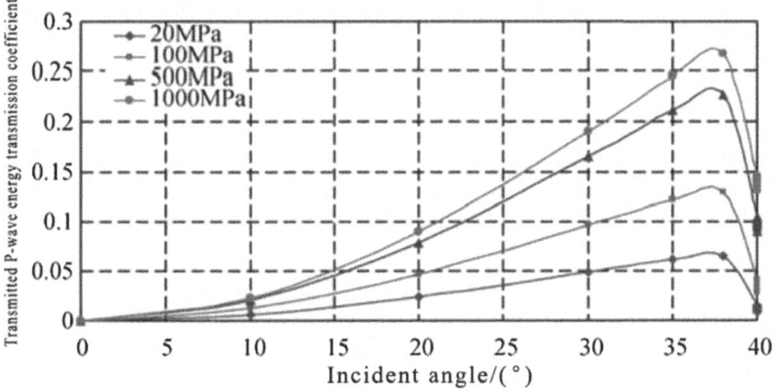

Fig. 4.59 Energy transmitted coefficient of transmitted *P*-wave

latter is in the range of 0°–38°, and it decreases with the increase of incident angle when the latter is in the range of 38°–40°.

4.3.4.4 The Influence of Elasticity Modulus of Cover Layer on Transmitted and Reflected Coefficients of the Structural Surface Energy

In order to study on the influence of elasticity modulus of cover layer on transmitted and reflected coefficients of the structural surface, this chapter now still takes the above example to explain, the elasticity modulus of the cover layer, respectively, are 20, 100, 500, and 1000 MPa, and the elasticity modulus of the bedrock is 30,000 MPa. See Figs. 4.60, 4.61, 4.62, and 4.63 for detailed results.

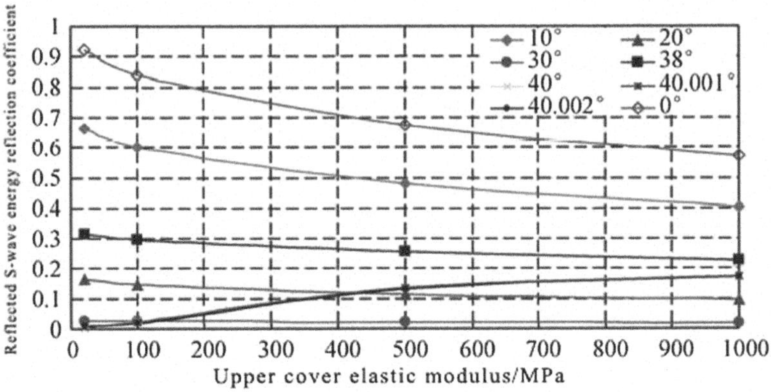

Fig. 4.60 Energy reflected coefficient of reflected S-wave

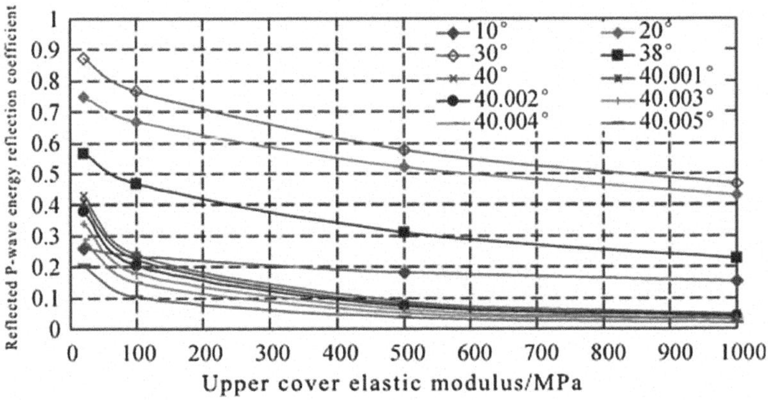

Fig. 4.61 Energy reflected coefficient of reflected P-wave

Through comprehensive analysis of Figs. 4.60, 4.61, 4.62, and 4.63, we can come to the following rules: The energy reflected coefficient of reflected S-wave and the energy reflected coefficient of reflected P-wave all decrease with the increase of elasticity modulus of the cover layer and finally reach the balance; the energy transmitted coefficient of transmitted S-wave increases with the increase of elasticity modulus of the cover layer and finally becomes stable. The energy transmitted coefficient of transmitted P-wave reaches its maximum when the thickness of cover layer meets 2000 MPa and slightly decreases afterward.

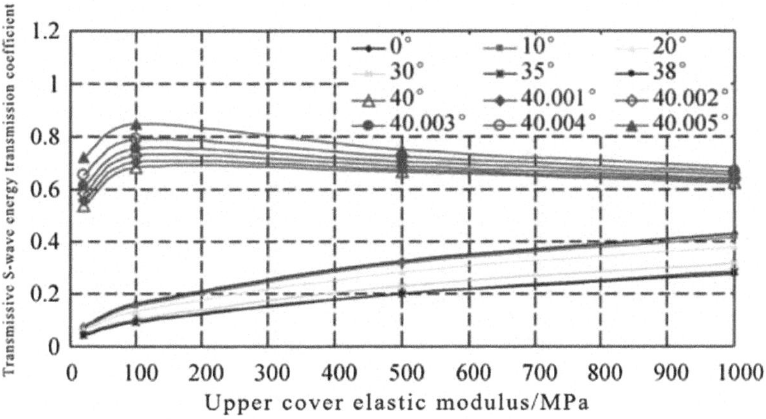

Fig. 4.62 Energy transmitted coefficient of transmitted *S*-wave

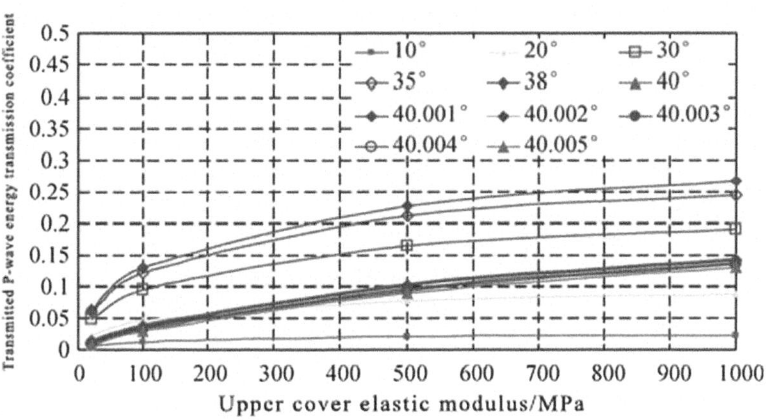

Fig. 4.63 Energy transmitted coefficient of transmitted *P*-wave

4.3.4.5 The Influence of Elasticity Modulus of Cover Layer on Transmitted and Reflected Coefficients of the Structural Surface Energy

In order to study on the influence of elasticity modulus of cover layer on total transmitted and reflected energy coefficients of the structural surface, this chapter now still takes the above example to explain, the elasticity modulus of the cover layer, respectively, are 20, 100, 500, and 1000 MPa, and the elasticity modulus of the bedrock is 30,000 MPa. See Figs. 4.64 and 4.65 for detailed results.

Through comprehensive analysis of Figs. 4.64 and 4.65, it can be known that the total reflected energy coefficients slightly decrease with the increase of incident angle when the latter is in the range of 0°–40° and reach its minimum when the

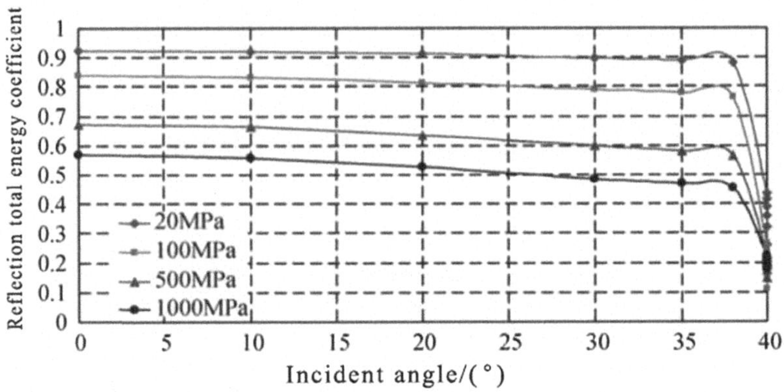

Fig. 4.64 Total reflected energy coefficients

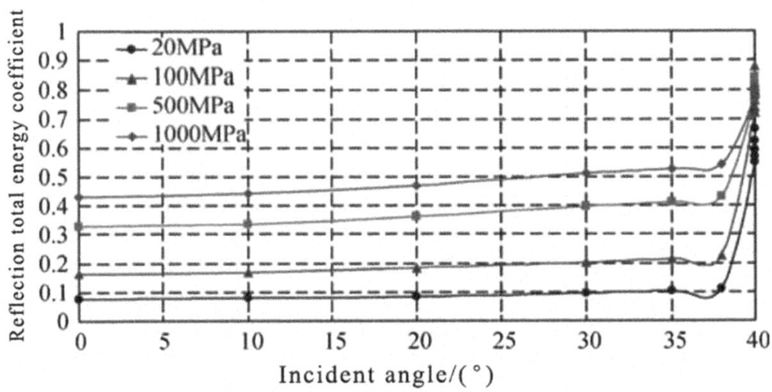

Fig. 4.65 Total transmitted energy coefficients

incident angle is 38°. Then, with the continuous increase of incident angle, the total reflected energy coefficients suddenly increase; the total reflected energy coefficients are precisely opposite to the total energy reflection, and with any incident angles, the sum of them is 1.

4.3.5 Advantage of Seismic Stability Time–Frequency Analysis of Bedrock and Overburden Slopes

In order to fully explain the advantage of seismic stability time–frequency analysis of bedrock and overburden slopes, this chapter selects sine wave whose frequency changes with time (see Figs. 4.66 and 4.67 for details) and uses time–frequency analysis method, pseudo-static method, and numerical analysis method to calculate

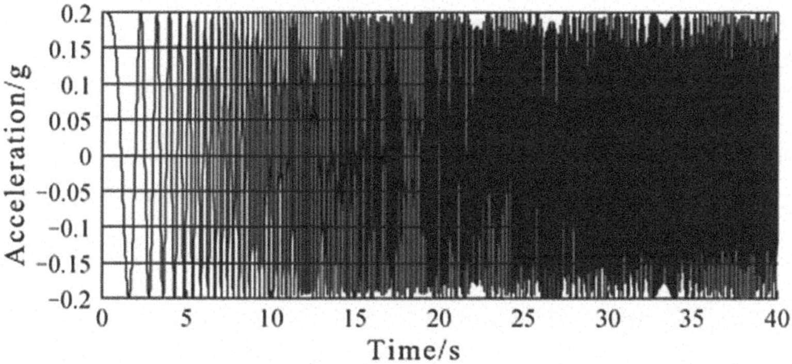

Fig. 4.66 Time history curve of sine wave

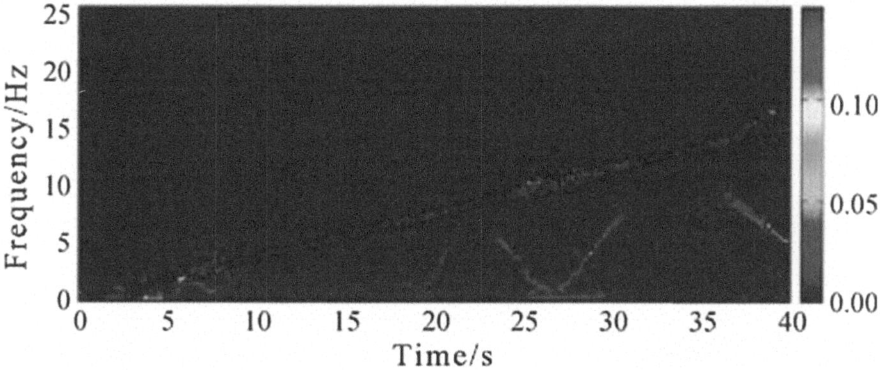

Fig. 4.67 Hilbert response spectrum of sine wave

Fig. 4.68 Slope analysis model

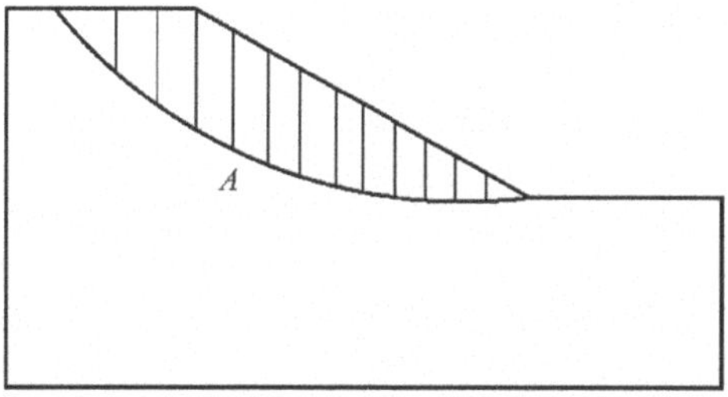

the slope stability (see Fig. 4.68 for analysis model). The concrete physical and mechanical indexes of bedrock, upper cover layer, and the sliding mass structural surface see Table 4.8. It is noteworthy that because the sliding block method could not concretely and quantitatively judge the seismic stability of slopes, this chapter will not take comparison of it.

Calculate the rest sliding shear stress of the left points by analogy and finally superimpose the results, and we can get the shear sling stress time history curve

Table 4.8 Physico-mechanical parameters of bedrock, upper cover layer, and the sliding mass structural surface

Model name	Unit weight	Shear wave speed	*P*-wave speed	Lame coefficient	Shear modulus
Bedrock	30 kN/m^3	826.19 m/s	3240 m/s	881.89 MPa	2047.24 MPa
Upper cover layer	19 kN/m^3	99.21 m/s	145.34 m/s	6.30 MPa	6.30 MPa
The structural surface	Normal stiffness	Shear stiffness	Internal friction angle	Tensile strength	Cohesive force
	4500 MPa	2300 MPa	32°	23 kPa	50

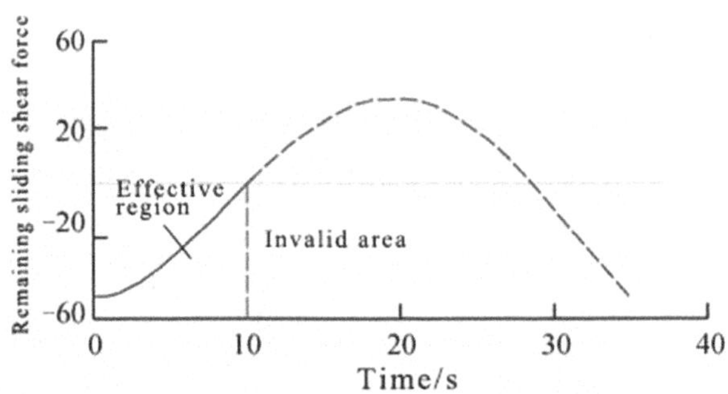

Fig. 4.69 Calculation results of time–frequency analysis method

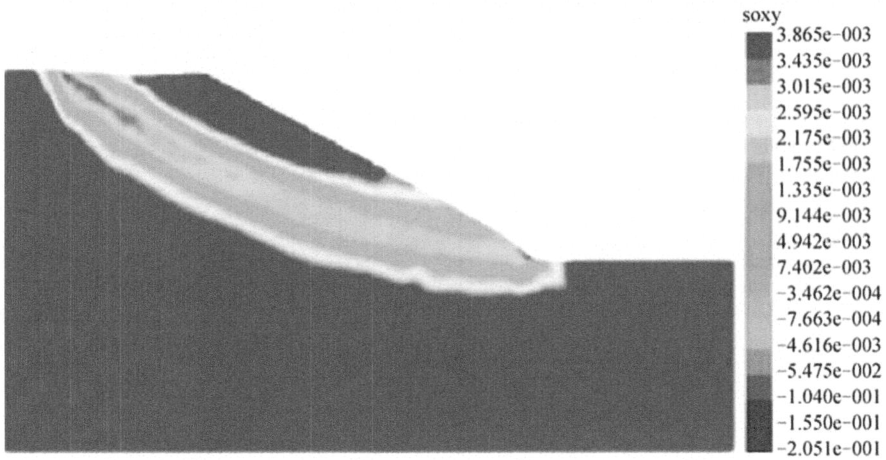

Fig. 4.70 Numerical calculation results (shear plastic strain figure)

of the whole sliding mass structural surface (see Fig. 4.69 for detailed results). See Fig. 4.70 for the calculation results of numerical analysis method.

At the same time, we select PGA = 0.2 g and calculate the rest shear sliding stress with pseudo-static method, and we can get the result is −47 kN, so the slopes are safe.

It can be known from Fig. 4.69 that when $T = 10.0$ s, the rest shear sliding stress of calculation results with time–frequency analysis method is negative and the slope has been damaged. While due to that the time–frequency analysis method in this chapter is derived through elastic wave theory which pays no attention to the nonlinear characteristics of rock and soil mass, so there still has calculation results when $T = 10.0$–35.0 s, but these results are invalid and can only display the rest shear sliding stress under different seismic wave frequency. Figure 4.70 shows that when the numerical calculation results display at $T = 9.3$ s, shear plastic zone occurs inside the slope, and the slope is damaged. Therefore, the results of time–frequency analysis method and numerical analysis method are basically the same. While the calculation results of pseudo-static method that shows the slope is safe and will not be damaged, which is opposite to the results of time–frequency analysis method and numerical analysis method. The above phenomenon may be caused by the following reasons: First, the natural frequency of the sliding mass is getting by the related formula proposed by Xu Jiyan and Liu Hongshuai et al. With the duration of input seismic wave, its frequency gradually increases and approaches the natural frequency of the sliding mass. The resonance effects become stronger while the rest shear sliding stress decreases. When $T = 10.0$ s, the instant frequency of seismic wave is 4.9 Hz, the rest shear sliding stress decreases to negative, the slope slides and failure occurs; second, for time–frequency analysis method and numerical analysis method, they cannot only consider the influence of PGA on slope stability, but can also consider the influence of frequency on the slope stability; while for pseudo-static method, it can only consider the influence of PGA on slope stability, which leads to difference among the above calculation results; lastly, time–frequency analysis method can pay no attention to the plastic strain and accumulative effect in the earthquake, which caused the result displacing time of time–frequency analysis method lagging behind of that of numerical analysis method.

Based on the above analysis, it can be known that the frequency of seismic ground motion has significant influence on the seismic stability of bedrock and overburden slopes, so we need to take it into consideration when calculating the seismic stability of bedrock and overburden slopes.

4.4 Motion Velocity of Different Phases After Slope Slides

After slides occur to the slopes, the motion velocity of the sliding mass directly influences the damage range of the slopes, which, therefore, becomes the study focus of many specialists and scholars. Thus, this section will, based on the previous research results, derive the motion velocity of different phases after slope slides so as to provide reference to the practical engineering.

4.4.1 Principle of Energy Failure

When input seismic wave arrives at the controlling structural surface, it will cause large amounts of reflection and transmission, and waveform conversion and energy transmission will be caused. Some energy of input seismic wave will be reflected by the structural surface and some will be transmitted inside the upper slope body. According to energy conservation theorem, this report takes that the energy-inducing slides to the slopes are equal to the energy of transmitted waves stored inside the slopes before structural surface failure. As the structural surface is still in elastic phase before slip failure and tensile failure occur to it, an overlying soil is still in its elastic phase. The overlying soil experiences Force F1 at T1, when the elastic energy inside the slope is E1, and then E1 is withdrawn, and the elastic energy turns into kinetic energy, which is stored outside the slope. Therefore, the energy stored inside the overlying soil is instant; i.e. the critical energy value causes structural surface failure that should be instant energy value. Due to that the overlying soil layer–bedrock structural surface satisfies the propagation theory of elastic waves; therefore, this section adopts energy formula of stress wave to solve the elastic energy stored inside the overburden layers in elastic phase.

As the displacement amplitude ratio of transverse wave equals to reciprocal of product of displacement potential amplitude ratio multiplying correspondent velocity ratio, i.e.

$$\frac{S^{21}}{S^{11}} = \frac{S_2^1}{S_1^1} \times \frac{V_{S1}}{V_{S2}} \tag{4.56}$$

In the formula, S_{21} and S_{11} represent the displacement amplitude ratio, S_{21} and $S_{11,}$ respectively, represent the displacement potential amplitude of transverse waves, and V_{S1} and $V_{S2,}$ respectively, represent the velocity ratio of transverse ratio. At the same time, it could be known from the energy calculation formula of stress wave that:

$$E'_{\text{in}} = \frac{1}{\rho_1 C_1} \int\limits_{T'_s}^{T'_s+\Delta T} \sigma_1'^2 \mathrm{d}t = E'_f + E'_t = E'_{fP} + E'_{fs} + E'_{tP} + E'_{tS} \tag{4.57}$$

In the formula,

$$E'_f = E'_{fs} + E'_{fp} = A'^2 \cdot E'_{\text{in}} + \frac{\tan\alpha_1}{\tan\alpha'_1} \cdot \left(\frac{B' \cdot V_{P1}}{V_{S1}}\right)^2 \cdot E'_{\text{in}} \tag{4.58}$$

$$\begin{aligned} E'_t = E'_{ts} + E'_{tp} &= \frac{\rho_2 \tan\alpha_1}{\rho_1 \tan\beta_1} \cdot C'^2 \cdot E'_{\text{in}} + \frac{\rho_2 \tan\alpha_1}{\rho_1 \tan\beta'_1} \cdot D'^2 \cdot E'_{\text{in}} \\ &= \frac{\rho_2 \tan\alpha_1}{\rho_1 \tan\beta_1} \cdot \left(\frac{C' \cdot V_{S2}}{V_{S1}}\right)^2 \cdot E'_{in} + \frac{\rho_2 \tan\alpha_1}{\rho_1 \tan\beta'_1} \cdot \left(\frac{D' \cdot V_{P2}}{V_{S1}}\right)^2 \cdot E'_{\text{in}} \end{aligned} \tag{4.59}$$

$$\sigma'_1 = \sqrt{\sigma_{zz}^2 + \tau_{zx}^2} \tag{4.60}$$

$$\sigma'_{ZZ} = \lambda_1(S_0^1 \cdot k_x^{11} \cdot \cos\alpha_1) - (\lambda_1 + 2\mu_1)(S_0^1 \cdot k_z^{11} \cdot \sin\alpha_1) \tag{4.61}$$

$$\tau'_{Zx} = \mu_1(S_0^1 \cdot k_z^{11} \cdot \cos\alpha_1 - S_0^1 \cdot k_x^{11} \cdot \sin\alpha_1) \tag{4.62}$$

Based on the above situation, it could firstly solve the critical displacement value S_0^{11} and IF ω_0^{11} of the correspondent seismic wave when slides happen to the controlling structural surface under the effects of single transverse wave, then the correspondent time T'_s of IF ω_0^{11} is the critical time of slides occurring to the controlling structural surface. It is noteworthy that the critical time T'_{s1} of shear failure occurring to the controlling structural surface might be different from the critical time T'_{s2} of tensile failure occurring to it, and if $T'_{s1} < T'_{s2}$, it takes that if shear failure occurs to the structural surface of sliding mass, then the calculated energy $E'_{\text{in}1}$ of input wave at this time is the critical energy value of shear failure that occurs to the controlling structural surface; with the continuing seismic motion afterward, tensile failure gradually happens to the controlling structural surface, and then the energy value $E'_{\text{in}2}$ at this time T'_{s2} by calculation is the critical energy value before the slides; on the contrary, if $T'_{s1} > T'_{s2}$, it takes that tensile failure happens to the sliding mass, and it will then get the calculated energy of input waves $E'_{\text{in}2}$ is the critical energy value when tensile failure happens to the controlling structural surface; with the continuing seismic motion afterward, there will be no energy import and, at this time, it can take that slides happen to the sliding mass.

4.4.2 Dynamic Mechanism of High and Steep Slopes Collapse Under Seismic Effects and Distribution Principle of Elastic Recoil Energy

4.4.2.1 Dynamic Mechanism of High and Steep Slopes Collapse Under Seismic Effects

This book takes that the collapse mechanism of slopes under seismic effects is as follows: At the beginning of seismic force or gravity, the sliding mass mainly produces elastic deformation and stores large quantity of elastic strain energy inside; when the instant input energy of the earthquake is over the critical energy allowed by the slope structural surface, shear and tensile failure points happen to it; with the continuing seismic motion, the failure points gradually develop to lock-fixed section, which finally causes gradual failure of the lock-fixed section and forms penetrating failure surface; at this time, the elastic strain energy stored inside the sliding mass is suddenly released, which causes elastic recoil to the sliding mass and that is further transformed into the kinetic energy of it, and then the sliding mass collapse out of the shear opening and forms sliding slopes.

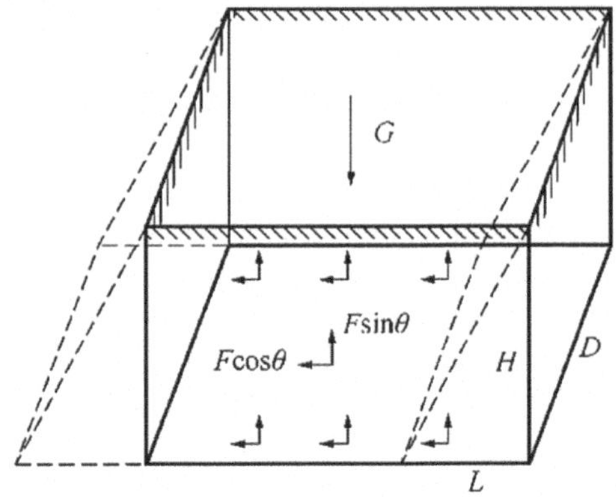

Fig. 4.71 Elastic mechanics analysis model

4.4.2.2 Distribution Principle of Elastic Recoil Energy

The slope mass will store part of the elastic strain energy inside under seismic effects, which is mainly caused by the constant squeeze of transmitted stress on slope mass and that makes the slope mass deformed. As the slope mass surface belongs to free end, it won't deform under seismic effects, and when failure happens to the controlling structural surface, it will move with it, so the movement of both are simultaneous. Therefore, this book sets up a three-dimensional elastomer as shown in Fig. 4.71, in which the position of the fixed boundary is equivalent to the slope mass surface, and the stress distribution surface is equivalent to the controlling structural surface of slope mass–bedrock. The length of the elastomer is L, width is D, height H, elastic modulus E, shear modulus G, density ρ, gravitational acceleration g, and the uniform load F acts on the slope top, with included angle θ between it and the interface, and the work of Fx, Fy and gravity on the elastomer are, respectively,:

$$W_y = (F \cdot \sin\theta)^2 \cdot LDH/2E; W_x = (F \cdot \cos\theta)^2 \cdot LDH/2G,$$
$$W_g = (\rho g H)^2 \cdot LDH/6E \tag{4.63}$$

Then the strain energy stored inside the elastomer is $W = W_x + W_y - W_g$, and when slip failure happens to the I-I interface, the elastomer only produces rigid displacement along X-direction, and the work of external force on the elastomer is:

$$W_y = (F \cdot \sin\theta)^2 \cdot LDH/2E; W_x = 0; W_g = (\rho g H)^2 \cdot LDH/6E \tag{4.64}$$

So the strain energy stored inside the elastomer is $W = W_y - W_g$. Therefore, if shear failure happens to the structural surface, slides happen to the slope mass, and according to energy conservation theorem, the ratio of elastic recoil energy W_{t1} with the total energy W is:

$$W_{t1}/W = 3E \cdot (F\cos\theta)^2/[3G \cdot (F\sin\theta)^2 + 3E(F\cos\theta)^2 - G(\rho g h)^2] \tag{4.65}$$

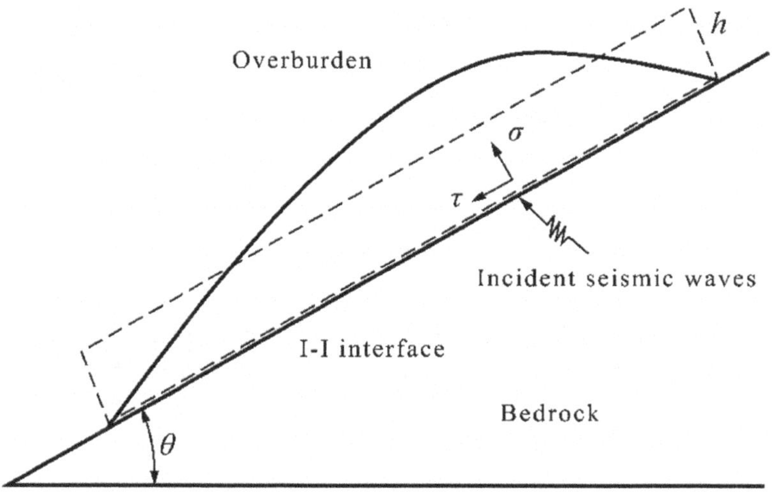

Fig. 4.72 Overlay equivalent model

If tensile failure happens to the elastomer, then the ratio of elastic recoil energy W_{t2} with the total energy W is:

$$W_{t2}/W = [3G\cdot(F\sin\theta)^2 - G(\rho gh)^2]/[3G\cdot(F\sin\theta)^2 + 3E(F\cos\theta)^2 - G(\rho gh)^2] \quad (4.66)$$

When the elastomer is tossed up, other elastic energy stored inside the elastomer will also be released, i.e. $W_{t2} = W$.

This book sets up model for high and steep cover layer slopes as shown in Fig. 4.72. As the geometric shape of cover layers is quite complex in practical engineering, so this book adopts method of equivalent cross-sectional area to calculate, which could ensure that the average stress value of the controlling structural surface stays stable and could also simplify calculation.

When the transverse wave is taken as input wave, the value of normal stress σ' and shear stress τ' of transmitted longitudinal wave and transmitted transverse wave can be seen, respectively, (4.67) and (4.68).

$$\sigma' = \lambda_2[S_0^4\cdot k_x^{41}\cdot\cos\beta_1 + S_0^5\cdot k_x^{51}\cdot\sin\beta_1'] + (\lambda_2+2\mu_2)[-S_0^4\cdot k_z^{41}\cdot\sin\beta_1 + S_0^5\cdot k_z^{51}\cdot\cos\beta_1'] \quad (4.67)$$

$$\tau' = \mu_2[S_0^4\cdot k_z^{41}\cdot\cos\beta_1 + S_0^5\cdot k_z^{51}\cdot\sin\beta_1' + S_0^5\cdot k_x^{51}\cdot\cos\beta_1' - S_0^4\cdot k_x^{41}\cdot\sin\beta_1] \quad (4.68)$$

It could be known from (4.67) and (4.68) that if tensile failure happens to the structural surface, then the ratio η_1 of elastic recoil energy W_t with the total energy W is:

$$\eta_1 = W_{t1}/W = 3E_2 \cdot (\tau')^2/[3G_2 \cdot (\sigma')^2 + 3E_2(\tau')^2 - G_2(\rho g h')^2] \quad (4.69)$$

If tensile failure happens to the structural surface, then the ratio η_2 of elastic recoil energy W_t with the total energy W is:

$$\eta_2 = W_{t2}/W = [3G_2 \cdot (\sigma')^2 - G_2(\rho g h')^2]/[3G_2 \cdot (\sigma')^2 + 3E_2(\tau')^2 - G_2(\rho g h')^2] \quad (4.70)$$

4.4.3 The Slope Velocity at Different Moment When Shear Failure Happens to the Controlling Structural Surface

4.4.3.1 Velocity Calculation at Slope Start-up Phase

When shear failure happens to the controlling structural surface, the elastic strain energy stored inside the overlying slope mass is converted into its initial kinetic energy, and at the same time, the energy value calculated in all formulas in Sect. 3.4.2 are all in unit area. Therefore, according to energy conservation theorem, the energy conservation formulas in the whole start-up phase is as follows:

$$\eta_1 \cdot E_t \cdot S = E_v = \frac{1}{2} m v_0^2 \quad (4.71)$$

In the formula, E_t represents the energy of transmitted waves, which should be solved according to Sect. 3.4.2; S represents the area of controlling structural surface; m represents the mass of the slope; v_0 represents the start-up velocity of the slope.

4.4.3.2 Velocity Calculation at Slope Stroke Phase

As the cohesive force c of the controlling structural surface does not exist at slope stroke phase, it won't be taken into consideration in the following energy conservation formulas. At the same time, the slope mass is still connected to the sliding bed in the sliding process, so the input wave still reflects and transmits at this place, and there is still part of energy transmitting inside the sliding mass. Thus, this book makes the following hypothesis: In the sliding process of the sliding mass, the energy transmitted coefficient is constant; when the sliding mass slides from Point A to Point B, the dip angle of the controlling structural surface and the mass of the sliding mass are stable, which are always α and m. In the sliding process, the gravity center drops from A_1 to A_2, with corresponding elevation dropping from h_1 to h_2; i.e. the sliding potential drops to h, and velocity of sliding mass at any time is V_L. Therefore, according to energy conservation theorem, the energy conservation formulas in the whole stroke phase is as follows:

$$\eta_1 \cdot E_{t后} + \frac{1}{2}mv_0^{\ 2} + mgh - mgh \cdot f_1 \cdot \cot\alpha = \frac{1}{2}mv_L^2 \qquad (4.72)$$

In the formula, represents the total energy of transmitted waves at any time when slides happen to the sliding mass; velocity of the sliding mass at any time is V_L; f_1 represents the kinetic friction coefficient of sliding mass structural surface.

4.4.3.3 Energy Conversion and Velocity Calculation at Slope Leap and Crash Phase

According to energy conservation theorem, the velocity of sliding mass after it crashed with the surrounding mountains is:

$$\frac{1}{2}mv_L^2 = U_c + \frac{1}{2}mv_2^2 \qquad (4.73)$$

In the formula, U_c represents collision energy loss, which mainly shows in dynamic fracture energy of rock mass within certain range of crash center of steep cliff on the other side and the further crashed clastic energy of earth and rock body in the sliding slope itself. Thus, it could be seen that when U_c is determined, the rebound fall speed V_2 of earth and rock body could be solved. Here, we adopt a simple method to solve V_2. As the collision process is quite complex, in the dynamics of high-speed rock landslides, it takes that the material property of rock body determines the energy attenuation coefficient after earth and rock body crashing with steep cliff on the other side to a large extent, which shows as recovery coefficient in numerical, namely adopting recovery coefficient to simplify the calculation of medium velocity after crash.

$$R = \frac{v_2}{v_L} \qquad (4.74)$$

In the formula, R could be divided into normal recovery coefficient and shear recovery coefficient by different directions. For slopes constituted by hard bedrocks, R is larger, on the contrary, R is smaller. According to the major experience of crash recovery coefficient of slope rock body domestically and abroad, we take $R = 0.8$, then velocity after crash is:

$$v_2 = R \cdot v_L \qquad (4.75)$$

4.4.4 The Slope Velocity at Different Moment When Tensile Failure Happens to the Controlling Structural Surface

Supposing that tensile failure happens to control structural surface, the elastic strain energy stored inside the overlying slope mass is partly converted into its initial kinetic energy, and at the same time, the energy value calculated in all formulas in Sect. 3.4.3 is all in unit area. Therefore, according to energy conservation theorem, the energy conservation formulas in the whole start-up phase is as follows:

$$\eta_2 \cdot E_t \cdot S = E_v = \frac{1}{2} m v_0^2 \tag{4.76}$$

In the formula, E_t represents the energy of transmitted waves, which should be solved according to Sect. 3.4.2; S represents the area of controlling structural surface; m represents the mass of the slope; v_0 represents the start-up velocity of the slope.

This book makes the following hypothesis: The mass of the slope stays unchanged, which is m constantly, and at the same time, as projection happens to the sliding mass, it is separated from the structural surface, so there will be no transmitted wave propagating into the sliding mass and the energy conservation formula that the sliding mass should satisfy at any moment T should be:

$$\frac{1}{2} m v_0^2 + mgH = \frac{1}{2} m v_L^2 \tag{4.77}$$

In the formula, H represents the elevation difference between projection points and the observation points, and at the same time, at any time, the horizontal projection distance S of the sliding mass is:

$$S = V_0 \cdot \cos\theta \cdot T$$

Bibliography

Cui Fangpeng. Study on slope collapse effect of seismic P-S wave time difference coupling action[doctoral dissertation] [D]. Doctoral dissertation from Geology and Geophysics Research Institute of Chinese Academy of Sciences, 2009.

Du Xiuli. Theory and method of engineering wave [M], Science Press, 2009.

DUNCAN J M. State of the art:limit equilibrium and fnite-element analysis of slopes [J]. Journal of Geotechnical Engineering, 1996, 22(7):577–596.

Hartzell S H, Carver D L, King K W. Initial investigation of site and topographic effects at Robin wood ridge, California [J]. Bull. Seis. Soc. Am. 1994, 84:1336–1349.

Huang N.E Shen Z. Long S R.et al. The empirical mode decomposition and Hilbert spectrum for nonlinear and non-stationary time series analysis [J]. Proc R. Soc Lond. 1998, 454:903–995.

Hu Guangtao, Dai Honglin. Landslide Dynamics [M]. Beijing: Geological Press.

Hu Yuxian, He xun. Fitting of artificial ground motion response spectrum considering phase spectrum[J]. Earthquake Engineering and Engineering Vibration, 1986, (2):37–51.

Hu Yuxian, Zhang Yushan, Liang Jianwen. Site liquefaction identification based on HHT [J], China Civil Engineering Journal, 2006, 39(2):66–77.

Hu Yuxian, Zhang Yushan, Liang Jianwen. Using HHT to identify the physical process of the liquefaction of the site from the vertical seismic acceleration records [J]. Earthquake Engineering and Engineering Vibration, 2003(3):1–11.

Kurita, R.R. and Rodriquez-Ovejero, L. A model with non-reflecting boundaries for use in explicit soil-structure interaction analyses, Earthq. Eng. Struct. Dynam, 2012, Vol. 8, pp: 361–374.

Lam. L, Fredlund. D.G. A general limit equilibrium model for three-dimensional slope stability analyses [J]. Canadian Geotechnical Journal, 1993, 30(6):905–919.

Liu Tielin, The accurate analysis of the causes of the whiplash effect and discussion on damping effect [J]. Journal of Liaoning Engineering Institute, 1997, 17(3):17–19.

Mao Yanlong, Hu Guangtao, Zhao Fasuo, et al. Mechanism of ground motion triggering landslide sliding [J]. Journal of Xi'an Engineering Institute, 1998, 20(4):47–50.

National Standards Compilation Group of the People's Republic of China. GB50011-2010, code for seismic design of buildings [S]. Beijing: China Architecture & Building Press.

National Standards Compilation Group of the Ministry of Electric Power of the People's Republic of China. DL/5073-2000, Code for seismic design of hydraulic structures [S]. Beijing: China Electric Power Press, 2000.

National Standards Compilation Group of the Ministry of Water Resources of the People's Republic of China. SL/386-2007, Code for design of slope in water conservancy and hydropower engineering [S]. Beijing: China Water & Power Press, 2007.

Qi Shengwen, Wu Faquan, Liu Chunling, Ding Yanhui. Engineering geological analysis of seismic slope stability [J]. Chinese Journal of Rock Mechanics and Engineering, 2004, 23(16):2792–2797.

Shang Jialan, Guo Hanyan. Effect of stress wave propagation on rock mass fracture. Proceedings of the Symposium on Protection Engineering, 1979.

Shi Chong, Zhou Jiawen, Ren Qiang, Zhou Xianqi. Ray theory solution of the elevation amplification effect of one side slope [J], Journal of Hohai University (Natural Science Edition), 2008, 36(2):238–241.

Tang Chun'an, Zuo Yujun. The slope of the earthquake in Wenchuan shallow fragmentation and ejection model and its dynamic explanation [C]. Investigation and analysis of seismic damage in Wenchuan earthquake, 2009.

WU XY. Examination of the Pseudo-static Limit Equilibrium Method for Dynamic Stability Analysis of Slopes [M]. Canadian: Canadian Geotechnical Conference. 1991.

Xu Qiang, Dong Xiujun. Genetic model of large scale landslide in Wenchuan earthquake [J], Earth Sciences-Journal of China University of Geosciences, 2011, 36(6):1134–1142.

Yang Changwei, Zhang Jianjing. Frequency analysis method for seismic stability of high and steep rock slope with the action of SV wave [J], Chinese Journal of Rock Mechanics and Engineering, 2013.

Yuan Jinke. Study on Structural characteristics of the rock mass of the slope and Post earthquake collapse identification system [D], doctoral dissertation from Chengdu University of Technology, 2012.

Zhang Zuwu, Yao Lingkan. Study on energy transfer and dissipation characteristics of seismic wave in the interface between soil and rock- the case of Wenchuan Ms 8.0 earthquake. Science of disaster, 26(1):5–9.

Chapter 5
Time–Frequency Analysis Theory of Seismic Stability of Retaining Structures

As lifeline engineering, highway and railway are the key steps of post-quake emergency rescue and reconstruction, which play a key role in reducing casualty and economic loss. The "5.12" Wenchuan earthquake in 2008 caused great loss to China. From the disaster rescuing process, it could be seen that the lifeline engineering—highway and railway engineering—in seismic disaster area showed up a great deal of inability in anti-seismic and disaster resistant work, which led to lots of transportation engineering restraining the emergency rescuing work immediately after the earthquake, especially in the severely affected counties and particularly severely affected counties such as Wenchuan, Beichuan, Mao County, and Qingchuan, as well as the breakdown of highway transportation in 254 severely affected counties, which resulted in delayed transportation of rescuing people, relief materials and medicines into the particularly severely affected areas like Yingxiu and Beichuan, and exerted great influence on rescue of disaster area.

There is no doubt that the transportation engineering damage caused by "5.12" Wenchuan earthquake is enormous, while at the same time, it offers a precious opportunity to summarize and improve the seismic techniques of highway and railway engineering in China. In order to improve the seismic design level of highway and railway retaining structures, it is quite necessary to carry out seismic damage research, failure mechanism analysis, and successful example summary to highway and railway in Wenchuan. Our research group conducted systematic research to the highway and railway retaining structures in Wenchuan seismic disaster area from January to June 2009 and complete statistical analysis to the collected data, and carried out shaking model test to the most severely damaged rigid retaining walls. On this basis, two kinds of representative retaining structures, namely rigid retaining walls and flexible retaining walls, are selected, and we adopted the time–frequency analysis theory of seismic motion to derive the seismic stability time–frequency analysis theory of two kinds of retaining structures with consideration to the influence of three major factors of seismic motion (amplitude, frequency, and duration).

C. Yang et al., *Three Dimensional Space-Time Analysis Theory of Geotechnical Seismic Engineering*, https://doi.org/10.1007/978-981-13-3356-9_5

5.1 Field Investigation Situation of Seismic Damage of Retaining Structures

5.1.1 Field Investigation Situation

After Wenchuan earthquake, transport agencies and railway agencies carry out seismic damage evaluation to the highway and railway in Wenchuan seismic areas. On the basis of sufficient data collection of previous emergency research and test and evaluation, our research group conducted systematic complementary research to particularly severe and severe disaster areas (Fig. 5.1), and the survey information includes four parts, namely on-the-spot survey of seismic work sites, feature data measurements of seismic work sites, sketch of seismic work sites, and image data of seismic work sites. The field survey method used is mainly, based on collecting the existing design materials, combining field complementary research, measurement, and detection. The detailed steps are as follows:

(1) Data collection: collecting the construction time, line level, and seismic fortification intensity of highways in particularly severe and severe disaster areas, as well as the original design and completion data of seismic retaining structures.
(2) Field Research Methods: using tools like measuring tape, steel measuring tape, compass, camera, video camera, GPS locator, laser distance meter, and hand level to gradually carry out research by route stake number to different routes and fill out the corresponding seismic damage questionnaire. When doing the measurements, take full advantage of digital cameras and video cameras to take photographs of the sites and make refined marks, arrangement, and storage; for some work sites attacked by the earthquakes, try to collect the personnel record who has been to the original damage field and its photos, videos, and other information.

Carry out survey, classification, and statistics of the retaining structures. The research contents are deformation characteristics under seismic effects of earthquakes

Fig. 5.1 Field investigation photographs

and secondary geological disasters of the retaining structures including gravitational retaining walls, reinforced retaining walls, slide-resistant piles, and anchored frame beams.

5.1.2 Explanation of Survey Routes

The Wenchuan earthquake caused disasters in three provinces of Sichuan, Gansu, and Shanxi, and our research group carried out seismic damage data collection and complementary research to the major highways and railways with a total length of 7081 km in the above three disaster areas.

The seismic damage research in Sichuan Province was mainly conducted within the intensity range of VII–XI, including administrative regions of Chengdu City, Deyang City, Mianyang City and Guangyuan City, as well as Ganxi and Aba Autonomous Region, as shown in Figs. 5.2 and 5.3. The survey routes are mainly the national roads and provincial roads in Sichuan Province as well as some township and county roads. The Sichuan survey route is over 40 sections in total with length of 2661 km. The national roads involve four routes of G108, G212, G213, and G317, the provincial roads include seven routes of S105, S106, S205, S210, S301, S302, and S303, and the railways include Guang Yue Railway, Baoji–Chengdu Railway, Cheng Wen Railway, and Chengdu–Dujiangyan Intercity Railway. The Shanxi survey route is mainly within the intensity range of VII–XI, which includes administrative regions of Hanzhong City, Baoji City, and Xianyang City. The survey route is over ten sections with a length of 4104 km, including four national roads of G108, G210, G310, and G316 and seven provincial roads of S104, S211, S212, S210, S306, S309, and Meitai section of Jiangmei Highway. For the Gansu Province survey area, it mainly carries out statistical research and analysis to the typical roadbed damage in Line 212 of the national road (from Section Dangchang to Section Guanzigou, with length of 316 km) and conducts selective analysis and arrangement to the seismic damage of 18 retaining structures with comprehensive field survey and complete data.

5.1.3 Seismic Damage Classification of Retaining Structures

Seismic Damage Classification is determined, on the basis of seismic damage work sites and existing seismic damage classification, through the comprehensive consideration of site performance aftershock. It is concretely divided into four grades of mild, moderate, severe, and damage, representing the damage degree of each type, and is represented as A, B, C, and D. The main feature description of the four seismic damage grades of retaining structures is shown in Table 5.1.

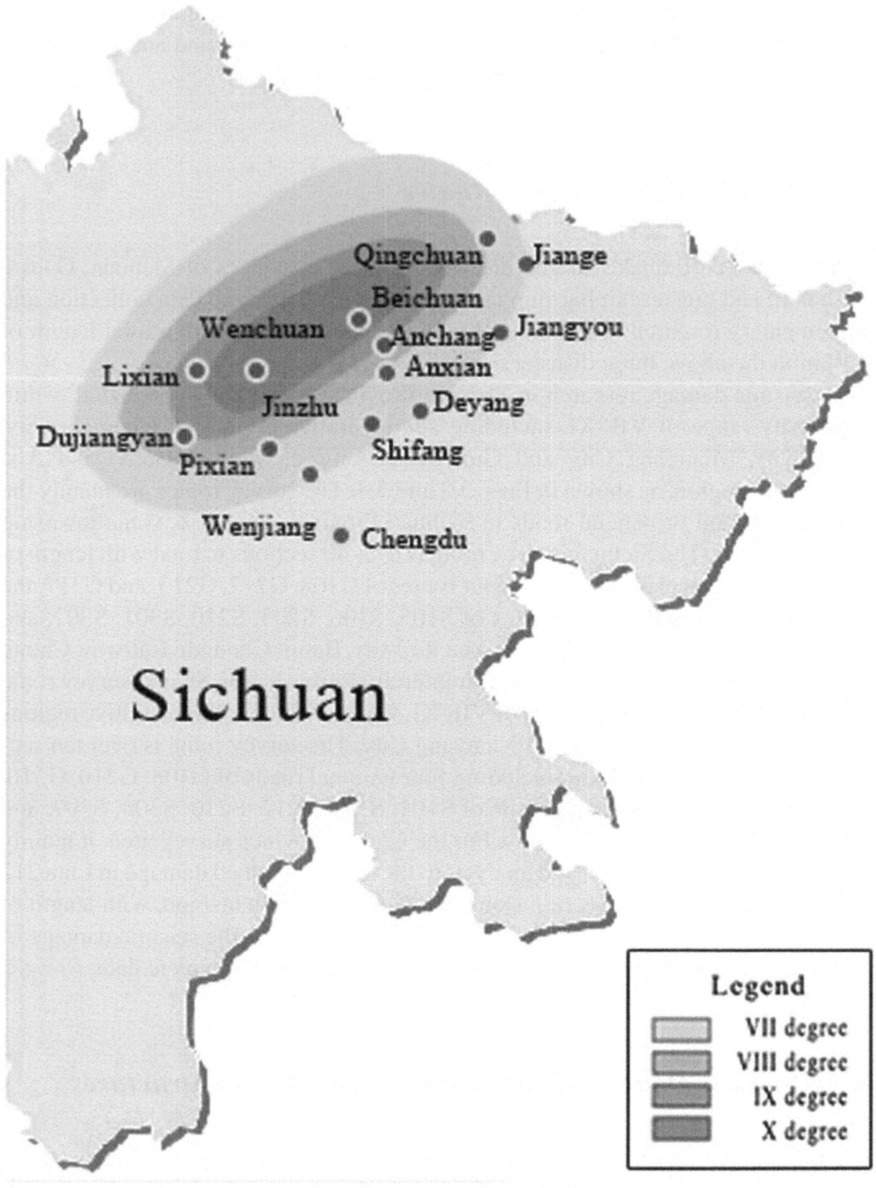

Fig. 5.2 Map of Wenchuan earthquake disaster area

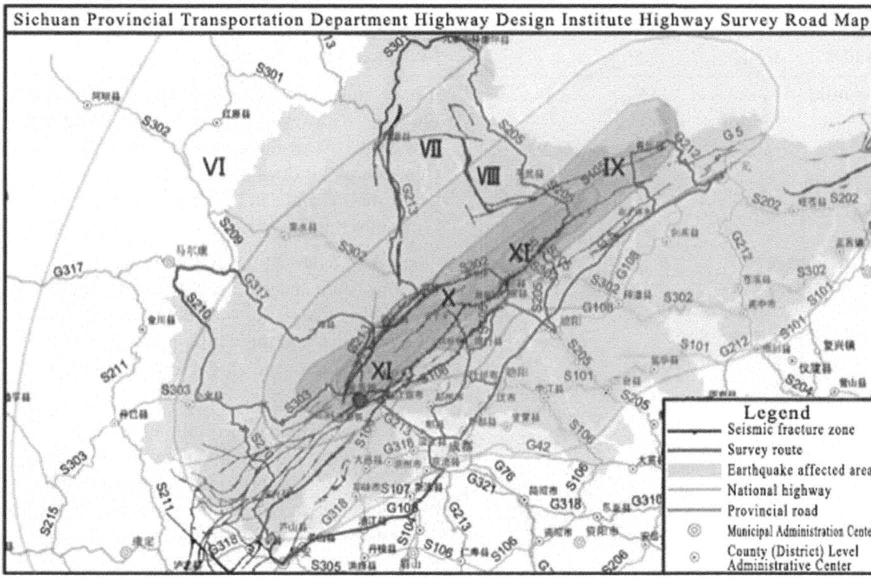

Fig. 5.3 Seismic damage situation of survey routes and retaining structures

5.2 Statistical Analysis of Field Investigation of Retaining Structure Seismic Damage

5.2.1 General Characteristics of Seismic Damage of Retaining Structures

The retaining structures refer to a kind of building structure that is used to support and stabilize the filled earth and slope body so as to prevent collapse and keep soil mass at the rear of the structure stable, and they mainly include two major kinds, namely the retaining walls and slide-resistant piles (sheet pile wall).

The retaining wall refers to the structure which is used to support the roadbed filled earth or slope body and to prevent deformation and instability of the filled soil and soil mass. Slide-resistant piles are the piles that penetrate the sliding slope into the sliding bed, which are used to resist slipping force of the sliding mass and stabilize the slopes. Figure 5.4 lists the sketch map of four commonly used retaining structures.

In Wenchuan earthquake, the seismic damage modes that happen to the retaining structures are several kinds of direct seismic damage, namely collapse, wall body snipping, overall tilt, toppling and structural surface deformation and cracks. Besides, burial and smash of the retaining structures caused by secondary geological disasters like landslides and collapse masses are also major seismic damage types.

Table 5.1 Classification features of seismic damage degree of retaining structures

Grades	Seismic damage features
A—mild	The seismic damage of retaining walls is not obvious, with seismic area (fracture length) less than 10% of the retaining wall area (length), and it does not lose the retaining performance and can be temporarily used without reparation
B—moderate	The seismic damage of retaining walls is obvious, with seismic area (fracture length) accounting for 10–30% of the retaining wall area (length), and the wall top displacement is obvious, which needs local reparation after earthquake
C—severe	The seismic damage of retaining walls is obvious, with seismic area (fracture length) accounting for 30–60% of the retaining wall area (length), and the wall top displacement is obvious, with the retaining wall area losing stability, which needs immediate reparation after earthquake
D—damage	The retaining wall of this section is completely damaged, and new retaining wall needs to be constructed

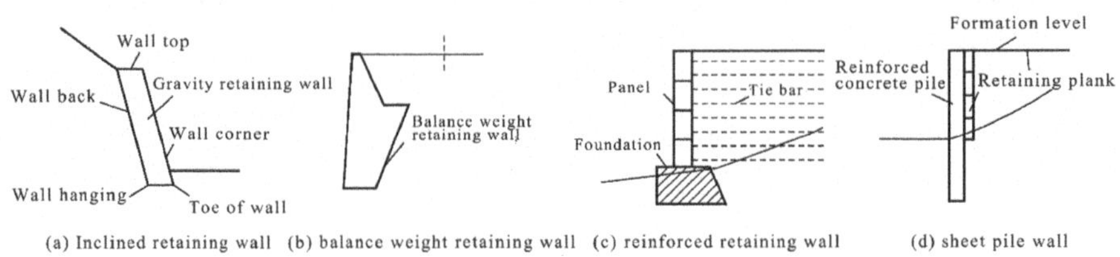

Fig. 5.4 Retaining wall types. *Note* Inclined retaining wall and balance weight retaining wall all belong to gravity retaining wall, and reinforced retaining wall and sheet pile wall are all flexible retaining structures

The seismic damage intensity of retaining structures is mainly in the range of IX–XI, with few in VII and VIII. By the structural modes of the retaining walls, the retaining structures of seismic damage could be mainly divided into three types of gravity retaining wall, reinforced retaining wall, and sheet pile wall. The research results show that most retaining structures that induce seismic damage are gravity retaining walls, while reinforced retaining walls and sheet pile walls, which belong to flexible retaining structures, induce less seismic damage, and they account for less than 2% (that is also related to small total number of reinforced retaining wall and sheet pile wall work sites). By the positions of retaining structures, they can be divided into cut slope walls and shoulder walls (upper retaining wall and lower retaining wall). From the perspective of statistical results, the seismic damage of retaining walls is mainly induced by cut slope walls.

From the perspective of retaining wall building method, seismic damage caused by retaining walls built with mortar rubble or rock blocks covers the most compared with retaining walls built with rubble concrete or pebble cement concrete. It is noteworthy that most highways in the mountain area are built with mortar rubble or rock blocks, and their large cardinal number and low construction quality are one of the basic reasons that lead to high damage percentage. In addition, although the total number of dry masonry retaining walls is few, their common damage is serious.

It could be seen that the seismic damage of highway and railway in the disaster areas is induced by both seismic motion and secondary geological disasters, and mortar masonry gravity retaining walls are the major structures that induced seismic damage, with an inclination to cut slope walls. The retaining structures in intensity of VII and VIII basically stand the earthquake test, and the anti-seismic performance of retaining walls built with mortar rubble or rock blocks is far below that of retaining walls built with rubble concrete or pebble cement concrete.

5.2.2 Seismic Damage Statistics

This section will explain the influence of total seismic damage trend of the retaining structures with respects to factors including intensity regions, retaining structure types, building methods, foundation conditions, direction of routes and fracture belt position and make classified statistics to the seismic damage types of retaining structures.

5.2.2.1 Statistic Analysis of Factors Influencing Seismic Damage of Retaining Structures

① Influence of route grades and intensity regions on seismic damage;

The number of seismic-damaged retaining structures is 375 in total. From Fig. 5.5, it could be seen that the seismic damage mainly concentrates in intensity region within the range of IX and XI around the fracture belt. The National Road 213 is from Dujiangyan to Yingxiu crossing the seismic center Yingxiu; thus, it is the most largely and severely damaged, far beyond other routes. In the surveyed routes, there are seven route sections which basically do not show obvious retaining structure seismic damage, and these routes have a small number of retaining structures and are far away from the fracture belts. Through classified statistics by route types, it can be seen that the design of township and county highways is different from that of provincial and national roads, and their retaining structures have low anti-seismic grades, which leads to severer seismic damage situation. National Road 212 and 213, with strike direction near the fracture belts, had a comparatively large number

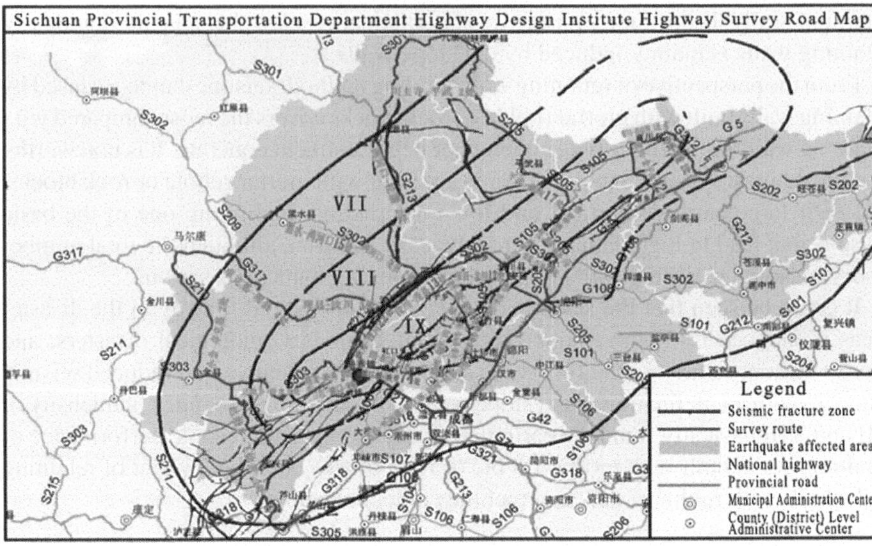

Fig. 5.5 General seismic damage situation of retaining structures

Table 5.2 Seismic damage number and percentage of retaining structures of all kinds of roads

Route type	Total number of seismic damage (N)	Percentage (%)	Mileage (km)	$R1 = N/L$
National roads	137	36.5	1614	0.10
Provincial roads	112	29.9	1608	0.17
Township and county roads	126	33.6	364	0.26

Table 5.3 Seismic damage number and percentage of retaining structures of all intensity regions

Intensity region	Total number of seismic damage (N)	Percentage (%)	Area (km^2)	$R2 =$ N/A
IX–XI	304	81.3	13,301	0.0219
VIII	13	3.5	27,786	0.0003
VII	48	12.8	84,449	0.0003
VI	9	2.4	–	–

of seismic damages; for example, Road 213 that is from Dujiangyan to Yingxiu was severely damaged as it passes through faults and seismic center. There are few provincial roads near high-intensity areas, so their seismic damage are comparatively low, but the percentage of these three kinds of routes are all over 30% (see Tables 5.2 and 5.3).

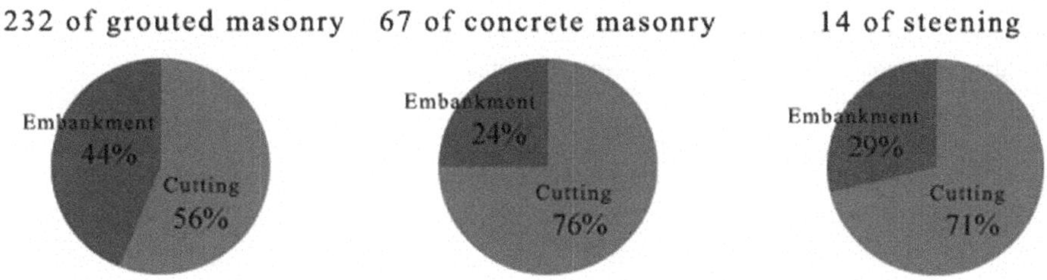

Fig. 5.6 Ratio of cut slope wall and embankment wall among all kinds of building methods

Within the intensity range from IX to XI, there are 19 affected routes, 304 affected retaining structures; in intensity range VIII, there are 6 affected routes, 13 seismic-damaged retaining structures; in intensity range VII, there are 6 seismic-damaged routes and 29 affected retaining structures. It could be seen from the distribution of seismic-damaged routes in intensity regions in the figure that no matter region area or route area in VIII intensity region is far less than that in VII intensity region. If it compares solely in seismic damage quantity, it is reasonable that seismic damage of retaining structures in VII intensity region is more than that in VIII intensity region, while if it calculates in unit seismic damage area (R2), the seismic damage in VII intensity region is almost equal to that in VIII intensity region. That shows that the seismic damage in VIII intensity area is more serious, which could also be seen from the field survey results.

② The influence of retaining structure types on seismic damage;

It could be known from the survey data that the seismic-damaged retaining structures are mainly gravitational cut slope walls. The survey finds out that some gravity retaining walls have been reinforced and form anchored frame retaining walls and anchored rod retaining walls and the like, which are all counted as gravitational retaining walls in statistics. According to the statistics, there are 371 seismic-damaged gravitational retaining walls, which cover 98.9% of the total number of seismic damage. In addition, there are four seismic-damaged reinforced retaining walls and slide-resistant piles (sheet pile wall) in total, and this is absolutely in relation to a small number of these kinds of retaining walls. Combining the analysis of anti-seismic mechanism of them in the following chapters, it could be generally taken that the flexible retaining structures have good anti-seismic performance.

Figure 5.6 shows that quantity of all types of seismic-damaged retaining structures. In the seismic-damaged retaining structures, 58.4% of which are cut slope walls and 41.6% are shoulder walls.

③ The influence of building methods of gravitational retaining walls on seismic damage;

The building methods of 313 seismic-damaged work sites have been in classified statistics, the damaged highway and railway retaining walls are mainly built with mortar rubble or concrete, and a few dry built retaining walls have also been found

Table 5.4 Table of building methods and seismic-damaged quantity of retaining walls

Building methods	Total damage number	Percentage (%)
Mortar rubble masonry	232	74.1
Pebble piece masonry	67	21.4
Dry masonry	14	4.5

during the survey, so the statistics are classified into three kinds, namely mortar rubble masonry, dry masonry, and pebble piece and concrete masonry. The results show that the major seismic-affected retaining walls are mortar rubble masonry, with 74% of the total seismic-damaged quantity, while the seismic-affected retaining walls built by pebble piece and concrete masonry cover 21%.

It is noteworthy that in mountain regions, highways and railways are generally not built with dry masonry, and most of them are built with mortar rubble masonry or pebble piece and concrete masonry. Therefore, under the circumstance of dry masonry base number far less that of the other two building methods, the 4.5% got by statistics cannot explain that the anti-seismic performance of dry built retaining walls is better than that of the other two, and it could be known through practical survey that the seismic damage of dry built retaining walls is the severest.

The seismic damage percentage corresponding to different retaining wall building methods is listed in Table 5.4. Among all the building methods, the percentage of cut slope wall-induced seismic damage and embankment wall-induced seismic damage are as shown in Fig. 5.6. It could be seen in Fig. 5.6 that, for any of the building method, the cut slope wall-induced seismic damage is severe, which may show that there is great uncertainty in the current study to determine the pressure of cut slope walls.

④ The influence of wall foundation conditions on seismic damage;

The statistics divides the foundation where the seismic-affected retaining structures locate into three kinds, namely soil foundation, rock foundation, and upper soil and lower rock foundation. By the foundation conditions of 219 work sites, the quantity and percentage of corresponding retaining structures obtained by statistics are as shown in Table 5.5. It could be known from the table that most seismic-affected retaining walls are located on soil or upper soil and lower rock foundations, and the number of seismic damage on rock foundation does not amount to 1/4 of the total number. The result shows that the wall foundation conditions exert great influence on its seismic damage. Figure 5.7 shows the seismic damage ratio of cut slope walls and embankment walls on soil or rock–soil foundations. It could be known from the figure that the seismic-affected cut slope walls also accounts for the most.

⑤ The influence of route direction and fracture belt relation on seismic damage of retaining walls;

Compare the route strike of seismic-affected retaining walls got by the survey with the straight line of fracture belt (see Fig. 5.8). It could be known from Fig. 5.8

Table 5.5 Quantity table of seismic-affected retaining walls on all kinds of foundations

Foundation conditions	Total damage number	Percentage (%)
Soil foundation	128	58.4
Upper soil and lower rock foundation	42	19.2
Rock foundation	49	22.4

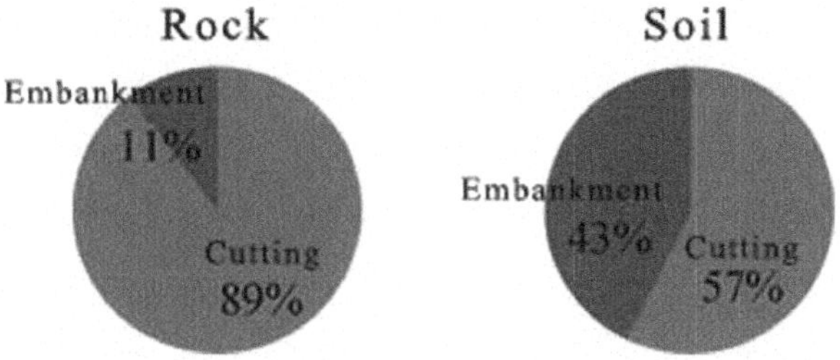

Fig. 5.7 Ratio pie of seismic-affected cut slope walls and embankment walls on soil and rock foundations

that when the retaining structure route is parallel to the fracture belt, the quantity of seismic-affected retaining structures is large, and with the increase in included angle between route strike and the fracture belt, the quantity of seismic-affected retaining structures gradually decreases; i.e., the normal in free face direction of the retaining structures is vertical to the fault; thus, the seismic damage is serious.

The damage trend of cut slope walls and shoulder walls is basically the same, but due to sample numbers at individual intervals, there is the difference between the two.

In the figure, abscissa represents the included angle between the strike direction of the retaining walls and the straight line of the fracture belt, ordinate represents a number of seismic-affected retaining walls, and the curve represents trend line got from regression analysis of data in each group.

⑥ Classification and mechanism analysis of the seismic-affected retaining walls.

With the field survey data, classified statistics is conducted to the failure type of seismic-affected retaining walls. The seismic damage of the retaining walls is firstly divided into two kinds, namely the damage directly induced by the earthquake and damage induced by the secondary disasters, in which the damages directly induced by the earthquake are mainly shown as collapse, snipping, tilt, wall surface deformation, and fracture. Damage induced by the secondary disasters is mainly burial of retaining structures by slope collapse and smash of retaining structures by rockfalls.

(1) The collapsed seismic damage of retaining walls mainly shows as wall collapse, soil mass slip behind the walls, or slope slides behind the walls. Here, this kind of collapse includes two types, namely entire collapse and local collapse, and as there is a large quantity of local collapse, therefore, the total number of collapsed damage is big. The main reason for collapse is because the earth pressure of soil mass behind the wall suddenly increases under seismic effects, which is

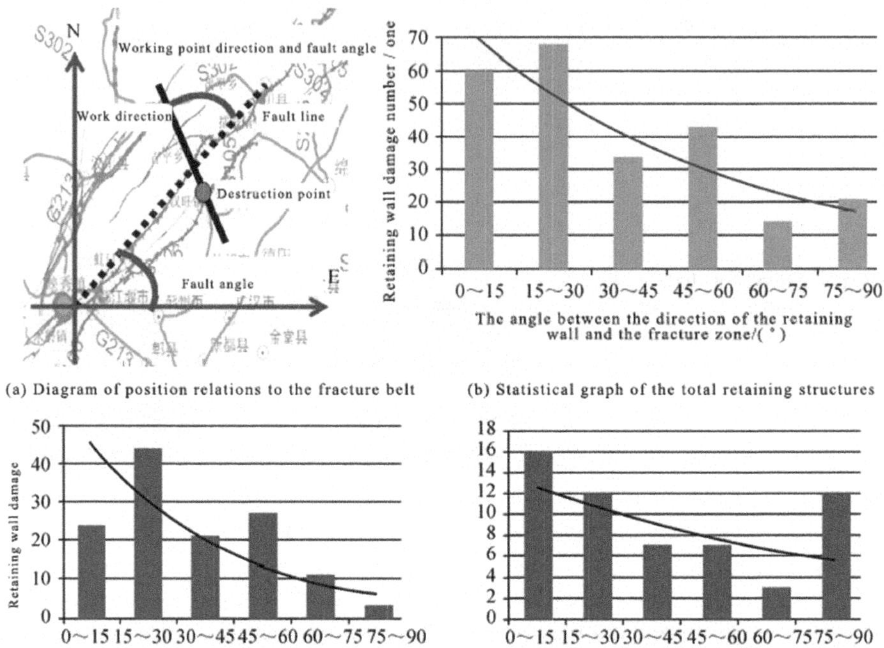

(a) Diagram of position relations to the fracture belt (b) Statistical graph of the total retaining structures

(c) Cut slope walls (d) shoulder walls

Fig. 5.8 Relation between seismic-damaged work sites and the fracture belts

beyond the anti-slip or anti-inclination ability and causes collapse. The other reason for this is because the slope slides behind the wall directly destroy the retaining walls.

(2) The snipping seismic damage mainly shows that the wall body is cut off which makes the upper part protruding. This kind of seismic damage is mainly caused by insufficient local shear strength of wall body, which mainly happens to mortar rubble-built walls and dry built walls.
(3) Deformed and cracked seismic damage mainly shows cracks and ballooning of the wall body. The main reason is that the soil pressure behind the wall increases beyond the shear ability of the wall body, which thus causes cracks and ballooning to it. In general, this kind of seismic damage mainly happens to mortar rubble and dry-laid rubble-built walls or block stone retaining walls.
(4) The tilted seismic damage mainly shows outward inclination of the retaining wall and displacement on the wall top. Which mainly caused by soil pressure behind the wall increases or sedimentation of soil foundation in the earthquake.
(5) The buried seismic damage mainly shows that the whole retaining wall body is buried by crushed stones or soil mass. As the slopes behind the wall go through seismic damage like slides and collapse, the slipped soil mass or crushed stones bury all the retaining wall.

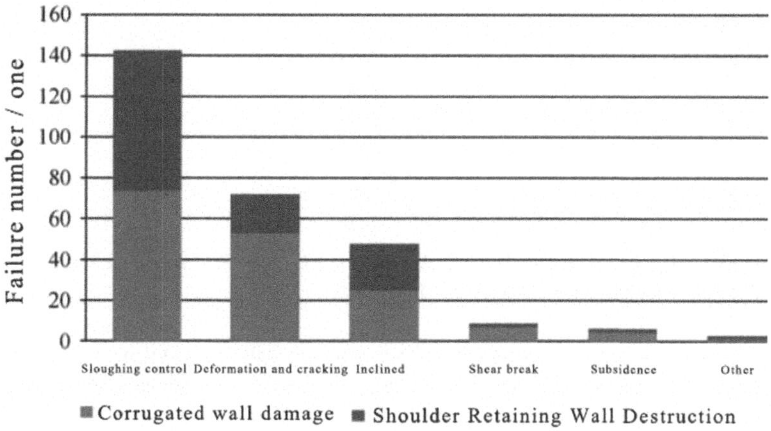

Fig. 5.9 Seismic damage distribution with direct seismic action

(6) The cracked seismic damage mainly shows the damage caused by crushed stones slip from cut slopes, which belongs to secondary disaster.

In addition to the above seismic damages, there are also other kinds of seismic damage, including faulting of slab ends of retaining wall construction joints (2) and falloff of wall panel (1).

The statistic results show that 89% of the seismic-damaged retaining walls found out in the survey belong to damage directly induced by earthquake, and the other 11% of seismic damage is induced by secondary disasters. As the survey time is late compared with the emergency rehabilitation time, most collapsed slopes have already been cleared; thus, the percentage of seismic-damaged retaining structures induced by secondary disasters will definitely exceed 11%.

With direct seismic action, the retaining structures mainly come across damages of collapse, deformation and cracks, and tilt. The other types of seismic damage and their quantity are shown in Fig. 5.9. Figure 5.9 shows that collapse, deformation and cracks, and tilt of the wall body are the major disasters.

5.2.2.2 Statistic Analysis of Factors Influencing Seismic Damage of Cut Slope Retaining Walls

The cut slope walls and shoulder walls are at different locations, with different heights (generally the shoulder walls are higher than the cut slope walls); thus, their failure features under seismic effects are different from each other. This section will, respectively, count and analyze the gravitational cut slope walls and shoulder walls which count for 98.9% of the retaining structure quantities in Wenchuan seismic region, including wall height, masonry method, geological conditions as well as the respective detailed situation of seismic damages.

Table 5.6 Height range of seismic-damaged retaining walls

Wall height range (m)	Percentage (%)
0–2	33.1
2–4	42.3
4–6	13.4
6–8	8.5
Above 8	2.8%

Note The percentage represents the ratio of damaged number of cut slope walls within the height range with the total damaged number of cut slope walls

Table 5.7 Statistics of height of seismic-damaged retaining walls from Dujiangyan to Yingxiu

Wall height range (m)	Total length of retaining walls (m)	Length of damaged retaining walls (m)	Length ratio	Total number of retaining walls	Number of damaged retaining walls	Quantity ratio
≤4	1248.917	564.9	0.45	51	11	0.22
4–6	791.582	481.6	0.61	23	10	0.43
6–8	820.45	546	0.67	13	10	0.77

① The influence of walls height on seismic damage;

Within the height of cut slope retaining walls recorded by the survey, the maximum height is 17 m, the minimum height is 0.3 m, and the average height is 3.26 m. The statistical results show that the seismic-damaged number of cut slope retaining walls with a height less than 4 m in the sample accounts for 3/4 of the total sample of the seismic-damaged retaining walls, which is related to the sample section distribution (Table 5.6).

In order to get further seismic-damaged rules, all the retaining walls and seismic-damaged retaining walls of National Road G213 from Dujiangyan to Yingxiu are counted and analyzed with the original design material (see Table 5.7, Figs. 5.10 and 5.11). The results show that the seismic-damaged cut slope walls and the length ratio (or quantity ratio) accounting for the total number of retaining walls increase with the height elevation, which is identical with the practical seismic damage. It could be known from this result that, with the increase in retaining wall height, the risk of being affected by seismic damage is also on the rise, which explains that the seismic earth pressure is closely related to the wall height.

② The influence of building methods on seismic damage;

The survey conducted classified statistics to the building methods of 147 work sites of the damaged cut slope retaining walls, which mainly includes three kinds, namely mortar rubble masonry, dry masonry, and pebble piece and concrete masonry. The detailed seismic-damaged number and percentage are shown in Fig. 5.12.

The survey results show that the seismic damage of retaining walls built with mortar rubble masonry is mainly of collapse and deformation and crack type; that

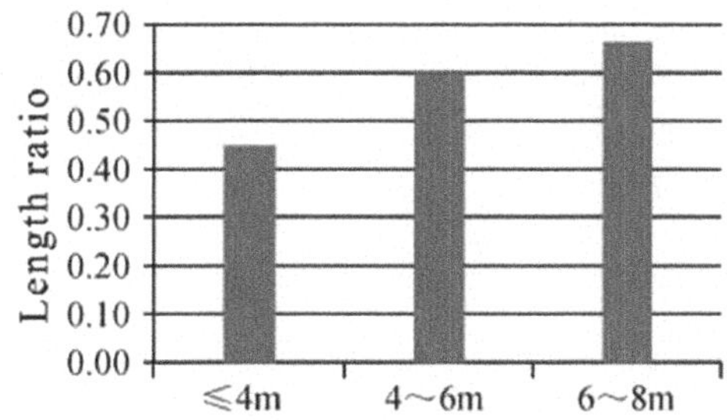

Fig. 5.10 Length ratio of retaining wall height in National Road G213

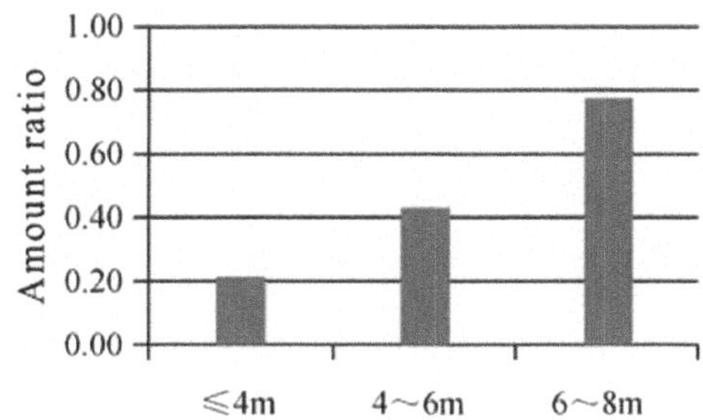

Fig. 5.11 Length ratio of retaining wall height in National Road G213

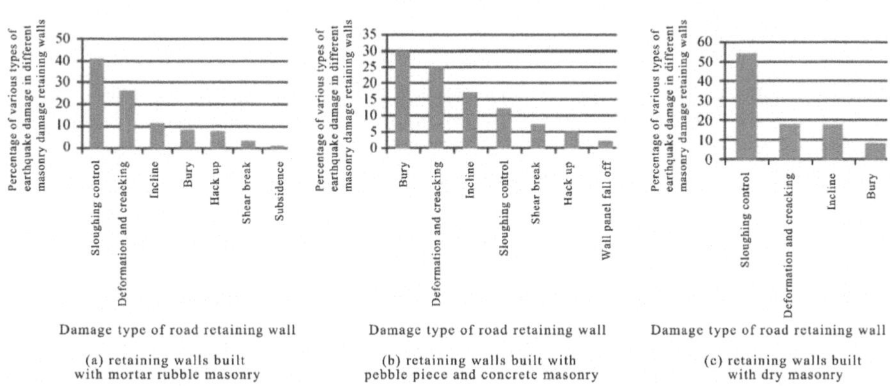

Fig. 5.12 Seismic damage-type distribution of retaining walls built with different masonry methods

of retaining walls built with concrete masonry is mainly of burial, deformation and cracks and tilt type; that of retaining walls built with dry masonry is mainly of collapse type.

With good integrity, the retaining walls built with pebble piece and concrete masonry rarely come across damage of collapse type, while most damages belong to local ballooning cracks and tilt damage. The integrity of retaining walls built with mortar masonry and dry masonry is poor, and most damages belong to collapse type. Figure 5.12 conducted detailed statistics on the type and number of seismic-damaged retaining walls built with different masonry methods.

Table 5.8 Foundation conditions and seismic damage distribution of cut slope retaining walls

Geological conditions	Total damage number	Percentage (%)
Soil foundation	71	54.6
Rock foundation	42	32.3
Upper soil and lower rock foundation	17	13.1
total	130	100

Note The percentage refers to the ratio of number of damaged retaining walls with different foundations with the total damaged number

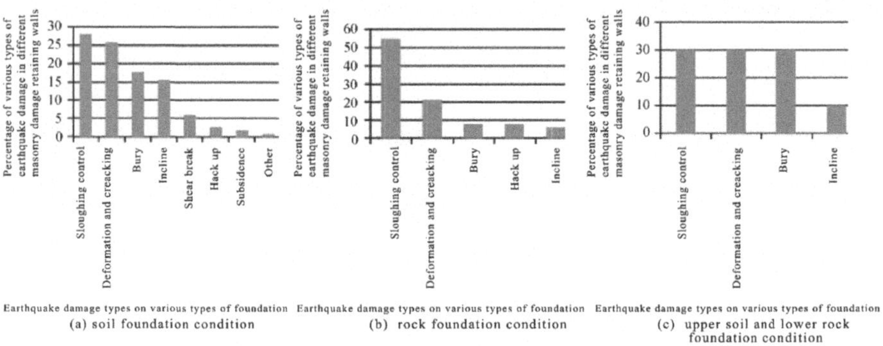

Fig. 5.13 Seismic damage types of cut slope retaining walls with different foundations

Abscissa represents the seismic damage type of retaining walls, and the ordinate represents the percentage of seismic damage of different types in damaged retaining walls built with different masonry methods.

③ The influence of wall foundation conditions on seismic damage.

Statistical classification was conducted on the geological conditions of 130 work sites of damaged cut slope walls, which is divided into three kinds, namely soil foundation, rock foundation, and upper soil and lower rock foundation. The detailed seismic damage number and percentage are as shown in Table 5.8.

It could be known from the table that the damage of cut slope retaining walls with soil and upper soil and lower rock foundations is more serious and that of retaining walls with rock foundations is less. The cut slope retaining walls mainly come across seismic damage of collapse, deformation and cracks, burial and tilt. With different geological conditions, the seismic damage types of gravitational cut slope walls are different and that of retaining walls with soil foundation are more complex, with almost ten kinds of damage (Fig. 5.13).

Abscissa represents the seismic damage type on different foundations, while ordinate represents the percentage of seismic damage types on this kind of foundation.

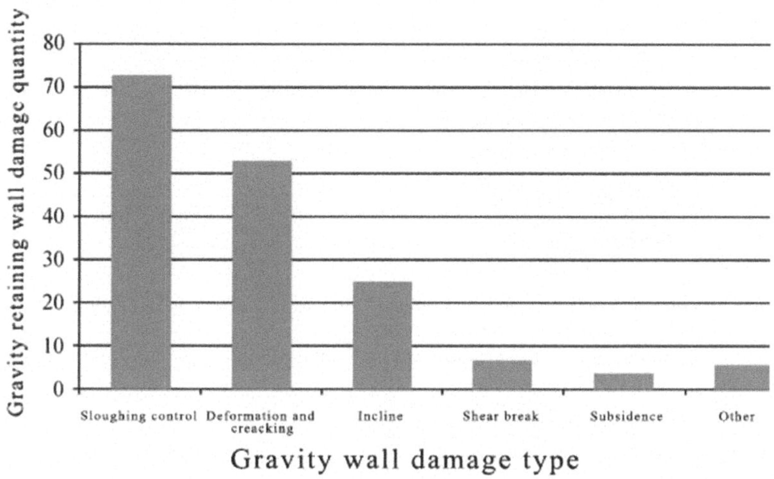

Fig. 5.14 Distribution map of seismic damage type and quantity of cut slope retaining walls (other types of earthquake damage means that the wall panel partially falls off)

5.2.2.3 Further Explanation of Seismic Damage Types of Retaining Walls

It conducted statistics to 163 of the seismic-damaged cut slope retaining walls in field survey. The results show that the seismic damage under direct seismic effects falls into collapse, deformation and cracks and tilt. As there is the difference in the shoulder walls, most burials of retaining walls caused by slope collapse that happen to cut slope retaining walls, and the seismic damage quantity and percentage are as shown in Fig. 5.14.

In the figure, abscissa represents the seismic damage type, while ordinate represents the quantity. Seismic damage of other types refers to that there is one falloff of wall panel.

① Collapse-type seismic damage;

Collapse-type seismic damage is one of the most common seismic damages of retaining walls in the survey and that is the same with gravitational cut slope retaining walls. For the 73 cut slope retaining walls coming across collapse-type seismic damage, that built with mortar rubble masonry accounts for 84%, that built with pebble piece and concrete masonry accounts for 7%, and that built with dry masonry accounts for 9%. As the integrity of mortar masonry and dry masonry is poor, their tendency to come across collapse is more obvious.

For this type of seismic-damaged retaining walls, that built with soil foundation accounts for 46%, that built with rock foundation accounts for 44%, and that built with upper soil and lower rock foundation accounts for 10%. As upper soil and lower rock foundation belongs to soil foundation, thus seismic damage happens on

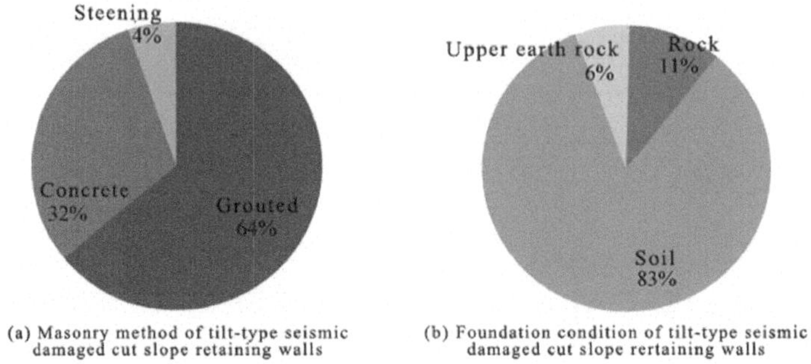

Fig. 5.15 Foundation condition and masonry method of the collapse-type cut slope retaining walls

Fig. 5.16 K1011 + 750 of Dujiangyan to Yingxiu section of National Road 213

soil foundation accounts for 56% of the total seismic damage and that is a big part (Figs. 5.15 and 5.16).

Gravitational collapse-type seismic damage of cut slope retaining walls (wall height is 3.2 m, mortar rubble masonry, repaired after shock) (Fig. 5.17).

Gravitational collapse-type seismic damage of cut slope retaining walls (wall height is 8 m, block stone masonry).

② Deformation and crack type seismic damage of retaining walls;

For the 53 seismic-damaged retaining walls coming across deformation and crack, that built with mortar rubble masonry accounts for 75%, that built with concrete masonry accounts for 21%, that built with dry masonry accounts for 4%, that built with rock foundation accounts for 24%, that built with soil foundation accounts for 61%, and that built with upper soil and lower rock foundation accounts for 15%. It can deduce the geotechnical properties of slopes through foundation types, and with soil foundation damage being the majority, it can be known that most deformed and cracked retaining walls are under soil slopes (Figs. 5.18 and 5.19).

Fig. 5.17 K1021+050 of Dujiangyan to Yingxiu section of National Road 213

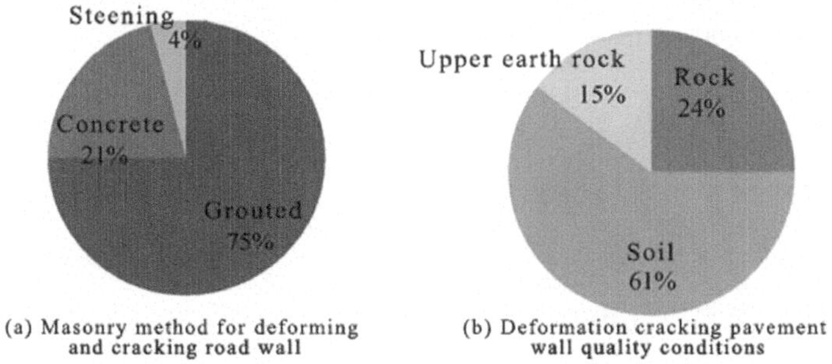

Fig. 5.18 Foundation condition and masonry method of the deformation and crack type cut slope retaining walls

Deformation and crack of gravitational retaining walls (wall height is 8.5 m, built with pebble concrete).

③ Tilt-type seismic damage of retaining walls.

For the 25 retaining walls coming across tilt-type seismic damage, that built with mortar rubble masonry accounts for 64%, that built with pebble piece and concrete masonry accounts for 32%, and that built with dry masonry accounts for 4%. In addition, that built with rock foundation accounts for 11%, that built with soil foundation accounts for 83%, and that built with upper soil and lower rock foundation accounts for 6% (Fig. 5.20).

The wall top displacement caused by tilt is in the range of 1–75 cm, with average displacement of 33 cm. The relation between the wall top displacement range and seismic damage quantity is as shown in Fig. 5.21.

Fig. 5.19 Longchi Tourism Highway of County Road XN16

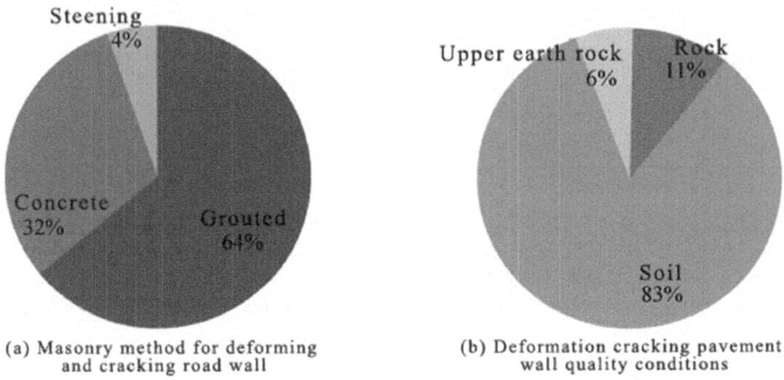

Fig. 5.20 Foundation condition and masonry method of the tilt-type cut slope retaining walls

5.2.2.4 Statistical Analysis of Influencing Factors of Seismic-Damaged Shoulder Retaining Walls

① The influence of walls height on seismic damage;

For the surveyed retaining walls, the maximum height is 15.5 m, the minimum height is 0.8 m, and the average height is 4.84 m, which is higher than that of the cut slope retaining walls (the average height of cut slope retaining walls is 3.26 m). The height of most seismic-damaged shoulder retaining walls is in the range of 2–8 m, which accounts for about 80% total quantity of seismic-damaged shoulder retaining walls. The relation of concrete height range and seismic damage quantity is as shown in Table 5.9.

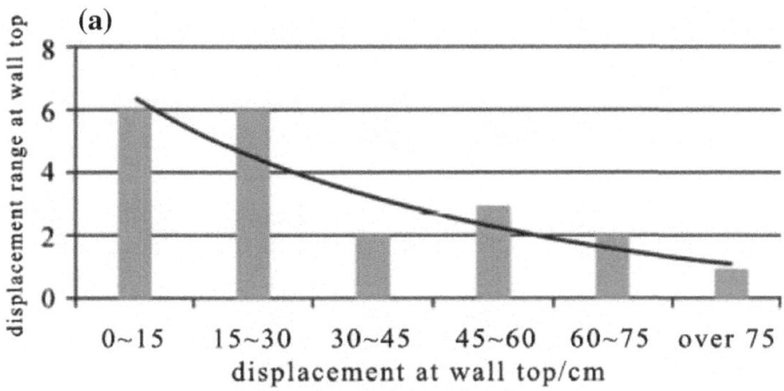

Fig. 5.21 **a** Distribution map of wall top displacement, **b** retaining wall tilt at K1021+840 of Dujiangyan to Yingxiu section of National Road 213 (the wall height is 8 m, and with mortar rubble masonry, the wall top displacement is 50 cm). Distribution map of wall top displacement and failure instance

Table 5.9 Height range percentage of seismic-damaged retaining walls

Wall height range (m)	Percentage (%)
0–2	2.4
2–4	24.4
4–6	34.1
6–8	22.0
8–10	9.8
above 10	7.3

Note The percentage refers to the ratio of number of damaged retaining wall with different heights with the total damaged number

In order to get further seismic-damaged rules, all the retaining walls and seismic-damaged retaining walls of National Road G213 from Dujiangyan to Yingxiu are counted and analyzed with the original design material. The results show that the length ratio (or quantity ratio) of seismic-damaged retaining walls with the total

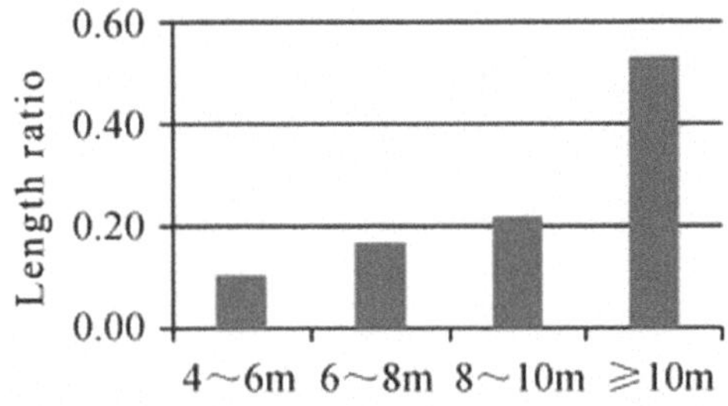

Fig. 5.22 Length ratio of shoulder retaining wall height from Dujiangyan to Yingxiu of National Road G213

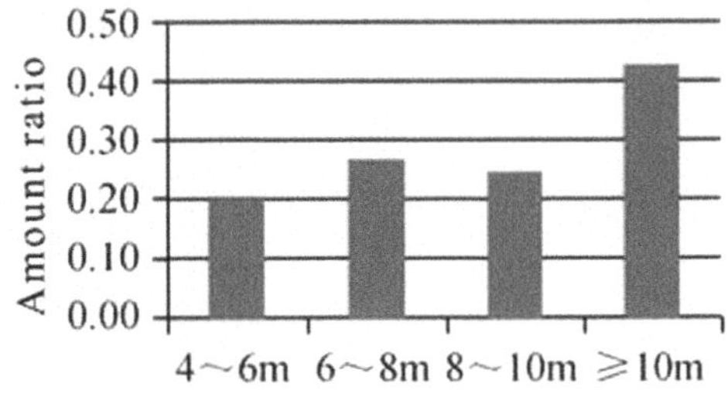

Fig. 5.23 Quantity ratio of shoulder retaining wall height from Dujiangyan to Yingxiu of National Road G213

Table 5.10 Statistics of height of seismic-damaged shoulder retaining walls from Dujiangyan to Yingxiu of G213

Wall height range (m)	Total length of retaining walls	Damaged length	Length ratio	total number of retaining walls	Damaged quantity	Quantity ratio
4–6	1743.86	187.6	0.11	25	5	0.20
6–8	2397.451	405	0.17	41	11	0.27
8–10	1259.413	280	0.22	16	4	0.25
≥10	428.266	228.5	0.53	7	3	0.43

number of retaining walls increase with the height elevation, which is identical with the practical seismic damage (Figs. 5.22, 5.23 and Table 5.10).

② The influence of building methods on seismic damage;

The survey conducted classified statistics to the building methods of 92 work sites of the damaged shoulder retaining walls, which mainly includes three kinds, namely mortar rubble masonry, dry masonry, and concrete masonry. The detailed seismic-damaged number and percentage are shown in Fig. 5.12 (Table 5.11).

The shoulder retaining walls built with mortar rubble masonry mainly come across collapse, tilt and deformation and crack seismic damage; that built with concrete masonry mainly come across collapse; that built with dry masonry mainly come across collapse. The seismic damage types that happen to retaining walls built with mortar rubble masonry are more than that happen to retaining walls built with concrete masonry. The integrity of retaining walls built with mortar rubble masonry is poor than that built with concrete masonry, which is one of the reasons that contributes

Table 5.11 Statistics of masonry method

Building methods	Total damage number	Percentage (%)
Mortar rubble masonry	78	84.8
Concrete masonry	10	10.9
Dry masonry	4	4.3

Note The percentage refers to the ratio of number of damaged shoulder retaining walls with different masonry methods with the total damaged number

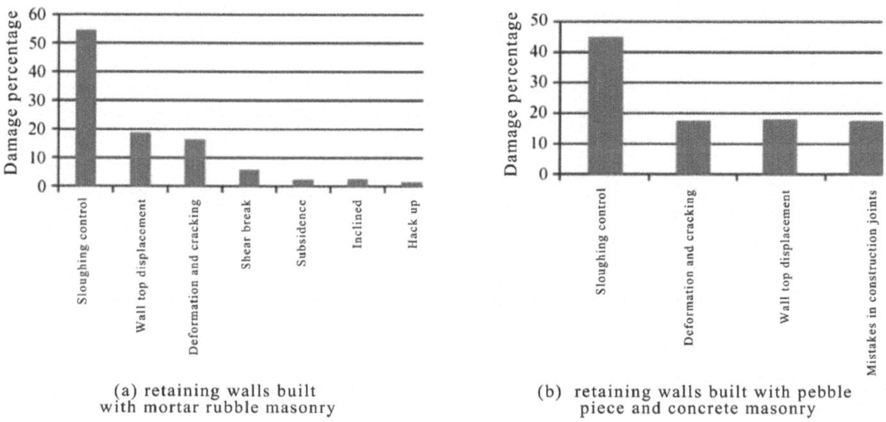

Fig. 5.24 Seismic damage-type distribution of retaining walls built with different masonry methods

to the seismic damage type complexity of mortar rubble-built retaining walls. The shoulder retaining walls built with concrete masonry did not come across damages like shearing and subsidence which may be because of their good integrity and active earth pressure produced by filled earth in front of the wall (Fig. 5.24).

In the figure, abscissa represents the seismic damage type, while ordinate represents the quantity.

③ The influence of wall foundation conditions on seismic damage;

Statistical classification was conducted to the geological conditions of 81 work sites of damaged shoulder walls, which is divided into three kinds, namely soil foundation, rock foundation, and upper soil and lower rock foundation. The detailed seismic-damaged number and percentage are shown in Fig. 5.25.

Soil foundation could come across seismic damage like liquidation under seismic effects, so the shoulder retaining walls built on this is easy to come across multi-type damages. The seismic damages that happen to retaining walls built on soil foundation are mainly collapse, deformation and crack, and tilt; that happen to retaining walls built on rock foundation are mainly collapse and shearing; that happen to retaining walls built on upper soil and lower rock foundation are mainly fall into four types, namely collapse, deformation and crack, burial and tilt. The detailed seismic-damaged number and percentage are shown in Fig. 5.25.

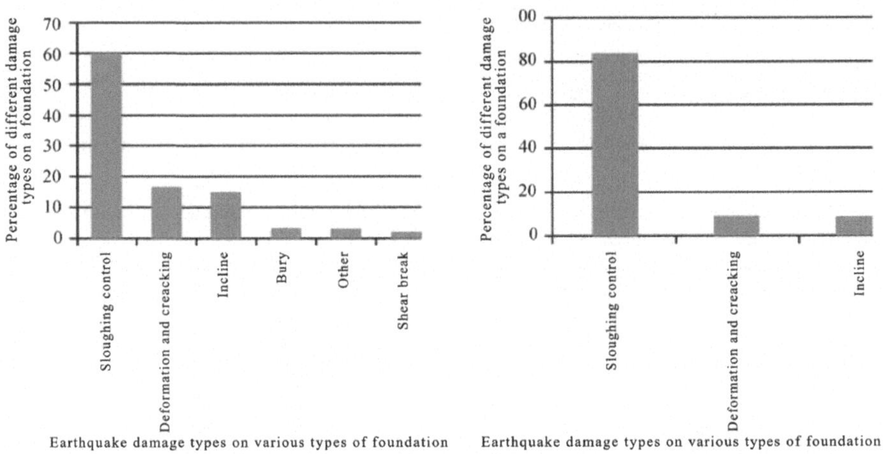

(a) Soil foundation condition (b) Upper soil and lower rock foundation condition

Fig. 5.25 Seismic damage distribution of retaining walls built on different foundations

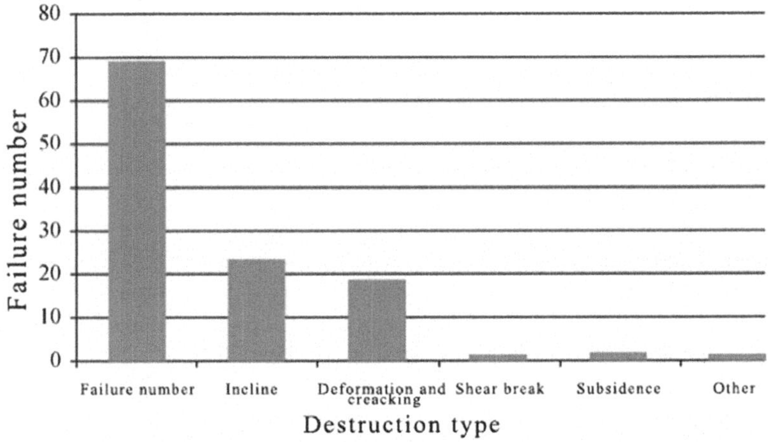

Fig. 5.26 Distribution map of seismic damage type and quantity of shoulder retaining walls

Abscissa represents the seismic damage type on different foundations, while ordinate represents the percentage of seismic damage types on this kind of foundation.

④ Further explanation of seismic damage types of retaining walls.

It conducted statistics to 117 of the seismic-damaged shoulder retaining walls in field survey. Seismic damages happening to shoulder retaining walls are mainly collapse, tilt and deformation and crack, in which shoulder retaining walls coming across collapse account for 59% of the total seismic-damaged retaining walls. The quantity and percentage of seismic damage are shown in Fig. 5.26.

In the figure, abscissa represents the seismic damage type of gravitational retaining walls, while the left ordinate represents the quantity, and the right ordinate represents the percentage.

5.3 Shaking Table Test Study on Seismic Stability of Gravitational Retaining Walls

Large-scale shaking table test is the important measure to study on the dynamic characteristics of constructions under seismic effects, and it is also an important method to conduct seismic damage mechanism analysis and overall anti-seismic performance evaluation, which has been widely applied to theoretical study and practice of seismic engineering. In order to further reveal the seismic damage mechanism of retaining walls, this chapter will, based on the previous work, combine the existing theory and experiment research result to conduct large-scale shaking table test that takes gravitational retaining walls as study object. The experiment includes two substrata: retaining walls with hard soil foundation and that with soft soil foundation (subgrade bearing capacity satisfies the design requirement). It firstly elaborates detailed experiment design and implement, then summarizes the wall displacement and soil pressure changing rules in the experiment process, compares soil pressure calculation results by seismic codes of different countries with the measured results, conducts focus research to the difference of anti-seismic performance of retaining walls built on different foundations, and finally proposes reasonable advice to the amendments of seismic design codes.

5.3.1 Large-Scale Shaking Table Test Design

5.3.1.1 Similarity Relation Design

In order to make the model test more authentically reflect the dynamic characteristics of the prototype, it must consider the similarity between the model and the prototype. Therefore, determination of the similarity relation between the prototype and the model is the key step in model design. Since 1970s, earthquake simulators have been successively built and applied domestically and abroad, and study of structural model similarity theory has been highly valued. As earlier scholars in this area, Harris and Lv Xilin explored and studied the model dynamic similarity theory for models on shaking tables, which laid foundation for the development of structural model test techniques. The similarity law and similarity law ignoring the gravity, which were built based on Buckingham π theory, have been widely applied to engineering dynamic model tests until today. This chapter will base on Buckingham π theory, adopt dimensional analysis method, and solve the similarity constant of dynamic response test model design of general embankments by considering the stress–strain constitutive relation between the model and the prototype (see Tables 5.12, 5.13 and 5.14).

5.3.1.2 The Physico-Mechanical Parameters of the Model Materials

Based on the preceding paragraphs, the experiment, respectively, adopts model soil and dry sand mixed by the similarity rate to simulate the field soil and embankment filling materials; every physical parameter of the model materials could be seen in Table 5.15.

5.3.1.3 Design of Retaining Wall Model

The test simulation carries out seismic design with intensity of VIII degree, and the prototype is the gravitational shoulder retaining wall with a height of 9.6 m. The size of test retaining wall model got by the geometric simulation rate is 1.6 m (height) × 1.5 m (width), with wall top width of 0.33 m, wall root width of 0.55 m, wall toe height of 0.204 m, and wall toe width of 0.102 m. The model is made of micro-concrete, and the filling material is made of dry sand. In order to study on the influence of foundation

Table 5.12 Similarity constant of embankment model test

Physical parameters	Similarity law	The similarity constant of embankment of 12 m
Geometric size L	C_L	6
Density of mass ρ	C_ρ	1
Input vibration acceleration A	$C_A = 1$	1
Modulus coefficient X	C_X	1
stress σ	$C_\sigma = C_\rho C_l$	6
Dynamic displacement u	$C_u = C_\rho^{1/2} C_l^{3/2} / C_X$	14.7
Dynamic strain ε	$C_\varepsilon = C_\rho^{1/2} C_l^{1/2} / C_X$	2.45
Vibration velocity of mass points $\dot{u}$	$C_{\dot{u}} = C_\rho^{1/4} C_l^{3/4} / C_X^{1/2}$	3.83
Acceleration of vibrating mass points $\ddot{u}$	$C_{\ddot{u}} = 1$	1
Time t	$C_t = C_\rho^{1/4} C_l^{3/4} / C_X^{1/2}$	3.83
Frequency f	$C_f = C_\rho^{-1/4} C_l^{-3/4} C_X^{1/2}$	0.261
Damping ratio ξ	$C_\xi = 1$	1
Friction angle φ	$C_\varphi = 1$	1

Table 5.13 Similarity constant of field soil model test

Physical parameters	Similarity law	The similarity constant
Length (m)	λ	6
Unit weight (kN/m^3)	1	1
Strain	1	1
Time (s)	$\lambda^{0.5}$	2.45
Stress (kPa)	λ	6
Acceleration (m^2/s)	1	1
Velocity (m/s)	$\lambda^{0.5}$	2.45
Displacement (m)	λ	6
Force (kN)	λ^3	216
Modulus of soil mass (kPa)	λ	6
Shear wave velocity (m/s)	$\lambda^{0.5}$	2.45
Cohesive force (kPa)	λ	6
Internal friction angle (°)	1	1
Frequency (Hz)	$1/(\lambda^{0.5})$	0.408
Flexural rigidity ($kN/m^2\ m^4$)	λ^5	7776
Axial rigidity ($kN/m^2\ m^2$)	λ^3	216
Angle of rotation	1	1 .
bending moment (kN m)	λ^4	1296

Table 5.14 Similarity constant of model test of gravitational retaining wall with a height of 9.6 m

Physical parameters	Dimension	Similarity relation	Similarity constant	Remarks
Length L	L	λ	6	Controlled variable
Density	ML^{-3}	λ_ρ	1	Controlled variable
Acceleration $\ddot{u}$	LT^{-2}	$\lambda_{\ddot{u}}$	1	Controlled variable
Velocity $\dot{u}$	LT^{-1}	$\lambda^{1/2}$	2.45	
Displacement u	L	λ	6	
Time t	T	$\lambda^{1/2}$	2.45	
Frequency ω	T^{-1}	$\lambda^{-1/2}$	0.408	

Table 5.15 Physical parameters of the model materials

Parameters	Hard soil foundation	Soft soil foundation	Fillings behind walls
Unit weight (kN/m^3)	20.26	18.6	17
Water content (%)	3.6	2.8	–
Internal cohesive force (kPa)	6.9	5	–
Internal cohesive force (°)	37.52	31	33

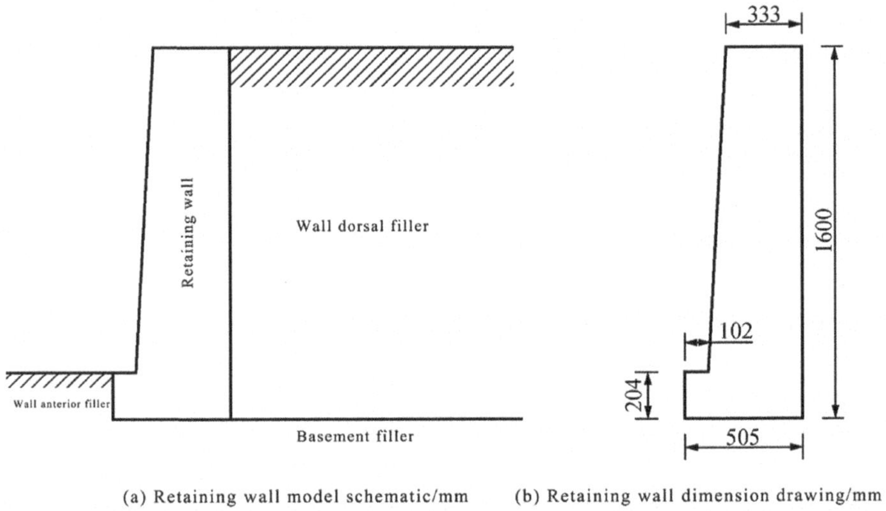

Fig. 5.27 Model diagram

conditions on anti-seismic performance of retaining walls, the test adopts both hard soil foundation and soft soil foundation. After subgrade model filling is loaded to the model container, it needs stratified compaction to ensure the compaction degree of backfilling foundation soil and the filling behind walls is identical with entitative engineering. The test model is shown in Fig. 5.27.

5.3.1.4 Design of Model Containers

In the shaking table test of geotechnical engineering, the typical model containers include rigid model container, laminar shear model container, and cylindrical flexible model container.

The feature of rigid model container is large overall stiffness and small lateral deformation of outer wall in vibration. The outer wall materials are mostly made up of steel plates, planks, or organic glass, and stable framework structure is added to enhance rigidity to lateral deformation of outer wall. This kind of model container has been widely applied to domestic and overseas shaking table test in earlier times. As the lateral deformation rigidity of outer wall is large, so the reflection of seismic wave on the border is intense, which together with changes to systematic vibrating form will bring certain error to the test results, which is the so-called boundary effect. To solve the above problem, flexible material is set up on the outer wall in test to reduce the boundary effect of containers. The generally adopted flexible materials are styrofoam and sponge. The property and thickness of flexible materials exert great influence on the boundary effect in the test. If the materials are thick and flexible, the boundary soil mass may not fit shear deformation feature; if they are thin and rigid, the absorbing ability is not enough, and the boundary reflection is still strong. In addition, it needs to consider the friction effect of outer wall and container bottom when designing the container structure. For the outer wall, some thin-film materials are generally adopted to reduce the friction effect of the outer wall to soil mass; for container bottom, materials like rubble with certain thickness are generally paved to increase friction and reduce relative displacement of container bottom and the soil mass.

Laminar shear model containers are taken as the ideal model container at present that could provide shear deformation conditions of soil mass. This kind of model containers is often constituted by piecing together several independent steel or aluminum rectangle or circular layered frameworks. A certain amount of rolling bearings is put between the layered frameworks to restrain vertical and lateral movement for one reason and to make the frameworks slides relatively in vibration direction so as to simulate shear deformation of soil.

Meymand developed the first cylindrical flexible model container, whose side wall is a piece of rubber membrane enclosed to cylinder, with its top end fixed by steel circular ring and bottom end fixed on the basal plates, which are mainly applied to the three-dimensional shaking table test for pile–soil interaction. Chen yueqing, Chen Guoxing, and Wang Zhihua also developed similar model containers on this basis and carried out a series of tests and researches on structure dynamic interaction.

As the test scale and loading strength require high of the container rigidity, this test, therefore, adopts top end-opening rigid model containers made of steel plates plus steel section plus organic glass, with inner-space size of 3.7 m × 1.5 m × 2.1 m (length × width × height), which is shown in Fig. 5.28. Styrofoam with a thickness of 30 mm is lined to back wall of filling in vibration direction to simulate the absorbing materials so as to reduce reflection of seismic waves on the border. As the organic glass is smooth, so it can eliminate the influence of frictional constraints of container lateral walls on tests. The container bottom materials are steel plates, which could produce large friction force after compaction and filling of field soil. In addition, the model container is sealed with steel plates and framework in input direction of dynamic load, so the relative movement between field soil and container bottom will be extremely small under the effects of dynamic loads.

Fig. 5.28 Model container

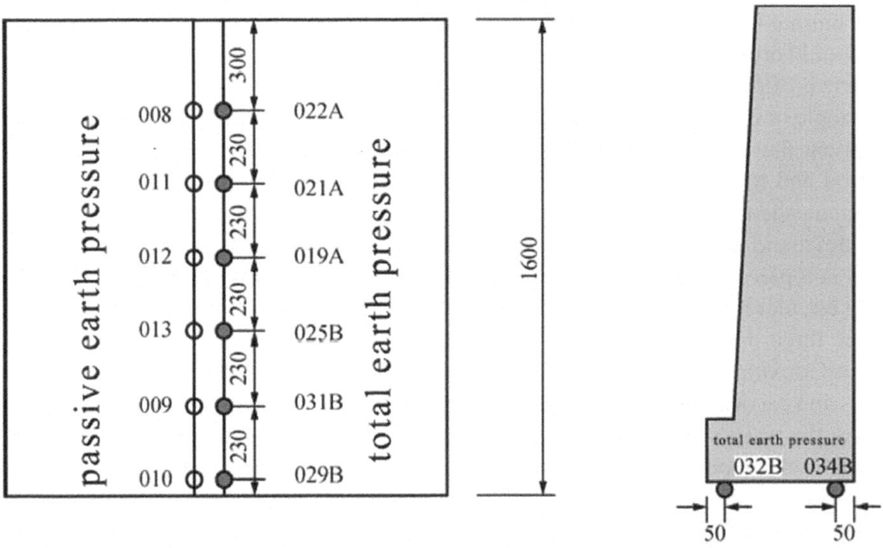

Fig. 5.29 Distribution of earth pressure sensors

5.3.1.5 Distribution of Sensors

The sensors adopted by the tests include earth pressure sensor, displacement sensor, and acceleration sensor, which are all set up on the longitudinal section between the retaining wall model and the filling. Sensor distribution of all substrata is shown in Tables 5.16, 5.17 and Figs. 5.29, 5.30, 5.31, and 5.32).

Table 5.16 Data collection and channel record

Accelerometer			Earth pressure meter			Displacement meter		
Self-defined number	Type	Corresponding channel self-defined number	Self-defined number	Type	Corresponding channel self-defined number	Self-defined number	Type	Corresponding channel self-defined number
1001	Horizontal	CH4	Total 032B	Strain type	CH41	Lower of gravity container	Differential	CH63
1002	Horizontal	CH5	Total 034B	Strain type	CH42	Upper of gravity container	Differential	CH64
1003	Horizontal	CH6	Total 029B	Strain type	CH43	Lower of gravity container	Guyed	CH67
1004	Horizontal	CH7	Total 031B	Strain type	CH44	Upper of gravity container	Guyed	CH68
1005	Horizontal	CH8	Total 025B	Strain type	CH45			
1006	Horizontal	CH9	Total 019A	Strain type	CH46			
1007	Horizontal	CH10	Total 021A	Strain type	CH47			
1008	Horizontal	CH11	Total 022A	Strain type	CH48			
1009	Horizontal	CH12	Dynamic 010	Piezoelectric	CH55			
1019	Horizontal	CH22	Dynamic 009	Piezoelectric	CH56			
1021	Vertical	CH24	Dynamic 013	Piezoelectric	CH57			
1022	Vertical	CH25	Dynamic 012	Piezoelectric	CH58			
1023	Vertical	CH26	Dynamic 011	Piezoelectric	CH59			
1024	Vertical	CH27	Dynamic 008	Piezoelectric	CH60			

Table 5.17 Data collection and channel record

Accelerometer			Earth pressure meter			Displacement meter		
Self-defined number	Type	Corresponding channel	Self-defined number	Type	Corresponding channel	Self-defined number	Type	Corresponding channel
1001	Horizontal	CH4	Total 027B	Strain type	CH33	Lower of gravity container	Differential	CH61
1002	Horizontal	CH5	Total 034B	Strain type	CH42	Upper of gravity container	Differential	CH62
1003	Horizontal	CH6	Total 028B	Strain type	CH35			
1004	Horizontal	CH7	Total 030B	Strain type	CH36			
1005	Horizontal	CH8	Total 026B	Strain type	CH37			
1006	Horizontal	CH9	Total 024A	Strain type	CH38			
1007	Horizontal	CH10	Total 020A	Strain type	CH39			
1008	Horizontal	CH11	Total 023A	Strain type	CH40			
1009	Horizontal	CH12	Dynamic 016	Piezoelectric	CH49			
1019	Horizontal	CH22	Dynamic 015	Piezoelectric	CH50			
1021	Vertical	CH24	Dynamic 014	Piezoelectric	CH51			
1022	Vertical	CH25	Dynamic 007	piezoelectric	CH52			
1023	Vertical	CH26	Dynamic 018	Piezoelectric	CH53			
1024	Vertical	CH27	Dynamic 017	Piezoelectric	CH54			

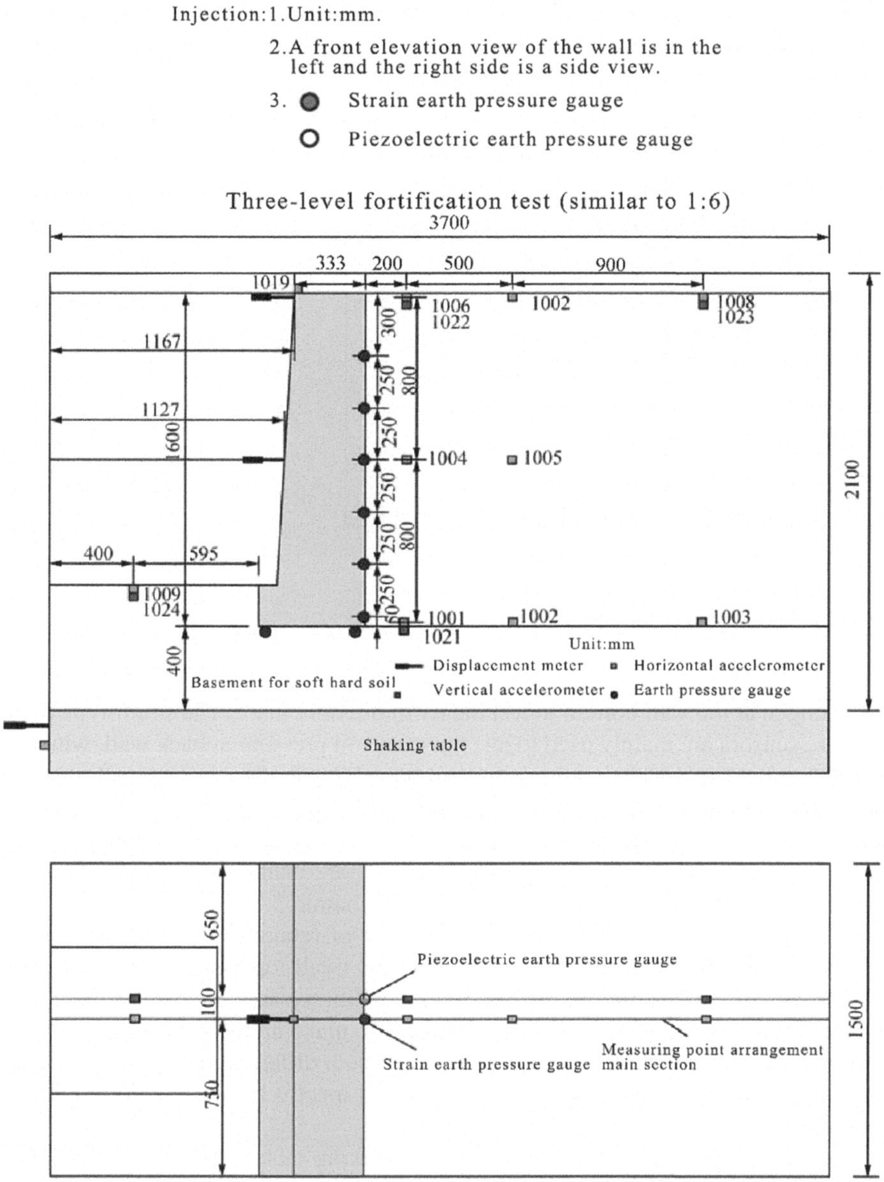

Fig. 5.30 Distribution of acceleration sensors and displacement sensors

The earth pressure sensors are set up on the contact surface between the gravitational retaining walls and the filling, with two rows along wall height and six in each row. One row is set up with strain-type earth pressure sensors, and the other is set up with piezoelectric earth pressure sensors; two strain-type earth pressure sensors

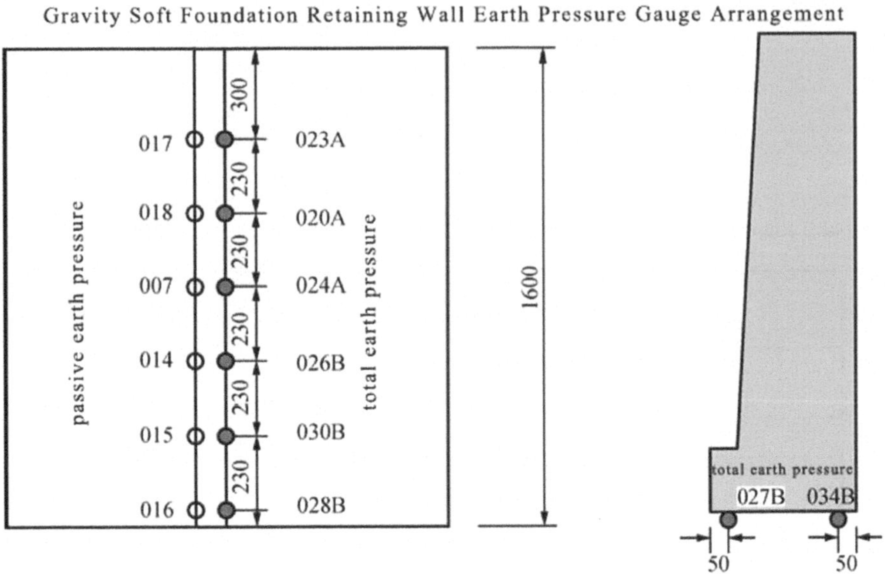

Fig. 5.31 Distribution of earth pressure sensors

are arranged at the wall bottom to test the earth pressure there. The strain-type earth pressure sensors are mainly used to test the total soil pressure at back wall, which is noted as "total," in which "total + serial number + B" is the large-range soil pressure meter that is buried at the middle–lower part and substrate of walls. "total + serial number + A" is the small-range soil pressure meter that is buried at middle–upper part of walls. Piezoelectric earth pressure sensors are mainly used to test dynamic earth pressure of back wall, which is noted as "dynamic."

In order to study on the displacement feature of retaining walls under seismic effects, it needs to record the wall displacements (wall top displacement and wall rotation angle) under different loading work conditions, and the displacement sensors are set up at wall top and inside the walls. To make up the measuring errors of apparatus and satisfy the test accuracy, it adopts both differential displacement sensors and guyed displacement sensors, which are, respectively, noted as "differential" and "guyed."

The acceleration sensors are arranged on shaking tables, filling at back walls, filling in front of walls and wall tops to detect the acceleration on the table surface so as to study on the acceleration dynamic response characteristics of filling at back walls, filling in front of walls and the retaining walls. The acceleration sensors could be divided into "horizontal" and "vertical" type by test direction.

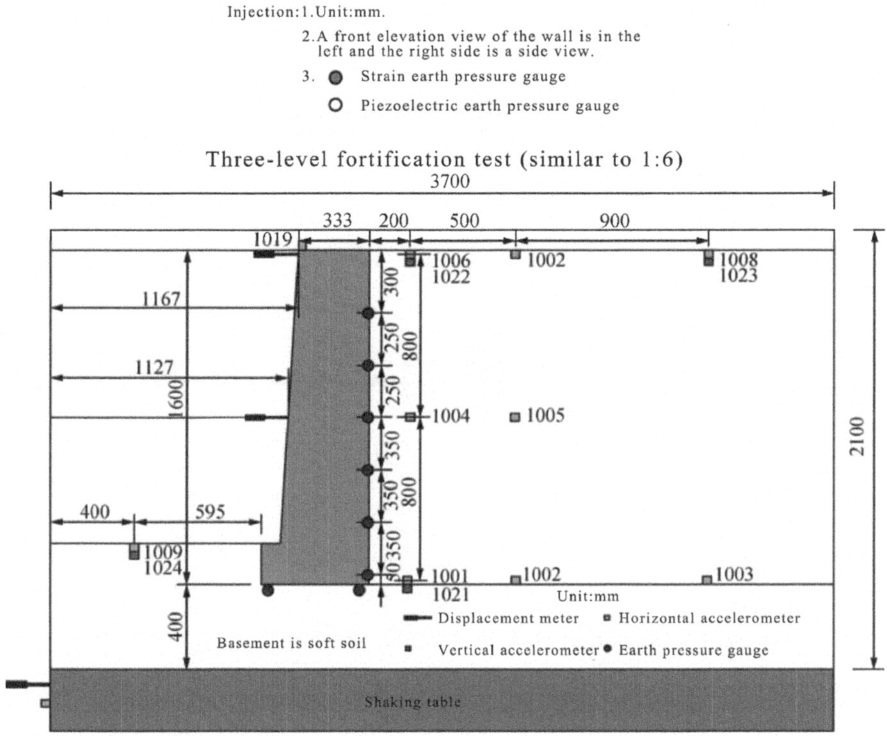

Fig. 5.32 Distribution of acceleration sensors and displacement sensors

5.3.1.6 Loading Conditions of the Experiment

According to the field survey and results of seismic risk evaluation, it adopts the existing digital seismic record of Wenchuan earthquake which has similar conditions to the model field. The duration is 180 s with 40 s of strong earthquake, and the PGA of E-W, N-S, and U-D components are, respectively, 957.4, 655.8, and 853.8 cm/s^2. The experiment adopts E-W and U-D components as the input loads for X-axis (horizontal, vertical route direction) and Y-axis (vertical, vertical route direction). Figures 5.33 and 5.34 display the time-history curve of E-W and U-D components of Wenchuan earthquake wave compacted by the simulation law, and the duration is 65.3 s. It carries out wave shape and peak value adjustments to the compacted Wenchuan seismic waves and conducts tests of simulated small earthquakes, middle earthquakes, and severe earthquakes to retaining walls in sequence. The loading order of the test is as follows: white noise→ 0.1 g→ 0.2 g→ 0.4 g→ 0.7 g→ 0.9 g until obvious damage occurring to the retaining walls.

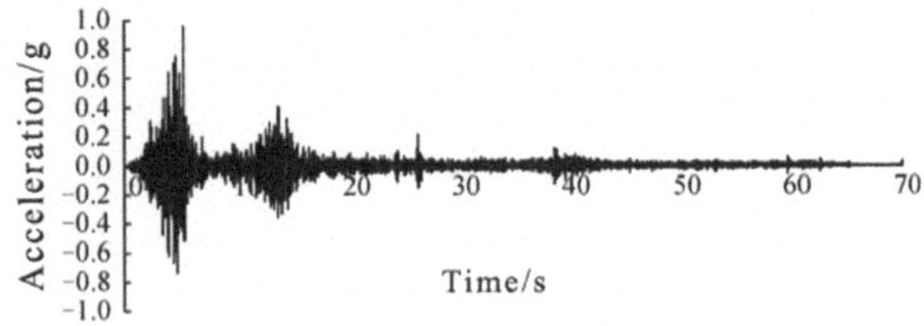

Fig. 5.33 E-W components of compacted Wenchuan seismic waves

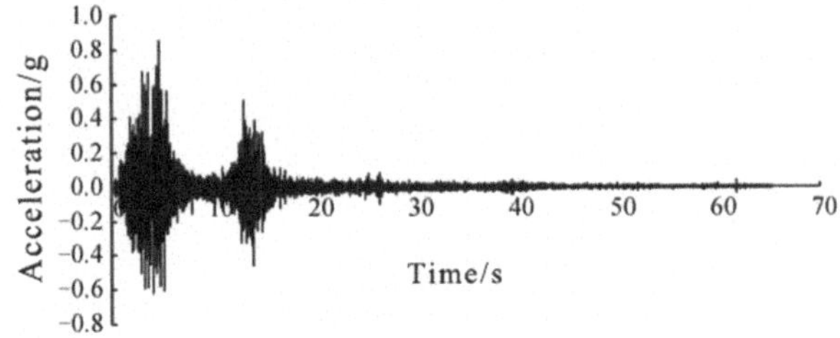

Fig. 5.34 U-D components of compacted Wenchuan seismic waves

5.3.2 *Implementation of Shaking Table Test*

5.3.2.1 General Situation of Data Collection System

This experiment adopts the BBM data collecting system with 128 channels, and its maximum fiducial error is less than or equal to 0.5%. Sensor signal conditioner is connected to electric charge transfer to convert the voltage signal, and its maximum fiducial error is less than or equal to 1%. Data collecting, signal monitoring, and online analyzing are conducted at the same time (see Fig. 5.35).

5.3.2.2 Experiment Flow

After the preparation work, retaining wall models, model containers, substrates, and embankment filling behind walls are all sent to the work site before formal beginning of the tests. The following paragraphs are a brief introduction to model test flow by order.

① Installation of displacement sensor frameworks;

To ensure normal operation of displacement sensors at wall top and middle wall, the displacement sensors must be fixed to the model containers with displacement sensor frameworks, whose rigidity and natural frequency must be ensured to a certain extent so as to make sure that there will be no relative movements between displacement sensors and model containers in loading. The displacement sensor frameworks are made of channel steel of 100 mm and are assembled to the model containers by the design size. The displacement test apparatus after installation is shown in Fig. 5.36.

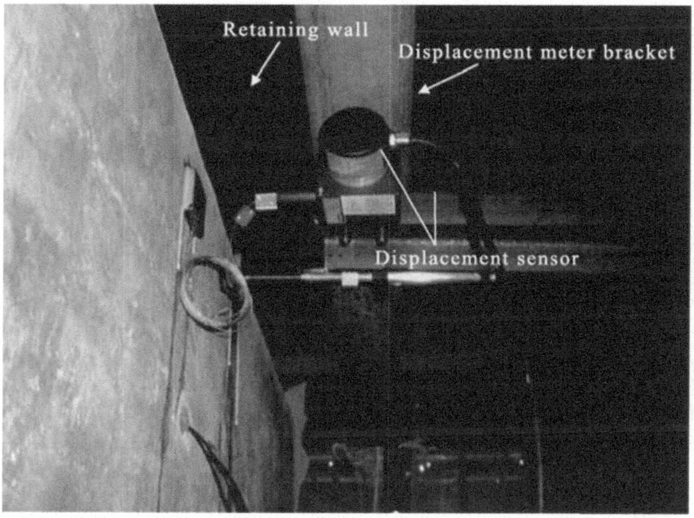

Fig. 5.36 Displacement test apparatus

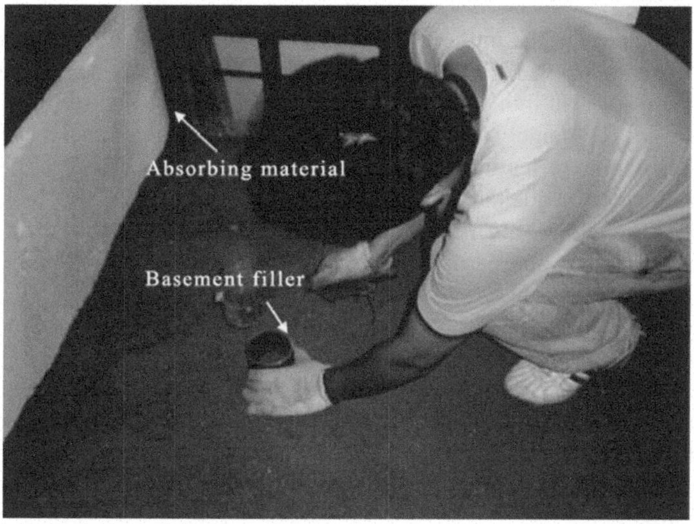

Fig. 5.37 Sample selection of basal filling

② Boundary process of model containers;

To avoid vibration wave reflecting at the model container border, it should eliminate the most of "boundary effect" influence on the experiment. Wave-absorbing materials of 30 mm thickness were added to the end walls in the vibrating direction, and foam was used as wave-absorbing materials (see Fig. 5.37).

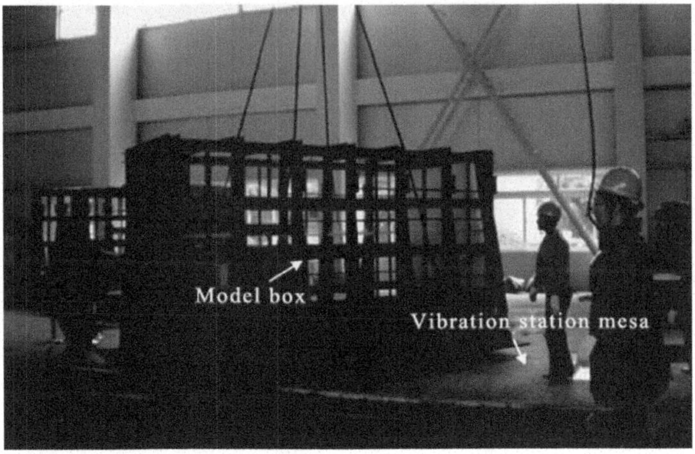

Fig. 5.38 Hoisting assembly of model containers

③ Compaction of basal filling;

In order to study on the influence of foundation conditions on anti-seismic performance of retaining walls, the foundation earth could be divided into hard soil foundation and soft soil foundation by the degree of hardness and softness. The test adopts three indexes, namely the capacity, cohesive force, and internal friction angle, to control the degree of hardness and softness of foundations. Therefore, the loose filling loosely paved inside the container needs to be compacted layer by layer, and it needs to ensure all indexes meet the regulated standards through testing the earth capacity and selecting some earth sample to make shear experiment. Figure 5.37 displays that the test researchers are selecting samples to the basal filling after layered compaction.

④ Hoisting assembly;

After completion of basal filling compaction, the model containers need to be assembled to the shaking tables, and the retaining wall models with earth pressure sensors assembled need to be hoisted inside the model containers (see Figs. 5.38, 5.39 and 5.40).

⑤ Stuffing of back wall basal filling;

After hoisting assembly of model containers and retaining walls, it needs to conduct layered stuffing of back wall basal filling, which mainly adopts quartz sand matched by the simulation rate. To simulate the compaction degree of the basal filling, it needs to compact the loosely paved quartz sand to specified density to meet the compaction standard. In order to more intuitively observe the seismic response (slides, sedimentation, fracture surface, etc.) of back wall basal filling from the container side, colorful sand with a thickness of 1 cm is loosely paved to every 30-cm-thick back wall filling (see Fig. 5.41). Given that the compression face of earth

Fig. 5.39 Retaining wall models

pressure sensors is relatively weak, and construction strength, roughness of filling as well as human factors all exert influence on the accuracy of earth pressure sensors, so to protect the earth pressure sensors from damage, when stuffing the back wall filling, it needs firstly stuffing fine sand around the earth pressure sensors, and carefully compacting their surrounding filling until the filled earth of the layer is compacted (see Fig. 5.42).

⑥ Embedding of acceleration sensors;

Before embedding of acceleration sensors, it needs firstly fixing them inside the insulated boxes, and ensure that the weight of acceleration sensors and the boxes are the same with earth of same volume. Then, seal the insulated boxes where acceleration sensors are fixed, and lay them horizontally to ensure complete contact of box bottom with filling so as to improve the test accuracy. See Fig. 5.43. The acceleration sensors on wall top are directly fixed with epoxy resin and 502 glue, which are shown in Fig. 5.44.

⑦ Earth stuffing in front of walls and embedding of acceleration sensors;

When the back wall filling is stuffed to the middle wall, in order to avoid large displacement in stuffing of back wall filling and in compaction, the retaining wall should be buried to a certain depth, and filling in front of walls should be to wall toe height and be compacted. To test the acceleration of free field, acceleration sensors should be buried in the filling in front of walls with certain distance to the retaining walls.

Fig. 5.40 Hoisting assembly of retaining walls

Fig. 5.41 In-situ colorful sand fabrication

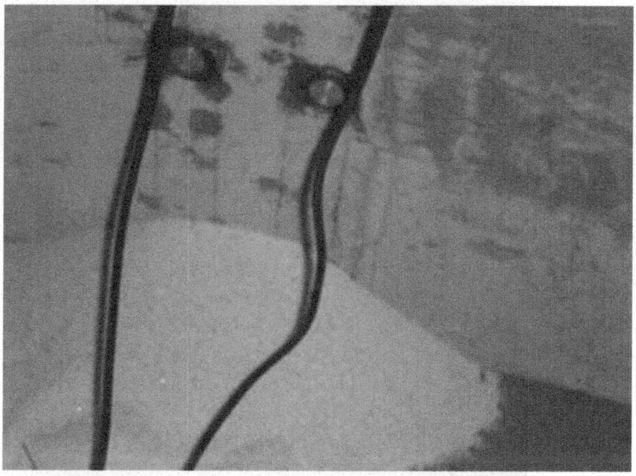

Fig. 5.42 Protection of earth pressure sensors by fine sand

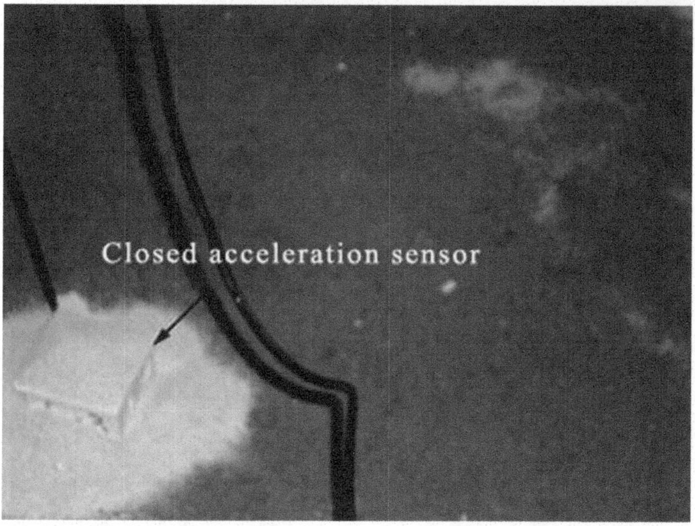

Fig. 5.43 Acceleration sensors

⑧ Installation of displacement sensors;

The displacement testing positions are wall top and middle wall, and each position is set up with two types of sensors, namely differential and guyed displacement sensors. The test proves that test results of differential displacement sensors are better than that of guyed displacement sensors. Figure 5.45 displays that the test people are installing displacement sensors.

⑨ Connecting sensors with the data collecting system;

When back wall filling, filling in front of wall and sensor arrangement is all finished, and the sensors are connected to the data collecting system and collect

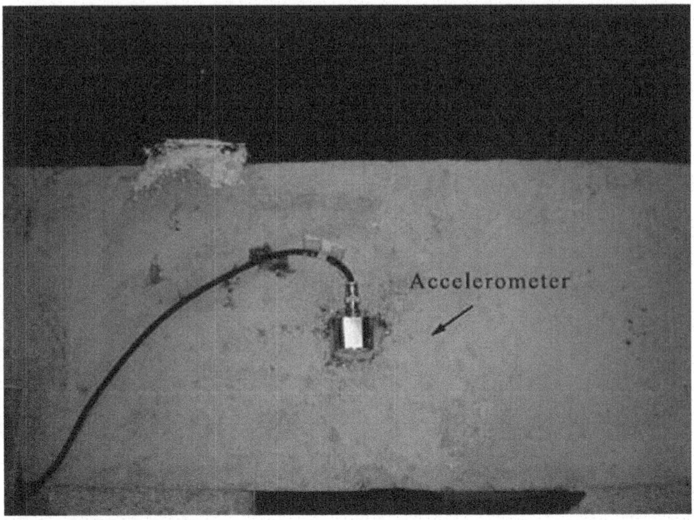

Fig. 5.44 Acceleration sensors at wall top

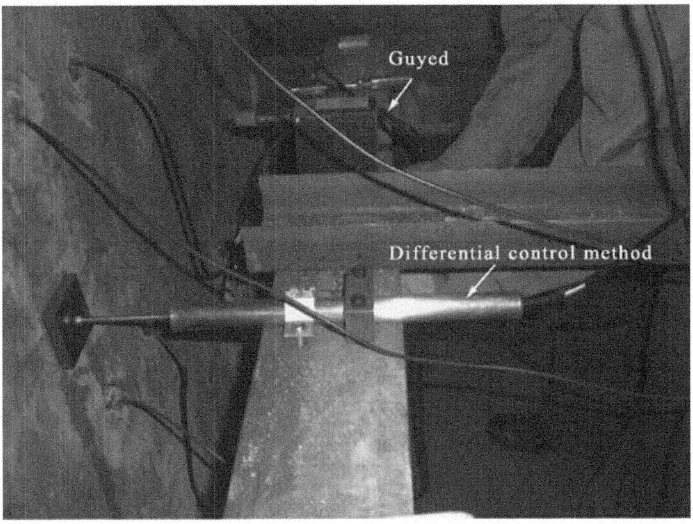

Fig. 5.45 Installation of displacement sensors

electric signals like acceleration earth pressure and displacement measured by the sensors. Figure 5.46 displays the overall view of gravitational retaining wall test model. As the relative movement of model containers with sensor wires in vibration will cause interference signals, so in order to avoid its influence on the accuracy of data collection, the test adopts seismic isolation materials such as foam or cotton cloth to wrap and fix the wires, which is as shown in Fig. 5.47.

Fig. 5.46 Overall view of gravitational retaining wall model

Fig. 5.47 Wrapping of wires with seismic isolation materials

⑩ Camera orientation.

Before the beginning of the test, high-speed digital cameras are installed on the side and top of model containers to constantly observe the vibration and damage of retaining wall models (see Fig. 5.48).

Fig. 5.48 Installation of cameras

5.3.3 *Statistical Analysis of Shaking Table Test*

5.3.3.1 Retaining Wall Displacement

① Necessity to study the retaining wall displacement;

The seismic damage research results of Wenchuan earthquake show that collapse, deformation and cracks, and overturning/slides, the three damage types, account for 82.6% of the total damage number. From the perspective of masonry method, most collapse damages are mortar rubble masonry, and for gravitational retaining walls built with concrete, it mainly happens with tilt damage or slide damage under seismic effects. With the development of anti-seismic techniques of transport infrastructures and people's rising requirements of anti-seismic performance targets, high-grade highways and railways have paid more and more attention to the seismic performance of retaining walls, which also raises demands on materials and specifications of wall body. Under this circumstance, compared with retaining walls built with mortar rubble masonry, gravitational retaining walls built with concrete are more widely applied in seismic areas, which makes a study on seismic mechanism of gravitational retaining walls built with concrete having more engineering significance and shows the importance of conducting model tests.

Under seismic effects, the displacement failure modes are mainly divided into slides, rotation around wall toe, rotation around wall top, slides coupling with rotation. The seismic damage research shows that, in Wenchuan earthquake, the cases in which retaining walls are taking rotation around wall top as damage mode are rare, and most damage modes are slides, rotation around wall toe, and slides coupling with rotation. For retaining walls coming across these kinds of damage modes, it produces obvious non-recoverable displacement to walls after earthquake, with residual displacement at wall top being the most serious. It shows that, for gravitational retaining walls, residual displacement at wall top could be taken as one of the most important indexes to judge the seismic performance of retaining walls. Given this, Zhang Jianjing et al. firstly proposed a quantitative index to judge the seismic performance of retaining walls, displacement index (represented as symbol δ in this article), which is defined as the ratio of wall top displacement aftershock with wall height. Figure 5.49 displays two seismic damage work sites of retaining walls in Dujiangyan to Yingxiu section of National Road 213, and wall top of both comes across obvious displacements.

In addition, performance design has always been paid high focus on in rock and soil seismic engineering. Displacement, as the most intuitive parameter to judge the seismic damage of gravitational retaining walls, effectively reflects the seismic performance of retaining walls, which makes seismic design of gravitational retaining walls based on displacement the subject that is most worthy of research and discussion in performance design at present. Therefore, based on large-scale shaking table test, to explore the displacement changing rules of gravitational retaining walls under seismic effects and to summarize the relation between wall displacement features and anti-seismic performance of retaining walls are not only of non-negligible engineering significance, and what is more is that it provides supports for improving anti-seismic level of retaining structures and developing new seismic design concept.

② Research methods of retaining wall displacement;

The main purpose of conducting large-scale shaking table test is to analyze the seismic mechanism of gravitational retaining walls through representation of its damage mechanism and features under seismic effects. By test observation, the model retaining walls, under input seismic effects, could effectively simulate the damage mechanism and features of field retaining walls, and the displacement mode of retaining walls under seismic effects is translational motion coupling with rotation around wall toe, i.e. RBT motion mode, which is shown in Fig. 5.50.

In Fig. 5.50, the total displacement of wall top is Δ, the translational displacement is $\Delta 1$, the rotational displacement is $\Delta R = \Delta 1 + \Delta R$, and the rotation angle of wall is θ. For the convenience of research, this book introduces translational displacement percentage $\Delta 1/\Delta$, rotational displacement percentage $\Delta R/\Delta$, and displacement index $\delta(\Delta/H)$to comprehensively analyze the anti-seismic performance of gravitational retaining walls. Displacement percentage reflects the displacement mode of retaining walls in earthquakes, and $\Delta 1/\Delta$ is in proportion to the sliding displacement of retaining walls; $\Delta R/\Delta$ is in proportion to the overturning displacement of retaining walls; δ reflects the displacement amplitude of retaining walls aftershock, which judges the anti-seismic performance of retaining walls with quantitative methods. If there are no special instructions, the displacements of retaining walls in this book are all aftershock residual displacements.

(a)

Central maximum displacement 50cm

(b)

Fig. 5.49 **a** Case ① of Dujiangyan to Yingxiu section of National Road 213 (δ = 6.67%), **b** Case ② of Dujiangyan to Yingxiu section of National Road 213 (δ = 6.67%)

③ Data analysis of retaining wall displacement.

In the shaking model test, input seismic grades, seismic wave features, geometric shape of retaining walls, and foundation conditions are all major factors that could

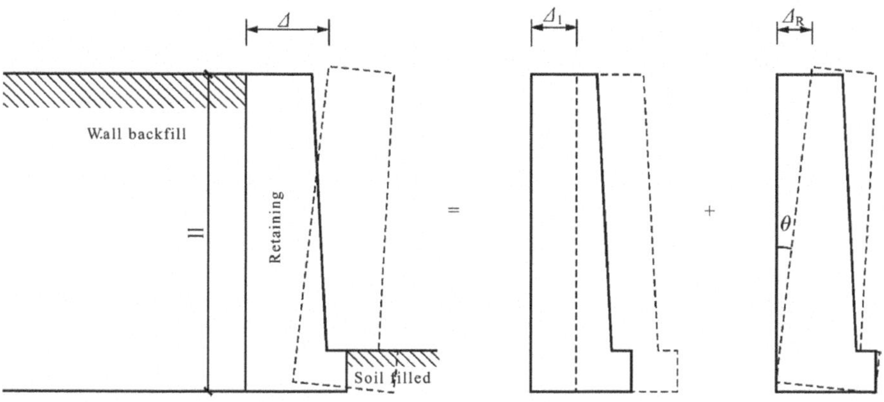

Fig. 5.50 Distribution modes of retaining walls

Table 5.18 Retaining wall displacement index δ

Seismic coefficient k_h	Hard earth road shoulder (%)	Soft earth road shoulder (%)
0.1	0	0
0.2	0.01	0.01
0.4	0.27	0.57
0.7	1.99	4.28
0.9	4.53	–

Table 5.19 Retaining wall rotation angle θ

Seismic coefficient k_h	Hard earth road shoulder	Soft earth road shoulder
0.1	0	0
0.2	0.006	0.005
0.4	0.132	0.253
0.7	0.972	2.012
0.9	1.927	–

affect retaining wall displacement. In order to deeply study on the displacement features of retaining walls, the experiment observes and records the retaining wall displacement under different working conditions and analyzes and summarizes the influence of foundation conditions on retaining wall displacement features and its changing rules. The concrete analysis data could be seen in Tables 5.18, 5.19, 5.20, 5.21 and Figs. 5.51, 5.52, 5.53, and 5.54.

It could be known by observing Figs. 5.51 and 5.52 that:

Firstly, the wall displacement is in proportion to the seismic motion acceleration, and the increasing amplitude of displacements of retaining walls under different seismic intensities is different.

When the seismic motion acceleration is less than 0.2 g, there is hardly any displacement to the retaining wall and the displacement index is less than 0.01%;

Table 5.20 Translational displacement percentage $\Delta 1/\Delta$

Seismic coefficient k_h	Hard earth road shoulder	Soft earth road shoulder
0.1	–	–
0.2	1.69%	17.65%
0.4	14.64%	22.50%
0.7	14.62%	18.06%
0.9	25.80%	–

Table 5.21 Rotational displacement percentage $\Delta R/\Delta$

Seismic coefficient k_h	Hard earth road shoulder	Soft earth road shoulder
0.1	–	–
0.2	98.31%	82.35%
0.4	85.36%	77.50%
0.7	85.38%	81.94%
0.9	74.20%	–

Note When seismic ground motion acceleration is 0.9 g, the displacement of shoulder retaining wall on soft earth foundation is so large that exceeds the maximum range of displacement sensors, which, thus, is represented as "–". As the retaining walls are designed by anti-seismic intensity of VIII degree, and when seismic ground motion acceleration is 0.1 g, which is within the anti-seismic intensity, the retaining walls hardly have any displacement. Thus, it is represented by "–"

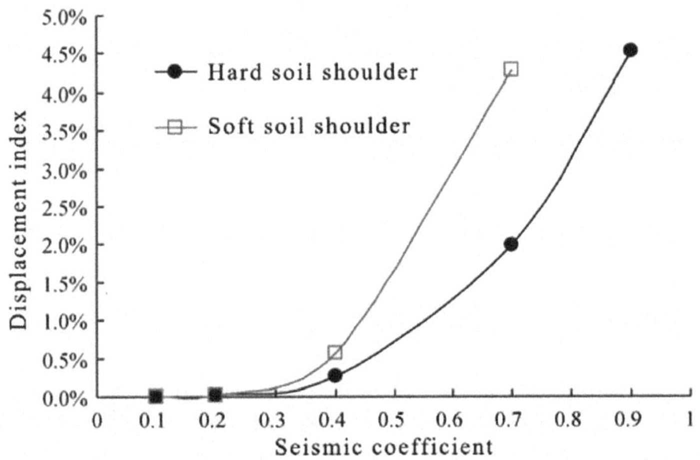

Fig. 5.51 Changing map of displacement indexes with different foundation conditions

when the seismic motion acceleration reaches 0.3 g, there is tiny displacement to the retaining wall and the maximum displacement index approaches 0.15%; when the seismic motion acceleration is more than 0.3 g, there is obvious displacement to the retaining wall, and it can be known from the tangency changes of displacement

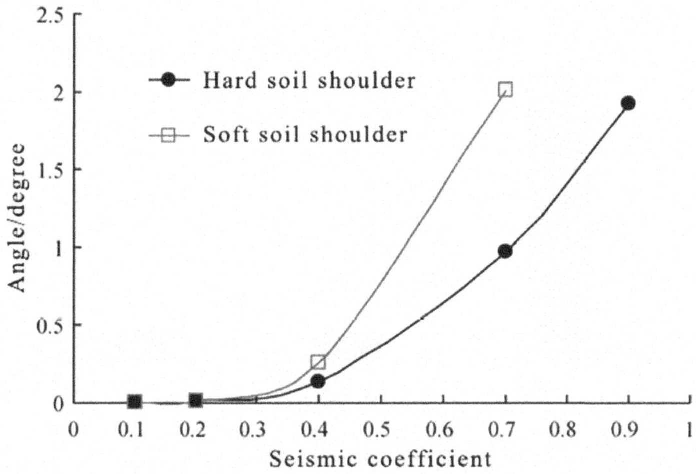

Fig. 5.52 Changing map of wall rotation angles with different foundation conditions

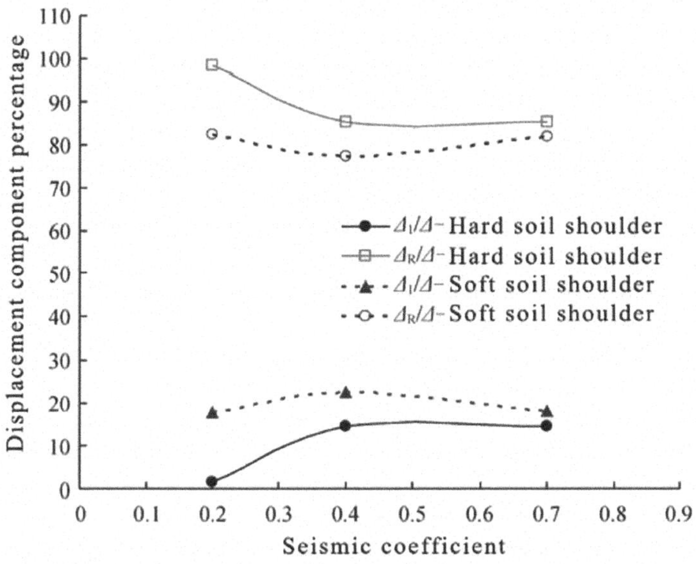

Fig. 5.53 Changing map of slides and rotational components percentage

index curve that the increasing amplitude of wall top displacement increases with the augmentation of seismic motion acceleration.

Secondly, the displacement of retaining walls is affected by the foundation conditions. Under the same situation, the displacements of retaining walls on soft earth foundations are larger than that of retaining walls on hard earth foundations, and the influencing extent is rising with the increase in seismic motion acceleration.

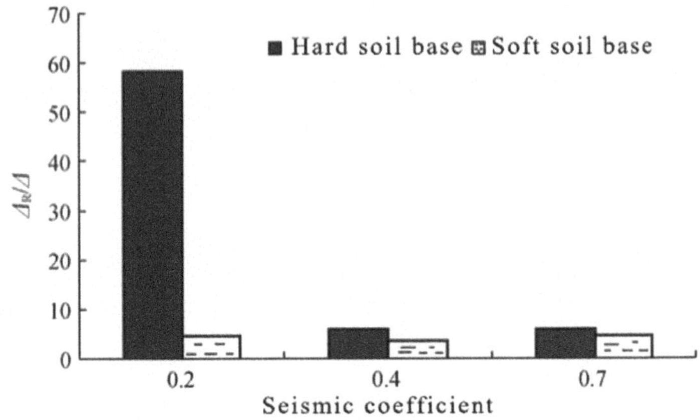

Fig. 5.54 Changing map of rotational components/sliding components ($\Delta R/\Delta 1$)

When the seismic motion acceleration is less than 0.3 g, the influence of foundation conditions on total displacement of retaining walls is not obvious. When the seismic motion acceleration is 0.2 g, the displacement indexes are all 0.01%, while when the seismic motion acceleration is 0.3 g, the displacement indexes on soft earth foundations are all approaching 0.15%, which are slightly higher than that on hard earth foundations. It explains that the anti-seismic design of retaining walls in regions with seismic intensity under 8° could omit the influence of foundation conditions on total displacement of retaining walls. When the seismic motion acceleration is over 0.3 g, the influence of foundation conditions on total displacement of retaining walls rises significantly with the acceleration increase. When the seismic motion acceleration is 0.4 g, the displacement index of retaining walls on soft earth foundations is about 2.1 times of that of retaining walls on hard earth foundations, which explains that the anti-seismic design of retaining walls in regions with seismic intensity over 8° should emphasize the influence of foundation conditions on retaining wall displacement and should, respectively, consider the hard earth foundations and soft earth foundations.

It could be known by observing Figs. 5.53 and 5.54 that:

For retaining walls on hard earth foundations, when seismic acceleration reaches 0.2 g, the rotational components account for 98.31% of the total percentage and translational components only account for 1.69%, with rotational components being about 58.17 times of translational components; when seismic acceleration is over 0.2 g, the rotational components gradually decrease and translational components gradually increase; when seismic acceleration is in the range of 0.4–0.7 g, both rotational and translational components rarely change with the seismic coefficients, with rotational components being 5.84 times of translational components.

For retaining walls on soft earth foundations, when seismic acceleration reaches 0.2 g, the rotational components account for 82.35% of the total percentage and translational components account for 17.65%, with rotational components being about

4.67 times of translational components; when seismic acceleration is in the range of 0.2–0.4 g, the translational components are in slight increasing trend, and when seismic acceleration is 0.4 g, rotational components are 3.44 times of translational components; when seismic acceleration is in the range of 0.4–0.7 g, translational components decrease slightly, with rotational components being 5.53 times of translational components when seismic acceleration is 0.7 g.

Through the above analysis, it can come to the following conclusion:

Firstly, from the aspect of the total trend of displacement component percentage, the rotational components account for a large proportion, which shows that overturning is the major damage mode that the retaining walls coming across large displacement.

Secondly, under seismic effects, the influence of foundation conditions on retaining displacement mode is significant.

With the increase in seismic coefficients, the rotational displacement mode of retaining walls on hard earth foundations changes toward rotation coupling with slides. When the seismic intensity is less than 8°, the retaining wall displacements are nearly all caused by rotation, and the displacement mode is rotation; when the seismic intensity is between 8° and 9°, the translational components gradually increase, while the rotational components gradually decrease, and the displacement mode is rotation coupling with translation; when the seismic intensity is over 9°, the translational components stay at fixed value and the displacement mode is also rotation coupling with translation.

The influence of seismic intensity on displacement mode of retaining walls on soft earth foundations is little. Compared with retaining walls on hard earth foundations, the displacement mode of retaining walls on soft earth foundations is more likely to be rotation coupling with slides, and the softer the earth is, the more contribution to the total displacement of the retaining wall sliding displacement is.

5.3.3.2 Earth Pressure of Wall Back

The calculation of seismic earth pressure under seismic effects has always been taken as the core content of anti-seismic design of retaining walls. A lot of research works have been done on this subject, while there are still no satisfying results. Retaining walls and earth have constituted a nonlinear dynamic interactive system under seismic effects. The currently conducted theories and experiments have simplified the practical situation, and many important influencing factors are looking for a further study, such as nonlinearity of soil mass, permanent deformation of soil mass, seismic amplification coefficient, coupling effects of earth pressure and water pressure and residential strength of soil mass. Therefore, it is hard for the current theories to accurately and comprehensively describe the practical situation of damage mechanism and seismic earth pressure distribution. Thus, the study on seismic earth pressure still has a long way to go. This book, taking the shaking model test as a basis, mainly discusses the wall back earth pressure distribution under different working conditions and the changing rules of resultant force and resultant action points with seismic

Table 5.22 Total back earth pressure strength of retaining walls on hard earth foundations (kPa)

Distance of measured points to wall top (m)	Loading work conditions				
	0.1 g	0.2 g	0.4 g	0.7 g	0.9 g
0.30	1.34	1.98	1.81	2.25	1.93
0.55	3.18	3.43	3.50	2.92	2.85
0.80	3.12	3.82	6.72	6.72	11.21
1.05	2.71	2.98	4.51	6.41	9.44
1.30	7.54	6.65	7.34	12.21	19.45
1.55	6.74	6.01	5.00	8.59	10.04

Table 5.23 Total back earth pressure strength of retaining walls on soft earth foundations (kPa)

Distance of measured points to wall top (m)	Loading work conditions				
	0.1 g	0.2 g	0.4 g	0.7 g	0.9 g
0.30	0.05	0.04	0.03	0.03	0.03
0.55	4.99	4.90	3.86	5.25	4.56
0.80	6.46	6.87	7.64	9.72	12.48
1.05	4.64	5.01	9.13	11.12	13.88
1.30	3.94	4.18	7.84	18.63	36.31
1.55	5.34	6.29	6.76	16.79	36.87

intensity. It comparatively studies the seismic earth pressure features under different foundations, demonstrates, and analyzes the experiment results combining with seismic damage data of retaining walls in Wenchuan earthquake and anti-seismic codes of different countries, which further reveals the damage mechanism of retaining walls under seismic effects and provides solid foundation for the development of seismic earth pressure calculation theories and methods.

① Earth pressure distribution;

Tables 5.22, 5.23, 5.24, 5.25 and Figs. 5.55, 5.56 display the measured values of earth pressure strength and earth pressure distribution rules with different seismic motion accelerations.

It could be known by observing Figs. 5.55 and 5.56 that:

Firstly, the strength of wall back pressure under seismic effects is in nonlinear distribution along wall height, and the earth pressure strength is almost in proportion to seismic motion acceleration. From the aspect of the entire distribution rule of earth pressure, the bigger the acceleration is, the more likely the distribution shape of earth pressure approaches triangle.

Secondly, the changing speed of total earth pressure strength of all measured points is significantly influenced by retaining wall displacement. As displacement suddenly increases to the upper part of retaining walls, the earth pressure changes are slow, while displacement to the lower part of retaining walls is tiny, the earth

Table 5.24 Total dynamic back earth pressure strength of retaining walls on hard earth foundations (kPa)

Distance of measured points to wall top (m)	Loading work conditions				
	0.1 g	0.2 g	0.4 g	0.7 g	0.9 g
0.30	0.54	0.91	1.56	1.76	2.44
0.55	1.06	1.81	2.39	2.12	2.98
0.80	1.44	1.90	1.97	3.33	5.59
1.05	0.92	1.21	2.31	4.01	7.17
1.30	1.04	0.96	1.82	2.87	5.03
1.55	0.38	0.56	0.95	4.24	12.57

Table 5.25 Total dynamic back earth pressure strength of retaining walls on hard earth foundations (kPa)

Distance of measured points to wall top (m)	Loading work conditions				
	0.1 g	0.2 g	0.4 g	0.7 g	0.9 g
0.30	0.86	1.36	1.43	1.88	2.73
0.55	0.67	1.58	1.77	2.19	3.23
0.80	0.70	2.02	2.92	4.40	5.54
1.05	0.67	1.66	3.85	5.15	8.63
1.30	1.25	2.27	5.49	7.67	13.83
1.55	0.00	0.00	0.00	18.79	23.42

pressure increases quickly. There exists high residual stress region under 1/3*H* of wall height, and the closer it is to the wall foot residual stress region, the increasing amplitude of earth pressure becomes more obvious with seismic acceleration.

Thirdly, the earth pressure strength is related to the foundation conditions. Compared with retaining walls on hard soil foundations, under the same seismic acceleration, the earth pressure distribution area of retaining walls on soft soil foundations is bigger. In high-intensity region, the influence of foundation conditions on residual stress region is significant. When seismic acceleration is less than 0.4 g, the enlargement of residual stress region with seismic acceleration is not obvious; when seismic acceleration is over 0.4 g, the residual stress region suddenly enlarges with seismic acceleration. The softer the foundation earth is, the larger the residual stress region is.

② Resultant force of earth pressure;

For the convenience of conducting analysis to the earth pressure data, this book defines the following symbols: F is the total earth pressure, f is the dynamic earth pressure, K total = $F/\gamma H2$ and K dynamic = $f/\gamma H2$ are total earth pressure coefficient and dynamic earth pressure coefficient, η is the foundation influencing coefficient, η total = (F soft soil − F hard soil)/F hard soil, and η dynamic = (f soft soil − f hard

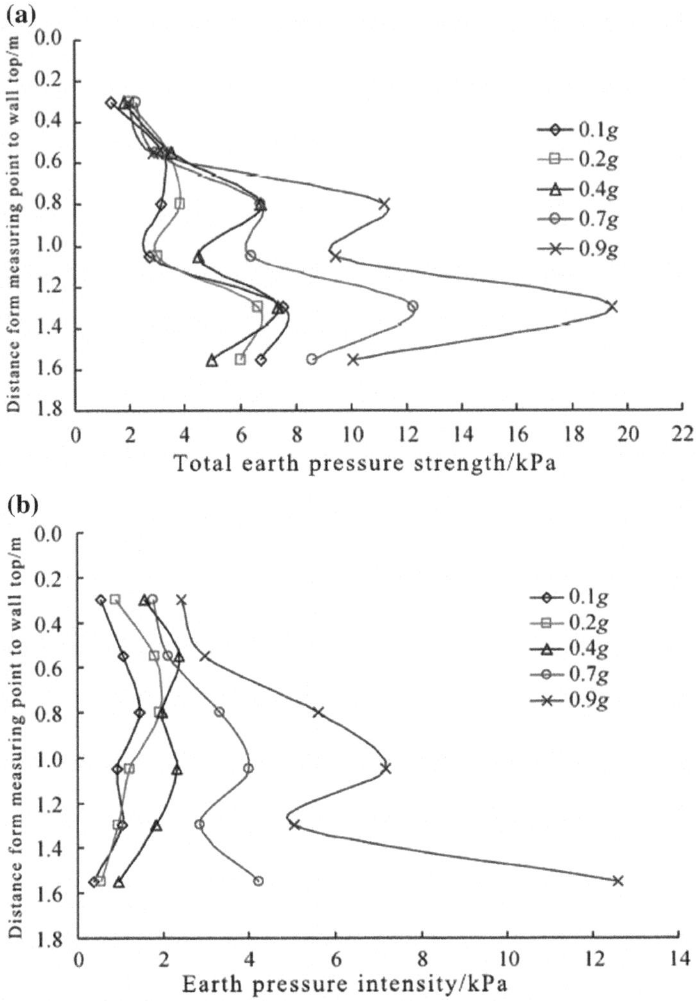

Fig. 5.55 **a** Distribution of total back pressure strength of retaining walls on hard soil foundation, **b** distribution of total dynamic back pressure strength of retaining walls on hard soil foundation

soil)/*f* hard soil, which represents the influence of different foundation conditions on earth pressure. Tables 5.26, 5.27 and Figs. 5.57, 5.58, 5.59, 5.60 show the changing rules of K_{total}, K_{dynamic} and η with seismic coefficient, and Fig. 5.60 shows the changing rules of total earth pressure coefficient, K_{total}, with displacement index δ (Tables 5.27 and 5.28).

It could come to the following conclusions by observing the curve changing rules:

Firstly, *F* and *f* are in proportion to seismic acceleration, and the increasing amplitude of earth pressure rises with seismic coefficient increase, which is as shown in Figs. 5.57 and 5.58.

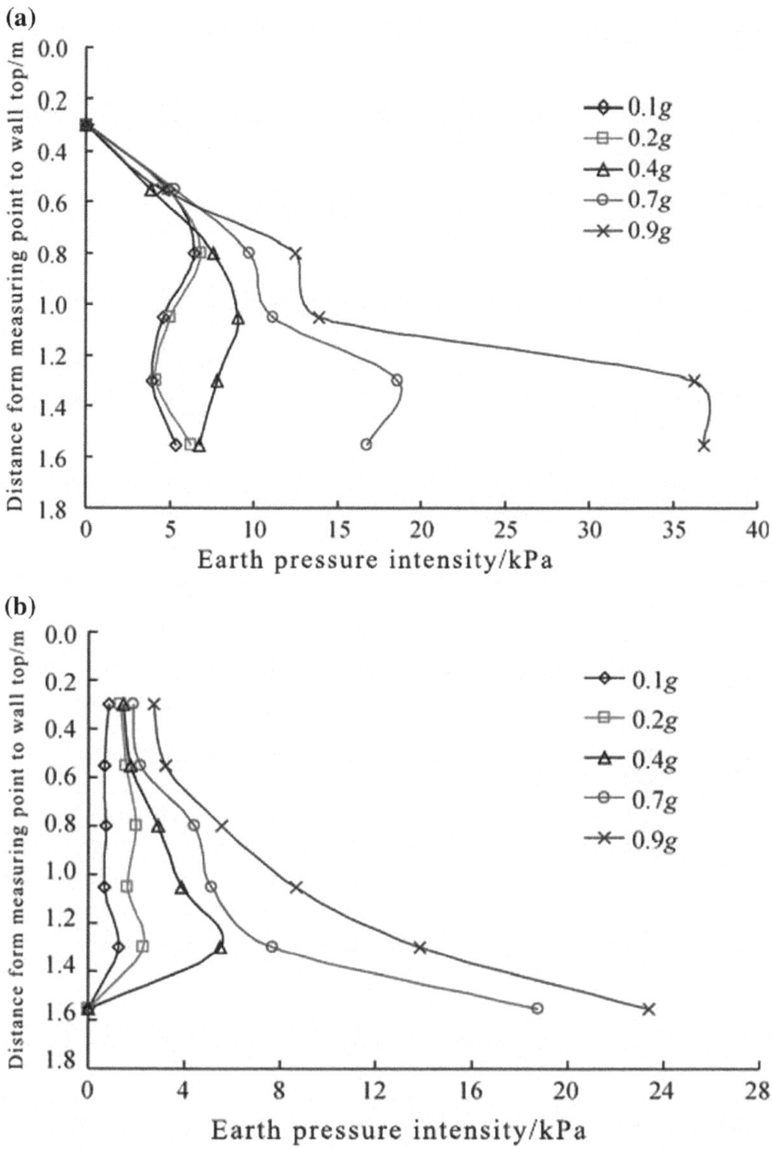

Fig. 5.56 **a** Distribution of total back pressure strength of retaining walls on soft soil foundation, **b** distribution of total dynamic back pressure strength of retaining walls on soft soil foundation

Secondly, F and f are affected by foundation conditions, and F and f of retaining walls on soft soil conditions are larger than that of retaining walls on hard soil conditions. The influence of foundation conditions on F is relevant to seismic inten-

Table 5.26 $F/\gamma H2$

Seismic coefficient	0.1	0.2	0.4	0.7	0.9
Hard soil foundation	0.126	0.130	0.155	0.206	0.293
Soft soil foundation	0.134	0.142	0.187	0.315	0.514

Table 5.27 $f/\gamma H2$

Seismic coefficient	0.1	0.2	0.4	0.7	0.9
Hard soil foundation	0.030	0.042	0.062	0.097	0.178
Soft soil foundation	0.024	0.051	0.090	0.188	0.277

Table 5.28 Foundation influencing coefficient η

Seismic coefficient	0.1	0.2	0.4	0.7	0.9
η_{total}	1.06	1.09	1.21	1.53	1.75
$\eta_{dynamic}$	0.81	1.24	1.45	1.95	1.56

sity, and when seismic intensity is small, the influence is little, while when seismic intensity increases, the influence also amplifies.

Figure 5.59 shows that when seismic coefficients are 0.1, 0.2, and 0.3, respectively, the corresponding η_{total} are 6.01, 9.35, and 15%, with average η_{total} near 10%, which demonstrates that when designing anti-seismic performance of retaining walls with seismic intensity under 8°, the influence of foundation conditions on earth pressure

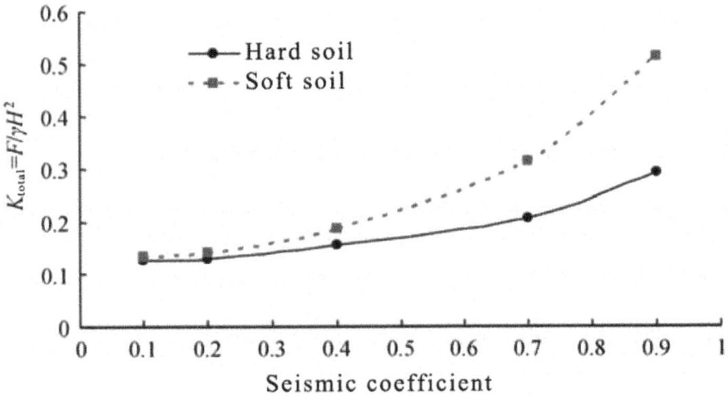

Fig. 5.57 Changing rules of K_{total} with seismic coefficient

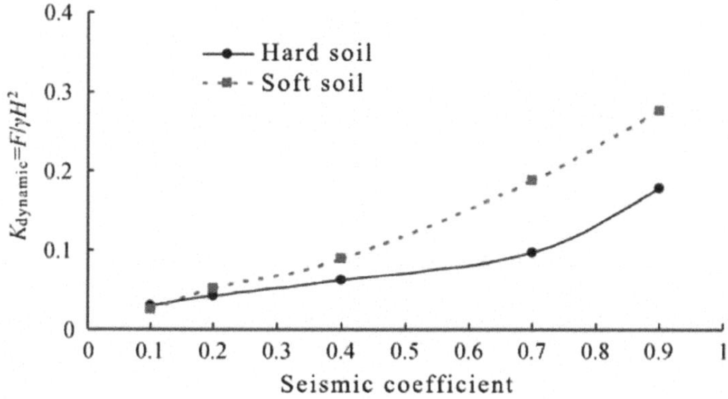

Fig. 5.58 Changing rules of K_{dynamic} with seismic coefficient

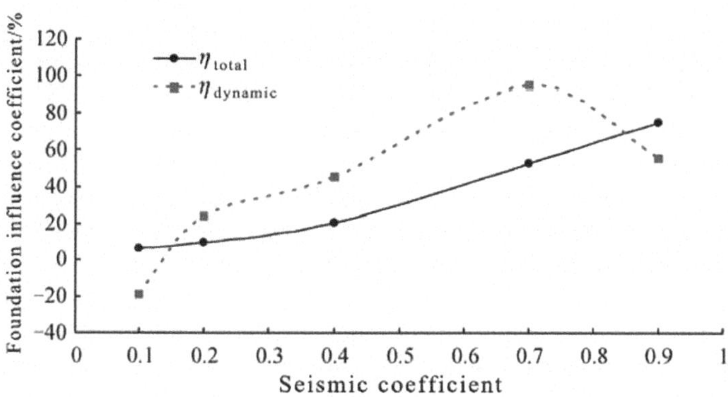

Fig. 5.59 Changing rules of η with seismic coefficient

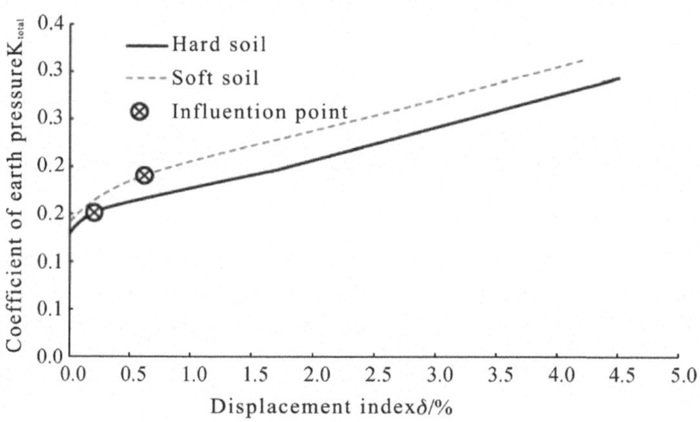

Fig. 5.60 Relation of K_{total} with δ

Table 5.29 Action points of earth pressure and resultant pressure of retaining walls on hard soil foundations

Seismic coefficient	0.1	0.2	0.4	0.7	0.9
F	0.34	0.38	0.39	0.34	0.33
f	0.46	0.49	0.48	0.40	0.35

Table 5.30 Action points of earth pressure and resultant pressure of retaining walls on soft soil foundations

Seismic coefficient	0.1	0.2	0.4	0.7	0.9
F	0.63	0.61	0.55	0.48	0.4
f	0.77	0.76	0.64	0.46	0.46

is so slight that it is within the acceptable range, so the earth pressure calculation could omit the influence of foundation conditions. With the increase in seismic accelerations, the influence of foundation conditions on earth pressure amplifies. When seismic coefficient is over 0.4, η_{total} is in linear increase with rise of seismic coefficient, and when seismic coefficient is 0.7 and 0.9, η_{total} are, respectively, 52.84 and 75.02%, which explains that for regions with seismic intensity under 8°, the earth pressure calculation of retaining walls should consider the influence of foundation conditions.

Thirdly, with the same displacement index, earth pressure of retaining walls on soft soil foundations is larger than that of retaining walls on hard soil foundations. At the preliminary period of displacement increase, the earth pressure increases linearly with the displacement when the latter exceeds a certain range.

Figure 5.60 shows that for retaining walls on hard soil foundations, the displacement index corresponding to the turning point of $K_{\text{total}} - \delta$ curve is nearly 0.3%, and when the displacement index is over 0.3%, K_{total} increases linearly with displacement index; for retaining walls on soft soil foundations, the displacement index corresponding to the turning point of $K_{\text{total}} - \delta$ curve is nearly 0.6%, and when the displacement index is over 0.6%, K_{total} increases linearly with displacement index.

④ Action points of earth pressure.

Tables 5.29 and 5.30, and Figs. 5.61 and 5.62, respectively, show the changing rules of action points of earth pressure, *h*/*H*, (*h* is the distance from the action point to wall heel, and *H* is wall height) with seismic coefficient under different loading conditions.

Combining the data in the figures and tables, it could come to the following conclusion:

① The resultant action point of *F* and *f* all decreases with the increase in seismic coefficient and that of *F* is lower than that of *f*. The bigger the seismic coefficient is, the closer the resultant action point of *F* and *f* are.

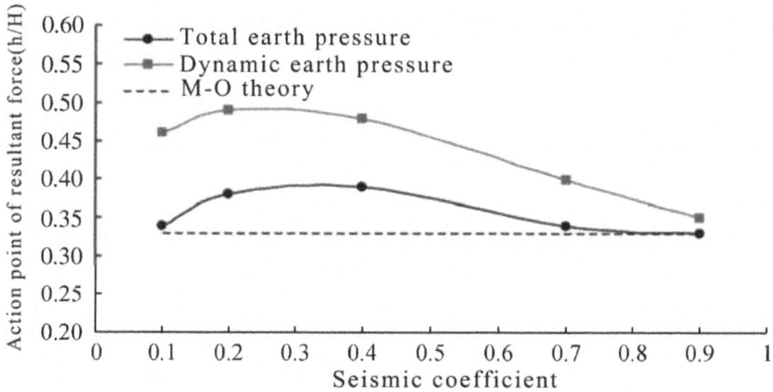

Fig. 5.61 Action points of earth pressure of retaining walls on hard soil foundations

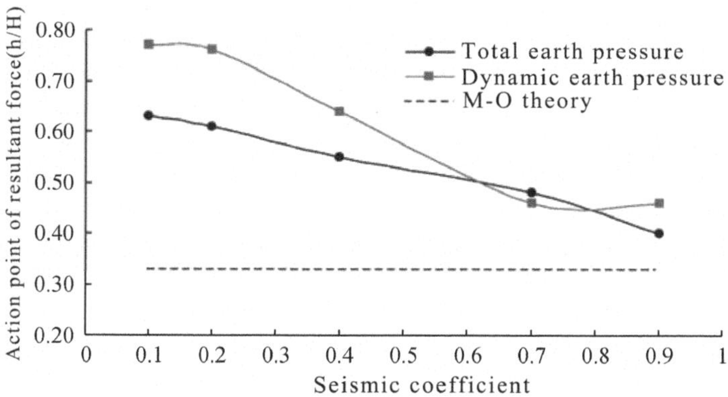

Fig. 5.62 Action points of earth pressure of retaining walls on soft soil foundations

② In general, when seismic coefficient is small, the measured resultant action points are higher than the 0.33H in M-O theory, and they decrease with the increase in seismic coefficient, while when seismic coefficient is big, the measured resultant action points near 0.33H.

③ The resultant action points are significantly affected by foundation conditions, and the softer the foundation soil is, the higher the resultant action points are. For retaining walls on hard soil foundations, the resultant action point of F is at 0.33H–0.39H, and the resultant action point of f is at 0.35H–0.49H; for retaining walls on soft soil foundations, the resultant action point of F is at 0.4H–0.63H, and the resultant action point of f is at 0.46H–0.77H.

5.3.3.3 Comparative Study of Measured Earth Pressure with Calculation Values by Seismic Codes of China, Europe, Japan, and New Zealand

The evaluation of design standards of existent seismic codes with shaking table test is of non-negligible engineering significance and scientific research value, and what is more is that it could help to deepen the recognition of seismic features and regularity of retaining structures in traffic engineering, to constantly complete the standards and specifications for seismic technology and solve the difficulties of it and to improve the seismic level and capability of transport infrastructures to the largest extent. Due to limits in technological standards and equipment, there are few large-scale shaking table test conducted in China for retaining walls in highly intensive regions, and due to lack of real seismic data of retaining walls in near field regions (with intensity of 9° or above), there are some blank fields for the existing seismic codes. The current seismic codes for highway and railway all omit the seismic design of retaining walls in regions with intensity of 9° or above. Seismic earth pressure has always been the core content of seismic design of retaining structures for the seismic codes of all countries. The difference of their calculation methods not only embodies the technical level of their country, such as seismic background, regional division, attenuation laws, but also relates to the earthquake fortification level and purpose, which, to a large extent, reflects the development of anti-seismic level of retaining structures. Therefore, this chapter will, on the basis of Chap. 3, verify and compare the earth pressure calculation results of retaining walls with intensity of 9° or below for seismic codes of four countries combining with the shaking table test results. Through comparison of the analysis results, it affirms and extends the calculation methods of seismic earth pressure in current seismic technological standards of China and provides supports for the improvement of seismic earth pressure designing framework of traditional retaining structures.

Table 5.31 and Fig. 5.63 show the comparing results of measured total earth pressure coefficients and that in seismic codes of different countries with different seismic coefficients. Table 5.32 and Fig. 5.64 show the difference between the calculated values and measured values of earth pressure by standards of China with different seismic intensity.

It could be known by observing Figs. 5.63 and 5.64 that:

Generally speaking, compared with other countries, the specified values of China are more close to the measured values.

For retaining walls on hard soil foundation, the specified seismic earth pressure calculated values are slightly higher than the measured values. When seismic intensity is 9° or below, the difference of specified calculated values to the measured values is in the range of 9.44–15.13%, with the average difference reaching 12%.

For retaining walls on soft soil foundation, the specified seismic earth pressure calculated values are slightly higher than the measured values for regions with seismic intensity of 8° or below, with average difference merely of 5%, which is in the acceptable error range in engineering design. When the seismic intensity reaches

Table 5.31 K_{total} of real measurement and in standards of different countries

Seismic coefficient	0.1	0.2	0.4
China	0.140	0.150	0.170
Europe—hard soil (kv↓)	0.190	0.220	0.280
Europe—hard soil (kv↑)	0.180	0.200	0.260
Europe—soft soil (kv↓)	0.194	0.224	0.298
Europe—soft soil(kv↑)	0.186	0.210	0.278
Japan	0.151	0.172	0.222
New Zealand	0.152	0.174	0.228
Measured value—hard soil	0.126	0.130	0.155
Measured value—soft soil	0.134	0.142	0.187

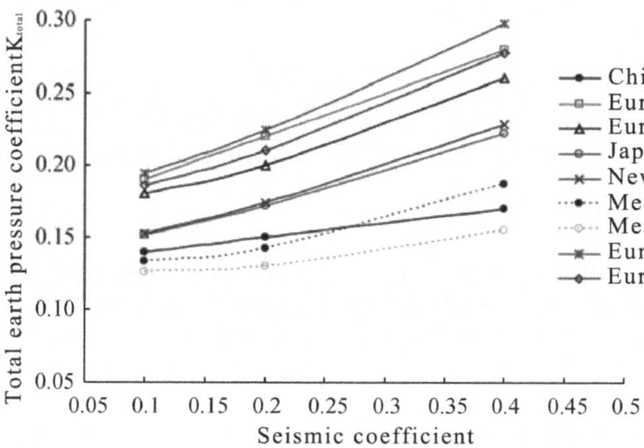

Fig. 5.63 Measured values and K_{total} by specified calculation

Table 5.32 Difference of measured values to specified values of China

Seismic coefficient	VII degree (%)	VIII degree (%)	IX degree (%)
Hard soil	10.98	15.13	9.44
Soft soil	4.69	5.29	−9.22

9°, the specified calculated values in China are obviously smaller than the measured values, with the difference reaching 9.22%.

The comparative results show that for retaining walls on hard soil foundation, the earth pressure calculation methods in existent seismic codes are relatively con-

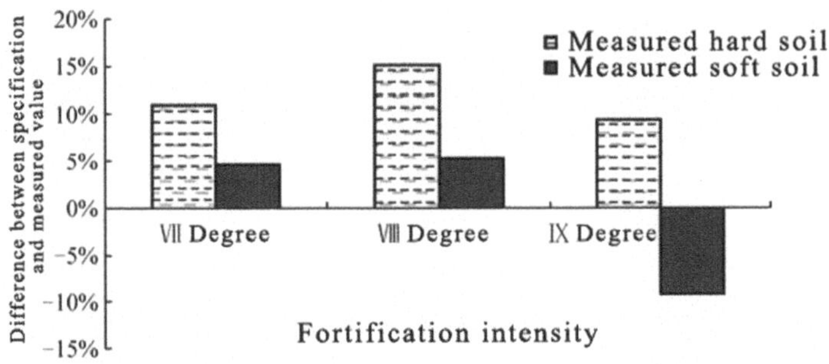

Fig. 5.64 Difference of measured values to earth pressure in codes of China

servative. For retaining walls on soft soil foundation, the earth pressure calculation methods in existent seismic codes are reasonable when seismic intensity is under 8° or below, while when seismic intensity is over 8°, the existent calculation methods are relatively dangerous. The above show that it should consider the influence of foundation conditions when using the existent specifications to calculate seismic earth pressure.

5.4 Time–Frequency Analysis Theory of Seismic Stability of Rigid Retaining Structures

The above field research results and shaking table test results fully explain that the rigid retaining walls are most widely applied in highly intensive regions of China and objectively reveal the current studies of seismic design of rigid retaining walls are still not mature enough. The core of seismic design of retaining walls is the earth pressure calculation. Up to now, the calculation methods of seismic earth pressure of rigid retaining walls could be roughly divided into three kinds: The first is limit equilibrium method, i.e. supposing that the relative movement of retaining walls and wall backfilling is large enough, the stress–strain of soil mass will reach its limit or damage state, and the representative results of this are the famous Mononobe–Okabe seismic earth pressure theory; the second is coordinated deformation method, i.e. considering the interaction of wall–soil and the features of real stress–strain of soil mass, for example, the work done by Scott, Jain and Scoot, Alampali, Elgamal, Siller, Veletsos, and Younan, and Richards, etc.; the third is finite element method, for example, the finite element method built by Clough and Duncan which considers the real stress–strain relation of soil mass. For these three methods, limit equilibrium method could only consider the influence of PGA on seismic earth pressure of retaining walls, but could not consider the influence of seismic wave frequency and duration; coordinated deformation method could not consider the influence of seismic wave frequency and duration, and its calculation accuracy is low in practical engineering, which is not convenient to use; although finite element method could consider the influence of seismic wave amplitude, frequency, and duration, it is complex to operate, especially for the contact surface part of wall–soil. However, a large amount of research results show that the duration and frequency of seismic wave exert significant influence on the seismic earth pressure of rigid retaining walls. Thus, it is quite necessary to conduct a study on time–frequency calculation methods of active earth pressure of rigid retaining walls under seismic effects.

According to the seismic damage research results of Wenchuan earthquake, it can be known that SV wave (elastic wave whose vibration direction and propagation direction are vertical to the slope cross section) is the most disastrous and causes the most serious damage. But SV wave is a kind of complex non-stationary signal, whose frequency feature changes with time, which causes lots of limits for the application of limit equilibrium method and coordinated deformation method in this aspect. Therefore, this book will, based on elastic wave theory and horizontal slice method, generalize the dynamic analysis models of rigid retaining walls, build force equilibrium differential equation of micro-unit, and use Hilbert–Huang Transform (HHT) to propose the time–frequency calculation methods of seismic active earth pressure of rigid retaining walls under the effects of SV wave. It concretely explains and verifies the application and correctness of this method by conducting shaking model test to the rigid retaining walls and then, through carrying out parameter study, deeply analyzes the superiority of time–frequency calculation method.

5.4.1 Time–Frequency Analysis Theory of Seismic Active Earth Pressure of Rigid Retaining Structures

For the above discussed problems of limit equilibrium method and coordinated deformation method, in order to fully consider the influence of three major factors of seismic motion (amplitude, frequency and duration) on the seismic active earth pressure of rigid retaining walls, this book proposes the time–frequency calculation method of seismic active earth pressure of rigid retaining walls with horizontal slice method and HHT. The derivation process is as follows: First, generalize the analysis model, conduct horizontal slice to the retaining structures with horizontal slice method, and solve the wave power generated in micro-units of each layer by seismic waves with HHT and elastic wave theory; second, build the force equilibrium differential equation of micro-units and further build the differential equation of seismic active earth pressure strength; at last, solve the theoretical solutions of seismic active earth pressure strength, resultant force, and action points. It is noteworthy that the nature of inertia force generated by soil mass under seismic effects is a simplified quantitative presentation of seismic waveform in soil mass. Therefore, this book takes the horizontal seismic inertia force as equivalent to the wave force generated by seismic waves in soil mass.

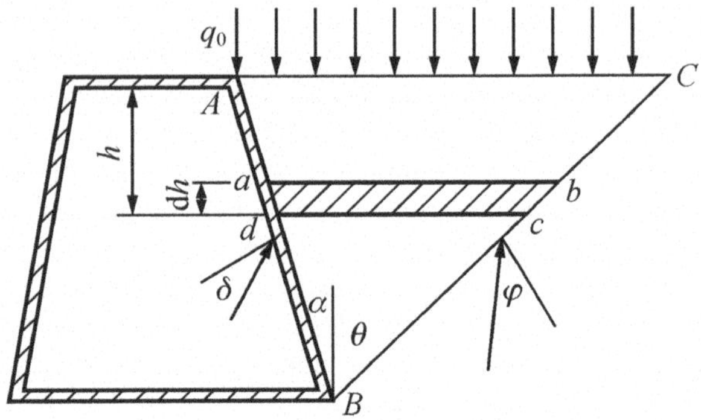

Fig. 5.65 Stress analysis model of rigid retaining walls

5.4.1.1 Formula Derivation of Time–Frequency Calculation Methods of Seismic Active Earth Pressure of Rigid Retaining Walls

① Basic assumption;

The derivation of time–frequency calculation methods of seismic active earth pressure of rigid retaining walls is based on the following assumption: ① the retaining structures are rigid body; ② the back wall filling is single, homogeneous, and isotropic; ③ when wall body drifts forward or backward, sliding soil wedge behind walls comes across slides at the surface between wall back and wall heel; ④ the seismic effects do not influence the fundamental mechanic performance of soil mass; ⑤ the material dump inside the filling behind walls consumes a large amount of seismic wave energy, which will finally cause the reflected seismic wave on the upper surface of filling bringing very weak seismic energy to propagate downward, so the formula derivation in this book will not consider the influence of reflected seismic wave on the upper surface of filling on micro-units.

② Generalized analysis model;

For rigid retaining walls, generalize the calculation mode of seismic active earth pressure, as shown in Fig. 5.65. Then, conduct horizontal slice to filling behind walls with horizontal slice method and select micro-units to study, as shown in Fig. 5.66. The parameters in Figs. 5.65 and 5.66 are specified as follows: The wall back dip angle of retaining walls is α, the internal friction angle is φ, the external friction angle δ, uniform overloading $q0$, active earth pressure strength Pa, soil strips self-weight dw, soil strips thickness dh, and q and $q + dq$ are, respectively, resultant of bilateral interslice forces of horizontal micro-units; r is the resultant of normal force and tangential force of sliding surface; FEH is the horizontal inertia force acted on soil slice. Figure 5.67 is the stress diagram of wave force acted on horizontal soil slices. It is noteworthy that the inertia force acted on the micro-units in this book is replaced by the wave force of seismic waves.

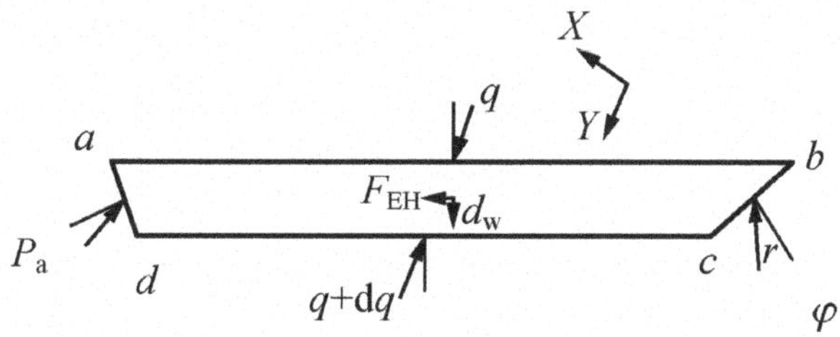

Fig. 5.66 Stress analysis model of horizontal micro-units

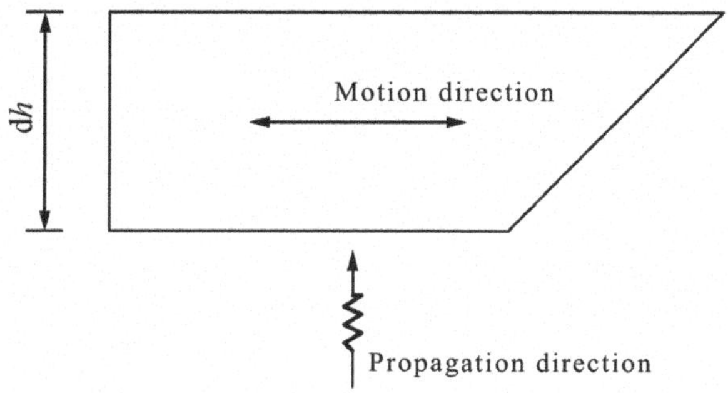

Fig. 5.67 Model of wave force acted on soil slices

③ Calculation of seismic active earth pressure strength.

By Fig. 5.65, it could be known that the geometric relations among different sizes of models are as follows:

$$\begin{cases} \overline{AB} = H/\cos\alpha;\ \overline{AC} = H\sin(\alpha+\theta)/\cos\alpha;\ \overline{BC} = H\cos\alpha/\cos\alpha\cos\theta \\ \overline{ad} = dh/\cos\alpha;\ \overline{ab} = (H-h)\sin(\alpha+\theta)/\cos\alpha\cos\theta \\ \overline{cd} = (H-h-dh)\sin(\alpha+\theta)/\cos\alpha\cos\theta\ \overline{bc} = dh/\cos\theta \end{cases} \tag{5.1}$$

The self-weight (neglecting second-order infinitesimal) of horizontal micro-unit *abcd* is:

$$dw = \frac{1}{2}\gamma S_{abcd} = \frac{1}{2}\gamma(\overline{ab}+\overline{cd})dh = \gamma(H-h)\frac{\sin(\alpha+\theta)}{\cos\theta\cos\alpha}dh \tag{5.2}$$

Select horizontal micro-unit *abcd* to build force equilibrium equation. Conduct stress analysis to horizontal micro-unit *abcd* according to the features of active earth pressure, as shown in Fig. 5.66.

(1) Build force equilibrium equation of *X* direction

$$-P_a\overline{ad}\cos(\alpha+\delta+\eta) + \gamma\overline{bc}\cos(\theta+\varphi-\eta) = 0 \tag{5.3}$$

(2) Build force equilibrium equation of Y direction

$$\begin{aligned} &-P_a\overline{ad}\sin(\alpha+\delta+\eta)+q\overline{ab}-(q+dq)\overline{cd}+dw/\cos\eta \\ &-\gamma\overline{bc}\sin(\theta+\varphi-\eta)=0 \end{aligned} \tag{5.4}$$

(3) Select moment balance for the midpoint of edge bc, and take counterclockwise direction as positive

$$\begin{aligned} &-P_a\overline{ad}\sin(\alpha+\delta)\frac{1}{2}(\overline{ab}+\overline{cd})+q\overline{ab}\left[\frac{1}{2}\overline{ab}\cos\eta-\frac{1}{2}\overline{bc}\sin(\theta-\eta)\right] \\ &-(q+dq)\overline{cd}\left[\frac{1}{2}\overline{cd}\cos\eta+\frac{1}{2}\overline{bc}\sin(\theta-\eta)\right]+\frac{dw}{\cos\eta}\cdot\frac{1}{4}\cdot(\overline{ab}+\overline{cd})\cos\eta=0 \end{aligned} \tag{5.5}$$

The potential function of all SV waves could be uniformly represented by elastic displacement. The details are as follows:

$$u(z,t)=U(z)\mathrm{e}^{i\omega t} \tag{5.6}$$

In the formula, $u(z, t)$ is the displacement function of micro-units; $U(z)$ is the displacement amplitude of micro-units; ω is the vibrating frequency of micro-units. Therefore, the inertia force generated by SV waves in micro-units is:

$$F_{EH}=\left|\mathrm{d}w\cdot\frac{\partial^2u}{\partial t^2}/g\right|=\mathrm{d}w\cdot\omega^2U(z)/g \tag{5.7}$$

Thus, the seismic angle η satisfies:

$$\tan\eta=F_{EH}/\mathrm{d}w=\omega^2U(z)/g \tag{5.8}$$

Substituting Formula (5.1) and (5.2) into (5.3)–(5.5), by simultaneous solution, it could get (neglecting second-order infinitesimal):

$$P_a=n_{a1}q \tag{5.9}$$

In the formula:

$$n_{a1}=\frac{\cos(\theta+\varphi-\eta)}{\cos\theta}\cdot\frac{\cos\theta\sin(\alpha+\eta)-\cos\alpha\sin(\theta-\eta)}{\sin(\alpha+\delta)\cos(\theta+\varphi-\eta)-\sin(\theta+\varphi)\cos(\alpha+\delta+\eta)}$$

Substituting Formula (5.9), (5.1) and (5.2) into Formula (5.4), through simplification, it can get (neglecting second-order infinitesimal):

$$\frac{\mathrm{d}q}{\mathrm{d}h}-\frac{A_1}{H-h}q=\frac{1}{H-h}\gamma \tag{5.10}$$

$$A_1 = 1 - \frac{\cos(\theta + \varphi - \eta)}{\cos\theta} \cdot \frac{\cos\theta\sin(\alpha + \eta) - \cos\alpha\sin(\theta - \eta)}{\sin(\alpha + \delta)\cos(\theta + \varphi - \eta) - \sin(\theta + \varphi)\cos(\alpha + \delta + \eta)} \cdot \frac{\cos\theta\sin(\alpha + \theta + \varphi + \delta)}{\cos(\theta + \varphi - \eta)\sin(\alpha + \theta)} \tag{5.11}$$

Using boundary condition: when $h = 0$, $q = q1/\cos\,\eta$ to solve Formula (5.10), it can get:

$$q = \left(\frac{\gamma}{\cos\eta}\frac{H}{1 + A_1} + \frac{q_1}{\cos\eta}\right)\left(\frac{H}{H - h}\right)^{A_1} - \frac{\gamma}{\cos\eta}\frac{H - h}{1 + A_1} \tag{5.12}$$

Substituting Formula (5.12) into (5.9), it can get the seismic active earth pressure strength is:

$$\begin{cases} P_a = m_{a1}(H - h)^{-A1} - m_{a2}(H - h) \\ m_{a1} = n_{a1}\left(\gamma\frac{1}{\cos\eta}\frac{H}{1+A_1} + \frac{q_1}{\cos\eta}\right)\cdot H^{A_1} \\ m_{a2} = n_{a1}\gamma\frac{1}{\cos\eta}\frac{1}{1+A_1} \end{cases} \tag{5.13}$$

④ Resultant force and action points of the seismic active earth pressure;

The resultant force of active earth pressure is:

$$E_a = \int_{o}^{H} \frac{P_a}{\cos\alpha}dh = \frac{1}{\cos\alpha}\left[\frac{m_{a1}H^{1-A_1}}{1 - A_1} - \frac{1}{2}m_{a2}H^2\right] \tag{5.14}$$

The distance of action point of resultant force of active earth pressure to wall foot is:

$$Z_{0a} = \frac{\int_0^H \frac{P_a(H-h)}{\cos\alpha}dh}{\int_0^H \frac{P_a}{\cos\alpha}dh} = \frac{\frac{m_{a1}H^{-A_1}}{2-A_1} - \frac{m_{a2}H}{3}}{\frac{m_{a1}H^{-A_1}}{1-A_1} - \frac{m_{a2}H}{2}}H \tag{5.15}$$

In summary, substitute n_{a1}, n_{a2}, m_{a1}, A_1, q_1 into Formula (5.16), and through simplification, they all can be uniformly written into the following form:

$$E_a = \frac{\gamma H^2}{2\cos\eta}\frac{\sin(\alpha + \theta)\cos(\theta + \varphi - \eta)}{\cos\alpha\;\cos\theta\sin(\alpha + \theta + \varphi + \delta)} + \frac{q_0 H}{\cos\eta}\frac{\sin(\alpha + \theta)\cos(\theta + \varphi - \eta)}{\cos\alpha\;\cos\theta\sin(\alpha + \theta + \varphi + \delta)} \tag{5.16}$$

In which, $\eta = \arctan(F_{EH}/dw) = \arctan(\omega^2 U(z)/g)$

⑤ Critical rupture angle of seismic active earth pressure.

According to the basic principle of seismic active earth pressure, for all the possible critical rupture angle θ, there is just one dip angle θ_{acr} that could lead to the maximum earth pressure. For Formula (5.16) it can be known that to solve the θ_{acr} that causes the maximum resultant earth pressure only if with $dEa/d\theta = 0$, but it generally cannot get the displayed solution in this way. Therefore, this book refers to the graphic methods in the references to solve the calculation formula for the critical rupture angle of seismic active earth pressure, and the details could be seen in Formula (5.17):

$$\tan\theta_{acr} = \frac{\sqrt{\frac{\sin(\alpha+\delta)\cos(\alpha+\delta+\eta)}{\cos\alpha\sin(\varphi-\eta)}} - \cos(\alpha+\delta)\sin(\alpha+\varphi+\delta)}{\cos(\alpha+\delta)\cos(\alpha+\varphi+\delta)} \tag{5.17}$$

Through the comprehensive analysis of Formula (5.17), it could be known that the solution of the critical rupture angle of seismic active earth pressure shall meet the following condition: $\alpha + \delta + \eta \leq 90°$, and $\varphi \geq \eta$, which is identical with the limit condition of Mononobe–Okabe formula.

5.4.1.2 Analysis of Seismic Wave Time–Frequency Effects

Based on the above derivation, it can be known that the time–frequency effects of seismic waves directly embody in the elastic displacement amplitude $U(z)$ and frequency ω of input seismic waves and indirectly embody in the solution of seismic angle in formula of active earth pressure. Therefore, this book proposes the following analysis process of seismic wave time–frequency effects: Firstly, break up seismic wave time–frequency into several of IMF (time history curve) through Empirical Mode Decomposition (EMD for short); secondly, carry out Channel Switch, i.e. switch multi-channel signal composed of multiple IMF into single-channel signal made up of single IMF; thirdly, solve the instant frequency of each channel, and draw the time–frequency curve of each IMF. Fourthly, bring each IMF and their instant frequency into Formula (5.6)–(5.8) to work out the results and superimpose the results, so we can get the seismic angle of rigid retaining walls under seismic effects. Lastly, substitute the relevant parameters into Formula (5.17), (5.13), (5.16) and (5.15) to, respectively, solve the critical rupture angle, strength, resultant force, and action points of seismic active earth pressure.

Based on the above discussion, the time–frequency calculation methods of seismic active earth pressure of rigid retaining walls derived by this book could well complement the shortcomings of limit equilibrium method and coordinated deformation method and could take good consideration of the influence of three major factors of seismic motion (amplitude, frequency, and duration) on the seismic active earth pressure of rigid retaining walls.

5.4.1.3 Practical Application of Seismic Wave Time–Frequency Effects

In order to more deeply and concretely explain the practical application of seismic wave time–frequency effects, this dissertation will, based on the analysis ideas of

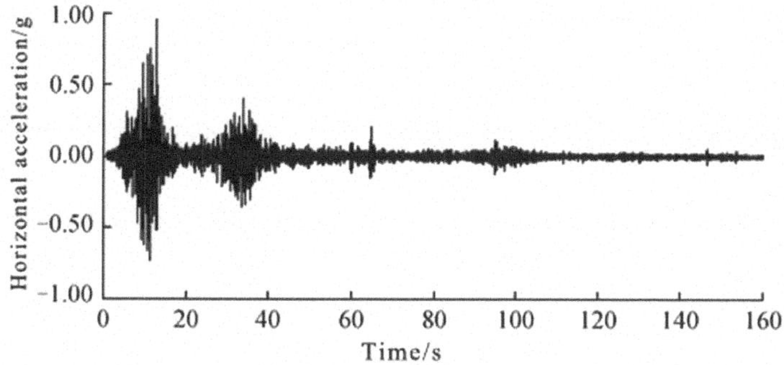

Fig. 5.68 Time-history curve of Wenchuan wolong seismic wave

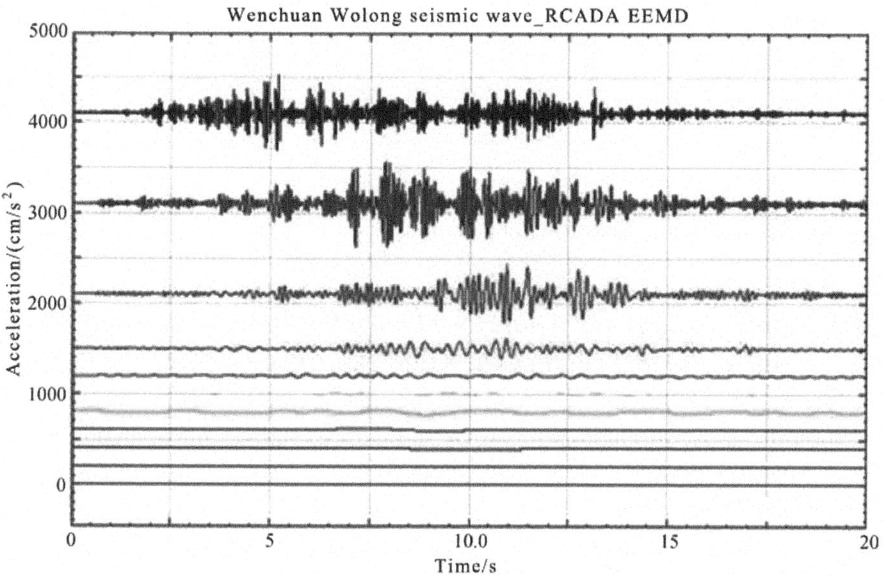

Fig. 5.69 Acceleration time-history curve of IMF

seismic wave time–frequency effects, conduct input wave time–frequency analysis and take the example of Wenchuan wolong seismic wave to explain. The details are as follows: Firstly, input the Wenchuan wolong seismic wave (see Fig. 5.68); secondly, carry out EEMD decomposition and get every IMF (see Fig. 5.69); lastly, solve the instant frequency of every IMF (see Fig. 5.5).

It is noteworthy that, due to that the duration of Wenchuan wolong seismic wave is a bit longer, so Figs. 5.69 and 5.70 only select calculation results of $T = 0$–20 s including peak seismic ground motion acceleration and demonstrate it clearly.

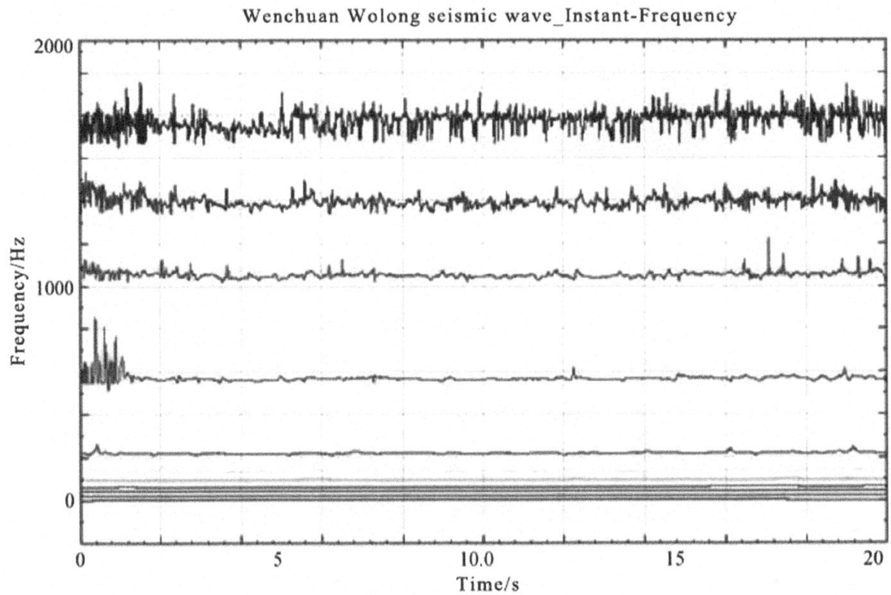

Fig. 5.70 Frequency time-history curve of IMF

5.4.2 Time–Frequency Analysis Theory Solution of Seismic Active Earth Pressure on Rigid Retaining Walls

Based on the above discussion, this book summarizes the solving approach of the time–frequency calculation methods of seismic active earth pressure of rigid retaining walls, which is as follows:

(1) Generalize the analysis model with a geometric shape, filling features, and foundation conditions of rigid retaining walls.
(2) Decide the design response spectrum curve of this region according to *Code for Seismic Design of Buildings*, and synthesize artificial ground motion time history with seismic ground motion time history fitting method so as to decide the seismic ground motion acceleration time history of this region.
(3) Decide time history curve of displacement based on seismic ground motion acceleration time history curve; work out IMF of seismic waves and the correspondent frequency time history curve with HHT.
(4) Substitute the above results into Formula (5.6) and (5.8) for solution, and the critical rupture angle of seismic active earth pressure of filling behind rigid retaining walls could be decided.
(5) Substitute the critical rupture angle and the relevant parameters into Formula (5.17), (5.13), (5.16) and (5.15) to, respectively, solve the critical rupture angle, strength, resultant force, and action points of seismic active earth pressure.

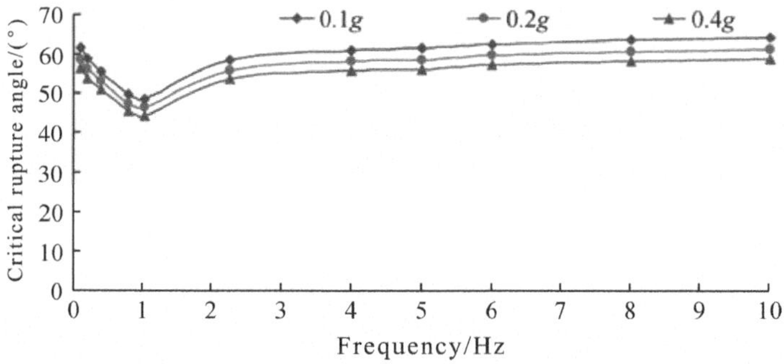

Fig. 5.71 Relation between the critical rupture angle and input wave frequencies

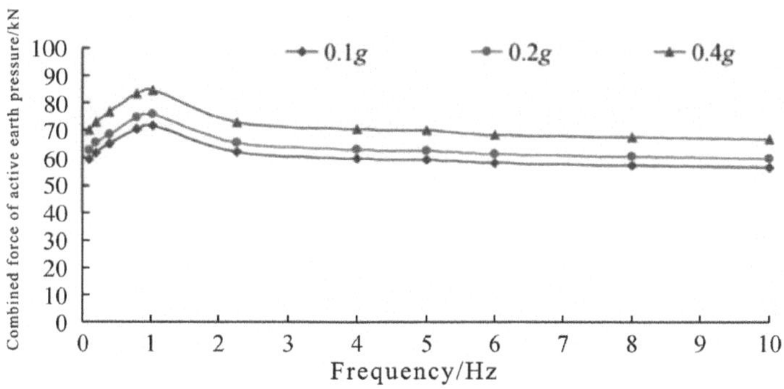

Fig. 5.72 Relation between the resultant force of seismic active earth pressure and input wave frequencies

5.4.3 Parameter Studies of Time–Frequency Analysis Theory of Seismic Active Earth Pressure on Rigid Retaining Structures

For the time–frequency calculation methods of seismic active earth pressure of rigid retaining walls, it conducts experiments of the influence of different loading frequencies on the critical rupture angle, strength, resultant force, and action points of seismic active earth pressure. The concrete samples are as follows: the rigid retaining wall height $H = 5$ m, unit weight of filling $\gamma = 17$ kN/m^3, cohesive force $c = 0$, the internal friction angle $\varphi = 33°$, the wall back is vertical, and the filling surface is horizontal and is without overload. Studying sine waves with PGA = 0.1, 0.2, and 0.4 g, and the frequency changes of sine waves are 0.1, 0.2, 0.4, 0.8, 1.0, 2, 4, 5, 6, 8, and 10 Hz. The detailed calculation results are as shown in Figs. 5.71, 5.72, 5.73, and 5.74.

Through the comprehensive analysis of Figs. 5.71, 5.72, 5.73, and 5.74, it can come to the following rules: (1) with the increase in seismic intensity, resultant

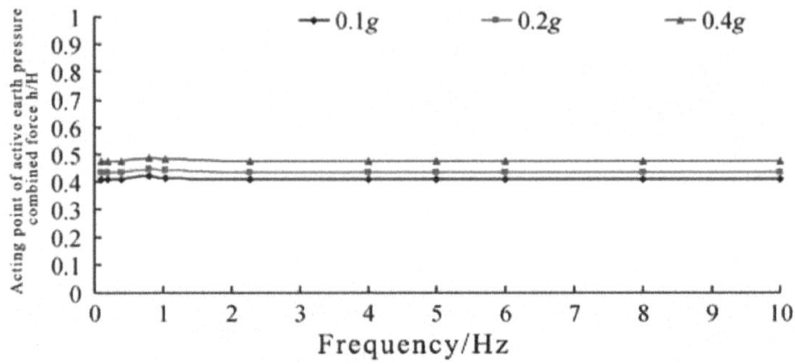

Fig. 5.73 Relation between resultant force action points and input wave frequencies

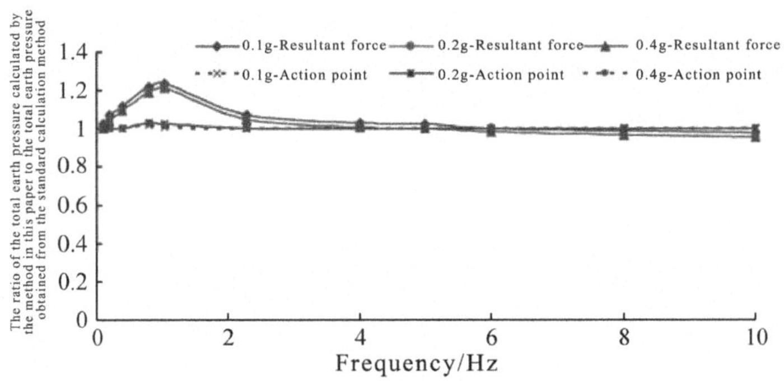

Fig. 5.74 Comparison of resultant force action points with standardized calculation results

force of seismic active earth pressure gradually augments, the critical rupture angle gradually decreases and its position gradually moves up under resultant force; (2) with the increase in input seismic wave frequency, the critical rupture angle and resultant force of seismic active earth pressure, respectively, distribute in "inverted saddle shape" and "saddle shape," and both of them reach their maximum value with $f = 1$ Hz. The reason for the above phenomenon may be that when the natural frequency f of this rigid retaining wall system is 1.32 Hz, the input wave frequency is 1 Hz, and the retaining walls come across resonance, which intensifies the seismic response of rigid retaining wall system and causes a decrease in the critical rupture angle and increase in resultant force of wall back seismic active earth pressure. (3) With the increase in input seismic wave frequency, the action points of resultant force of wall back seismic active earth pressure stay unchanged and are with little fluctuation. The reason may be that the seismic wave frequency mainly influences the wall back pressure level and slightly influences its distribution, which causes the position of action points basically stay unchanged with the input seismic wave frequency. (4) For rigid retaining walls, with different seismic intensity and input wave frequency, the ratio of resultant force of seismic active earth pressure calculated with the proposed method in this book with calculation results by the standardized method is in the range of 1.0–1.3, and the position of action points is basically the same, which fully explains that, by the existing codes that only pays attention to PGA but not input wave frequency, conducting seismic stability design to rigid retaining walls may reduce the seismic safety reserves of retaining walls.

With the above analysis, it can be known that input seismic wave exerts large influence on the seismic stability of rigid retaining walls, which should be taken into consideration of the anti-seismic design of rigid retaining walls.

5.4.4 Verification of Shaking Table Test Results of Time–Frequency Analysis Theory of Seismic Active Earth Pressure on Rigid Retaining Structures

To verify the correctness of time–frequency calculation methods for seismic active earth pressure of rigid retaining walls, our research group takes the rigid retaining structure with a height of 9.6 m on G213 as the prototype and carries out large-scale shaking table test according to the similarity system. The geometric similarity ratio adopted by the shaking table test is 1:6. Input the pressure wave of Wenchuan wolong seismic waves with time similarity ratio. The similarity materials are made up of quartz sand, barite powder, clay or sand and water with different proportions, and their volume similarity ratio is 1:1. The similarity ratio of internal friction angle and Poisson's ratio are all 1:1. The shaking table test adopts top end-opening rigid model containers made of steel plates plus steel section plus organic glass, with inner-space size of 3.7 m × 1.5 m × 2.1 m (length × width × height). Styrofoam with a thickness of 30 mm is lined to back wall of filling in vibration direction to simulate the absorbing materials so as to reduce reflection of seismic waves on the border. At the same time, it can get that, with the similarity system, the size of test retaining wall model is 1.6 m (height) × 1.5 m (width), with wall top width of 0.33 m, wall root width of 0.55 m, wall toe height of 0.204 m, and wall toe width of 0.102 m. The model is made of micro-concrete, and the filling material is made of dry sand. After subgrade model filling is loaded to the model container, it needs stratified compaction to ensure the compaction degree of backfilling foundation soil and the filling behind walls are identical with entitative engineering. In order to more accurately observe the distribution of earth pressure sensors of retaining walls, two rows of sensors are set up along wall height with six in each row. One row is set up with strain-type earth pressure sensors, and the other is set up with piezoelectric earth pressure sensors. The concrete physico-mechanical parameters are shown in Table 5.33, the shaking table test model is shown in Fig. 5.75, the distribution of sensors is shown in Fig. 5.76, and the pressure wave of Wenhcuan wolong seismic wave is shown in Fig. 5.77.

Based on the solving process of time–frequency calculation method of seismic active earth pressure of rigid retaining walls, this book will substitute the relevant parameters into Formula (5.15) and (5.16) to solve the resultant force and action points of seismic active earth pressure under the effects of Wenchuan wolong seismic

Table 5.33 Comparison of time–frequency calculation methods of resultant force and action points of seismic active earth pressure with shaking table test results

	Loading work condition (resultant force) (kN)				Loading work condition (resultant force) (h/H)			
	0.1 g	0.2 g	0.4 g	0.7 g	0.1 g	0.2 g	0.4 g	0.7 g
Test results	5.48	5.65	6.75	8.97	0.34	0.38	0.39	0.34
Calculation Results	5.16	5.25	6.19	8.22	0.33	0.355	0.36	0.37
Error (%)	5.84	7.08	8.30	8.36	2.9	6.6	7.7	-8.8

Note Error = (test results-calculation results)/test results * 100%

Fig. 5.75 Model of shaking table test

wave with PGA of 0.1, 0.2, 0.4, and 0.7 g. The concrete calculation results and shaking table test results could be seen in Table 5.33.

Through the comprehensive analysis of two tables, it could be known that the maximum error of resultant force and action points of seismic active earth pressure on rigid retaining walls are, respectively, 8.36 and −8.8%, and their minimum error are, respectively, 5.84 and 2.9%. All errors are less than 10%. Therefore, the time–frequency calculation results of the resultant force and action points of seismic active earth pressure on rigid retaining walls are basically identical with the shaking table test results, which fully explain that the calculation methods of seismic active earth pressure on rigid retaining walls are correct and credible.

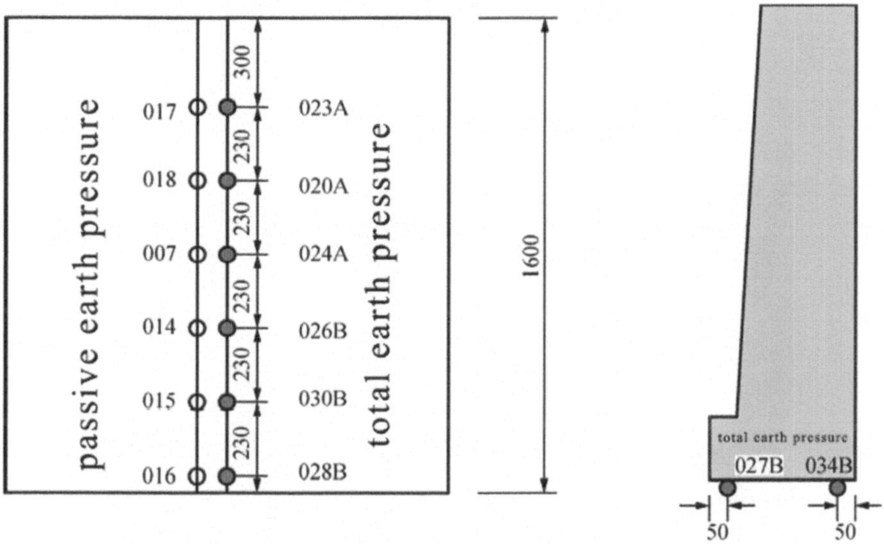

Fig. 5.76 Distribution of earth pressure sensors

Injection:1.Unit:mm.
2.A front elevation view of the wall is in the left and the right side is a side view.
3. ● Strain earth pressure gauge
○ Piezoelectric earth pressure gauge

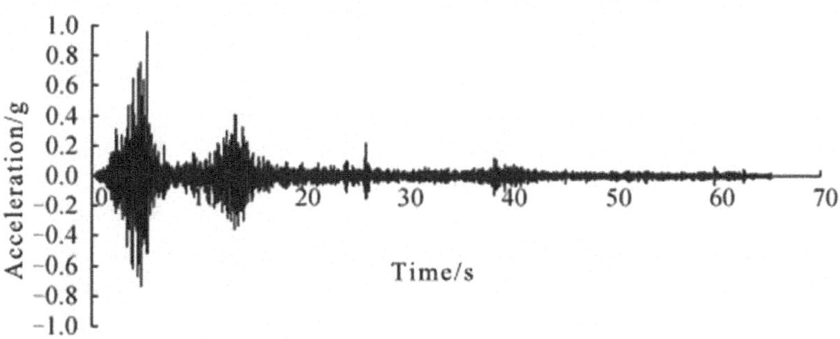

Fig. 5.77 Pressure wave of Wenchuan wolong seismic wave

5.5 Time–Frequency Analysis Theory of Seismic Stability of Reinforced Retaining Structures

In 5.12 Wenchuan earthquake, the performance of reinforced retaining walls was good on the whole, with only one coming across fracture of rigid plastic composite reinforced belt, which caused local damage to retaining walls (see Fig. 5.78). The result fully shows that the reinforced retaining walls are suitable to be applied in

Fig. 5.78 Physico-mechanical parameters of substrate and filling behind walls

seismic regions with high intensity. The usually adopted analysis methods for seismic stability of reinforced retaining walls are pseudo-static method, limit displacement method, and numerical analysis method, in which pseudo-static method could only consider the influence of PGA on seismic stability of retaining walls, but not the influence of seismic wave frequency and duration; limit displacement method could not consider the influence of seismic wave frequency and duration, and its calculation accuracy is low in practical engineering, which is not convenient to use.

Figure 5.78 Damage examples of reinforced retaining walls. Although numerical analysis method could carry out time–frequency analysis to seismic stability of reinforced retaining walls under seismic effects, it is complex in operation, especially for parts of lacing wire–soil contact surface, lacing wire–plate contact surface, and plate-filling contact surface. However, a large amount of research results show that the duration and frequency of seismic wave exert significant influence on the seismic stability of reinforced retaining walls. Thus, it is quite necessary to conduct study on time–frequency analysis methods of seismic stability of reinforced retaining walls under seismic effects.

According to the seismic damage research results of Wenchuan earthquake, it can be known that SV wave (elastic wave whose vibration direction and propagation direction are vertical to the slope cross section) is the most disastrous and causes the most serious damage, while SV wave is a kind of complex non-stationary signal, whose frequency feature changes with time, which causes lots of limits for the application of pseudo-static method, limit displacement method in this aspect. Therefore, this book will, based on elastic wave theory, generalize the dynamic analysis models of reinforced retaining walls, and for both ductile and non-ductile reinforced mate-

Fig. 5.35 BBM data collecting system with 128 channels

rials, it builds force equilibrium differential equation of micro-unit with horizontal slice method and uses Hilbert–Huang Transform (HHT) to propose the time–frequency calculation methods of seismic stability of reinforced retaining walls under the effects of SV wave. It concretely explains and verifies the application and correctness of this method by conducting shaking model test to the reinforced retaining walls and then through carrying out a comparative analysis on the calculation results of time–frequency analysis method, pseudo-static method, and numerical analysis method and analyzes the superiority of time–frequency calculation method.

5.5.1 Time–Frequency Analysis Theory of Seismic Stability of Reinforced Retaining Structures

For the above discussed problems of pseudo-static method and limit displacement method, in order to fully consider the influence of three major factors of seismic motion (amplitude, frequency, and duration) on the seismic stability of reinforced retaining walls, this book proposes the time–frequency calculation method for seismic stability of reinforced retaining walls with horizontal slice method and HHT. The derivation process is as follows: First, generalize the analysis model, conduct horizontal slice to the retaining structures with horizontal slice method, and solve the wave power generated in micro-units of each layer by seismic waves with HHT and elastic wave theory; second, build the force equilibrium differential equation of micro-units, and further build the differential equation of the entire sliding mass, so as to solve the total of tensile force of lacing wire and the reinforced material length of reinforced retaining walls. It is noteworthy that the nature of inertia force generated by soil mass under seismic effects is a simplified quantitative presentation of seismic waveform in soil mass. Therefore, this book takes the horizontal seismic inertia force as equivalent to the wave force generated by seismic waves in soil mass.

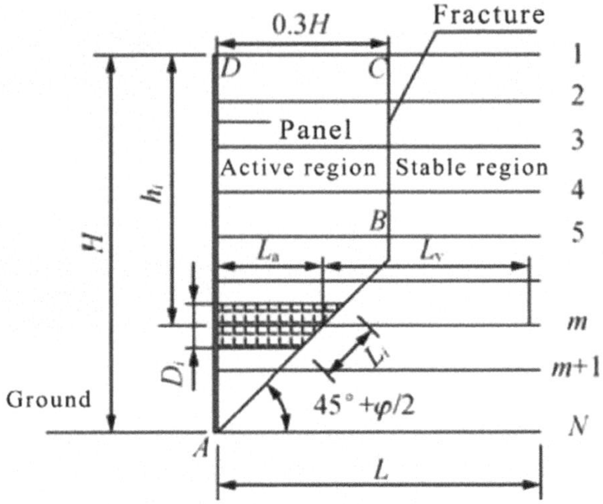

Fig. 5.79 Non-ductile reinforced materials

5.5.1.1 Formula Derivation of Time–Frequency Analysis Theory of Seismic Stability of Reinforced Retaining Structures

① Basic assumption;

The time–frequency analysis method derivation of seismic stability of reinforced retaining walls is based on the following assumption: ① According to the design codes of reinforced materials by the Ministry of Transportation, it can be known that two kinds of simplified fracture surface are adopted for reinforced materials with different modulus, which is as shown in Figs. 5.79 and 5.80; ② the filling behind walls is homogeneous and isotropic; ③ the material dump inside the filling behind walls consumes a large amount of seismic wave energy, which will finally cause the reflected seismic wave on the upper surface of filling bringing very weak seismic energy to propagate downward, so the formula derivation in this book will not consider the influence of reflected seismic wave on the upper surface of filling on micro-units.

② Generalized analysis model;

For reinforced retaining walls with both ductile and non-ductile reinforced materials, conduct horizontal slice to filling behind walls with horizontal slice method and select micro-units to study, as shown in Fig. 5.81. The parameters in Fig. 5.82 are specified as follows: Wi is soil strips' self-weight, and FN, i, FN, $i-1$, Ft, i, Ft, $i-1$ are, respectively, normal and shear interslice force on the upper and lower side of horizontal soil strips; *Ti* is the tensile force on the *i*th layer of reinforced material; FEH, i is the horizontal inertia force on the soil strips; αi is the dip angle between *i*th fracture surface of horizontal soil strip and the horizontal surface. Figure 5.82 is the stress diagram of wave force acted on horizontal soil strips. It is noteworthy that

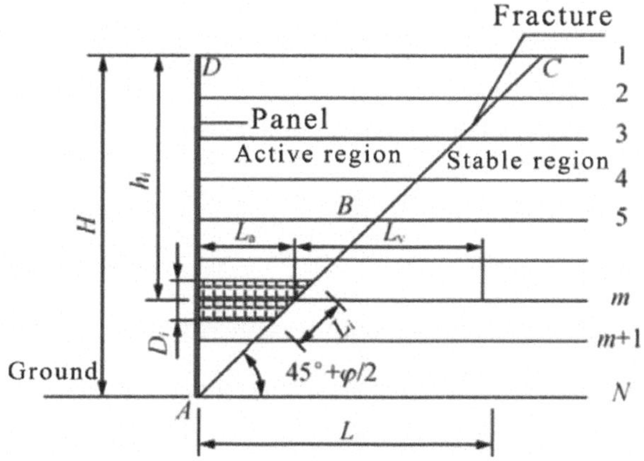

Fig. 5.80 Ductile reinforced materials

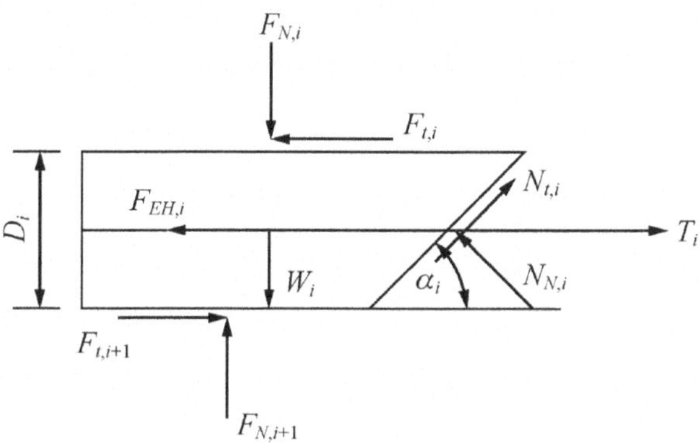

Fig. 5.81 Stress analysis model of horizontal soil strips

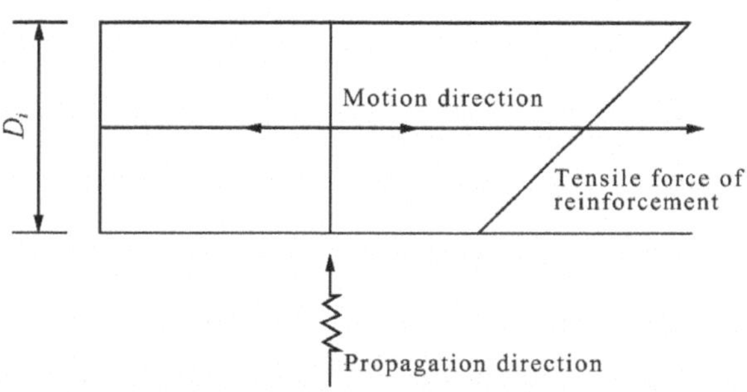

Fig. 5.82 Analysis model of wave power on horizontal soil strips

the inertia force acted on the micro-units in this book is replaced by the wave force of seismic waves.

③ Calculation of tensile force of reinforced materials.

According to Fig. 5.43, it can be known that when the ith horizontal soil strips reach force equilibrium, the sum of its vertical components is 0, i.e.:

$$F_{N,i+1} - F_{N,i} - W_i + N_{t,i}\sin\alpha_i + N_{N,i}\cos\alpha_i = 0 \tag{5.18}$$

$$N_{t,i} = cl_i + N_{N,i}\tan\phi \tag{5.19}$$

In the formula, c is the cohesive force of filling; φ *i* the internal friction angle of filling. Substituting Formula (5.19) into (5.20), it can get:

$$N_{N,i} = \frac{F_{N,i} - F_{N,i+1} + W_i - cl_i\sin\alpha_i}{\tan\varphi\ \sin\alpha_i + \cos\alpha_i} \tag{5.20}$$

When the entire sliding mass reaches force equilibrium, the sum of horizontal components is 0, i.e.

$$\sum_{i=1}^{n} T_i + \sum_{i=1}^{n} N_{t,i}\cos\alpha_i - \sum_{i=1}^{n} N_{N,i}\sin\alpha_i - \sum_{i=1}^{n} F_{EH,i} = 0 \tag{5.21}$$

Substituting Formula (5.19) and (5.20) into (5.21), it can get:

$$\sum_{i=1}^{n} T_i + \sum_{i=1}^{n} (cl_i + N_{N,i}\tan\phi)\cos\alpha_i - \sum_{i=1}^{n} \frac{F_{N,i} - F_{N,i+1} + W_i - cl_i\sin\alpha_i}{\tan\varphi\sin\alpha_i + \cos\alpha_i}\sin\alpha_i - \sum_{i=1}^{n} F_{EH,i} = 0 \tag{5.22}$$

The potential function of all SV waves could be uniformly represented by elastic displacement. The details are as follows:

$$u(z,t) = U(z)\mathrm{e}^{i\omega t} \tag{5.23}$$

In the formula, $u(z, t)$ is the displacement function of the *i*th micro-unit; $U(z)$ is the displacement amplitude of the *i*th micro-unit; ω is the vibrating frequency of it. Therefore, the inertia force generated by SV waves on the *i*th micro-units is:

$$F_{EH,i} = \left| W_i \cdot \frac{\partial^2 u}{\partial t^2}/g \right| = W_i w^2 U(z)/g \tag{5.24}$$

Thus, substituting Formula (5.24) into (5.22), it can get the total tensile force of ductile and non-ductile reinforced materials to keep inside stability of retaining walls are, respectively:

$$\sum_{i=1}^{n} T_i = \rho V w^2 U(z) - \frac{cH\cos\alpha}{\sin\alpha} - \frac{[\cos\alpha\tan\varphi - \sin\alpha][F_{N,0} + W - cH]}{\tan\varphi\sin\alpha + \cos\alpha} \tag{5.25}$$

$$\sum_{i=1}^{n} T_i = \rho V w^2 U(z) - 0.3cH + \frac{0.3(1-0.3\tan\alpha)\gamma H^2 - [c(1-0.3\tan\alpha)H]}{\tan\varphi} \times \frac{[\cos\alpha\tan\varphi - \sin\alpha][F_{N,0} + 0.045\gamma H^2\tan\alpha] - 0.3cH\tan\alpha]}{\tan\varphi\sin\alpha + \cos\alpha} \quad (5.26)$$

In Formula (5.25) and (5.26), $F_{N,0}$ is the overload of wall top; ω is the instant frequency of SV wave, and it changes with time for seismic waves; ρ is the intensity of filling behind walls; V is the volume of sliding mass *ABCD* behind wall; α is the dip angle between tilting part of retaining wall fracture surface with the horizontal surface.

Nondimensionalize the total reinforced materials needed by keeping inside stability of reinforced retaining walls $\sum_{i=1}^{n} T_i$ as parameter K, and take it as equivalent to the earth pressure coefficient in traditional design methods of retaining walls. Then, the expression of K is:

$$K = \sum_{i=1}^{n} T_i/\gamma H^2 \quad (5.27)$$

The tensile force on the ith layer of reinforced material is:

$$T_i = \gamma h_i D_i K \quad (5.28)$$

5.5.1.2 Analysis of Seismic Wave Time–Frequency Effects

Based on the above derivation, it can be known that the time–frequency effects of seismic waves mainly embody in the elastic displacement amplitude $U(z)$ and frequency ω of input seismic waves. Therefore, this dissertation the following analysis process of seismic wave time–frequency effects: Firstly, break up seismic wave time–frequency into several of IMF (time history curve) through Empirical Mode Decomposition (EMD for short); secondly, carry out Channel Switch, i.e. switch multi-channel signal composed of multiple IMF into single-channel signal made up of single IMF; thirdly, solve the instant frequency of each channel, and draw the time–frequency curve of each IMF. Lastly, substitute each IMF and their instant frequency into Formula (5.25) and (5.26) to work out the results and superimpose the results, so we can get the total of tensile force of lacing wire needed by the reinforced retaining walls.

Based on the above discussion, the time–frequency calculation methods of seismic stability of reinforced retaining walls derived by this book could well complement the shortcomings of pseudo-static method and limit displacement method and could take good consideration of the influence of three major factors of seismic motion (amplitude, frequency, and duration) on the seismic stability of reinforced retaining walls.

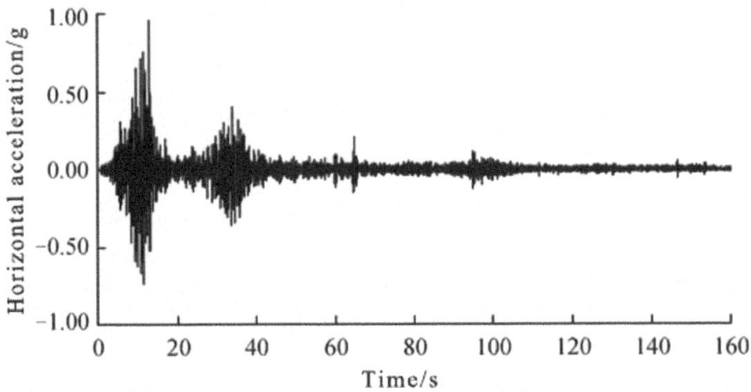

Fig. 5.83 Time history curve of Wenchuan wolong seismic waves

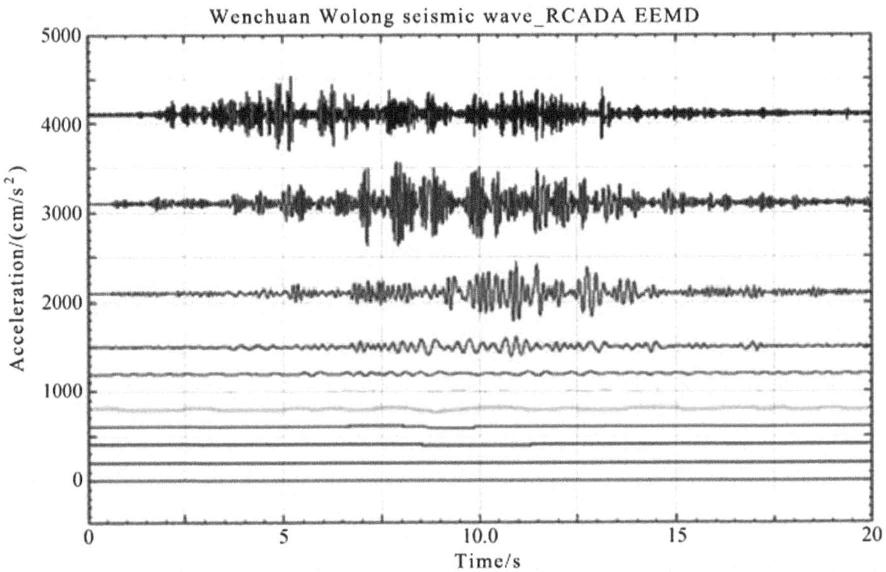

Fig. 5.84 Acceleration time-history curve of IMF

5.5.1.3 Practical Application of Seismic Wave Time–Frequency Effects

In order to more deeply and concretely explain the practical application of seismic wave time–frequency effects, this dissertation will, based on the analysis ideas of seismic wave time–frequency effects, conducts input wave time–frequency analysis, and take the example of Wenchuan wolong seismic wave to explain. The details are as follows: Firstly, input Wenchuan wolong seismic waves as shown in Fig. 5.83;

Secondly, carry out EEMD decomposition and get every IMF (see Fig. 5.84); lastly, solve the instant frequency of every IMF (see Fig. 5.85).

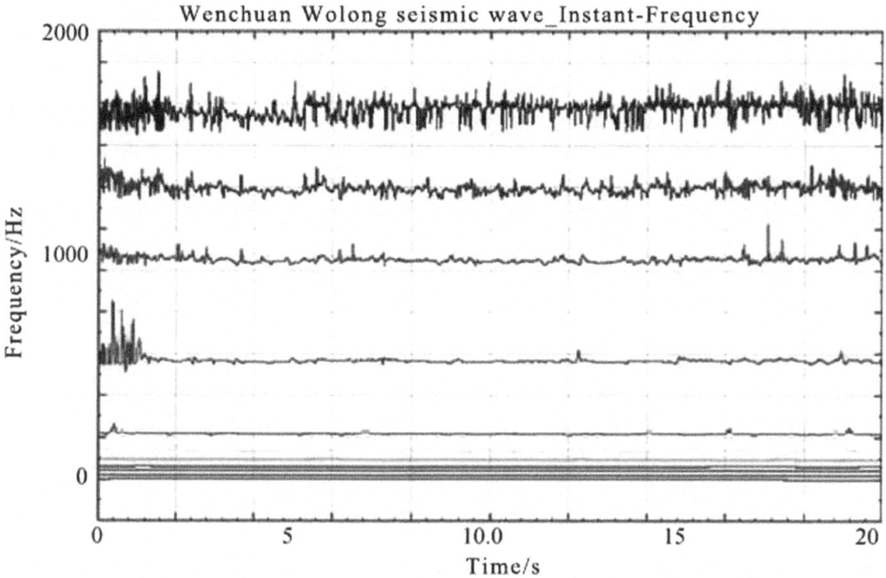

Fig. 5.85 Frequency time-history curve of IMF

It is noteworthy that, due to that the duration of Wenchuan wolong seismic wave is a bit longer, so Figs. 5.84 and 5.85 only select calculation results of T =0–20 s including peak seismic ground motion acceleration and demonstrate it clearly.

5.5.2 Solving Process of Time–Frequency Analysis Theory of Seismic Stability of Reinforced Retaining Structures

According to the above analysis, this chapter summarizes the solving process of time–frequency analysis method of seismic stability of reinforced retaining walls, which is as follows:

(1) Select plate type of reinforced retaining walls and type of filling and reinforced materials, and generalize analysis model combining with field conditions.
(2) Decide the position and lacing wire spacing of fracture surfaces according to the conditions like soil feature and reinforced material types.
(3) Decide the design response spectrum curve of this region according to *Code for Seismic Design of Buildings*, and synthesize artificial ground motion time history with seismic ground motion time history fitting method proposed by the reference so as to decide the seismic ground motion acceleration time history of this region.

(4) Decide time history curve of displacement based on seismic ground motion acceleration time history curve, work out IMF of seismic waves and the correspondent frequency time history curve with HHT.
(5) Substitute the above results into Formula (5.25) and (5.26) for solution, which could decide the total tensile force needed by the reinforced retaining walls, and combine with formula (5.27) to solve the tensile force on each layer.
(6) According to the calculation results of lacing wire tensile force on each layer, calculate the cross-sectional area of needed reinforced materials, and combining with the position of fracture surface, further calculate the lacing wire length on each layer.

5.5.3 Parameter Study of Time–Frequency Analysis Theory of Seismic Stability of Reinforced Retaining Structures

For time–frequency analysis method of seismic stability of reinforced retaining walls, conduct research of influence of seismic intensity, the internal friction angle as well as input wave frequency on parameter K under the condition of different reinforced materials. The concrete sample is as follows: The reinforced retaining wall height H is 5 m, unit weight γ of filling is 20 kN/m^3, cohesive force c is 0, and 18 layers of lacing wires are set up with the same spacing, with anti-pulling friction coefficient of 0.75, which studies sine waves with internal friction angle of filling $\varphi = 15°$, 20°, 25,° and 30° and PGA = 0.1 g, 0.2 g, 0.4 g and the frequency change of sine waves are 0.1, 0.2, 0.4, 0.8, 1.0, 2, 4, 5, 6, 8, and 10 Hz.

Based on the above condition, it calculates, respectively, parameter K with different reinforced material types, and the concrete calculation results could be seen in Figs. 5.86, 5.87, 5.88, and 5.89. It is noteworthy that K in Figs. 5.86, 5.87, 5.88, and 5.89 represents ductile reinforced materials and B represents non-ductile reinforced materials.

Through the comprehensive analysis of Figs. 5.86, 5.87, 5.88, and 5.89, it can come to the following rules: (1) With the increase in internal friction angle of filling, parameter K, the total tensile force needed by the reinforced materials gradually decreases; (2) with the increase in seismic intensity, K suddenly increases; (3) K of ductile reinforced materials is apparently higher than that of non-ductile reinforced materials, which may be caused by the difference of filling fracture surface behind walls; (4) parameter K distributes in saddle shape with increase in input seismic wave frequency and reaches its maximum with $f = 1$ Hz. The reason for the above phenomenon may be that when the natural frequency f of this reinforced retaining wall system is 1.16 Hz, the input wave frequency is 1 Hz, and the retaining walls come across resonance, which intensifies the seismic response of rigid retaining wall system, and causes increase in total tensile force needed by the reinforced materials. (5) For ductile and non-ductile reinforced materials, the K value calculated with the proposed method in this book and K value calculated by the standardized method

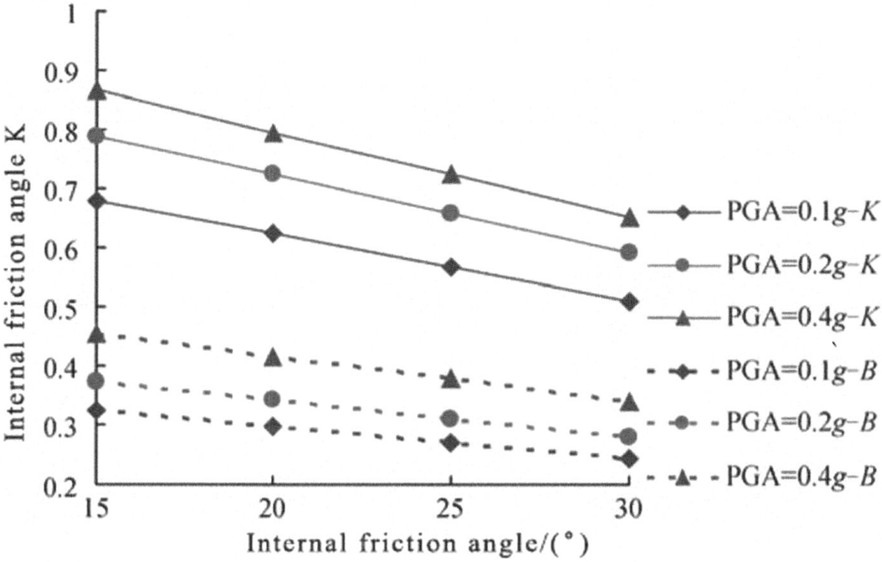

Fig. 5.86 Parameter K with $f = 0.2$ Hz versus internal friction angle of filling

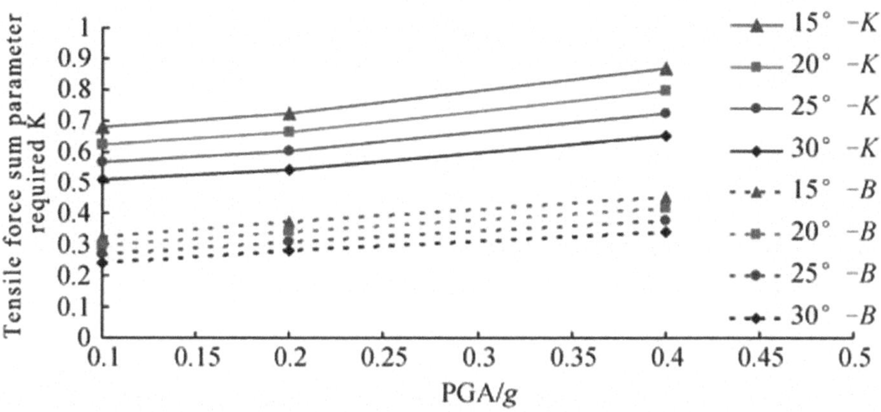

Fig. 5.87 Parameter K with $f = 0.2$ Hz versus PGA

are, respectively, in the range of 1.0–1.25 and 1.02–1.35, which fully explain that, by the existing codes that only pay attention to PGA but not input wave frequency, conducting seismic stability design to reinforced retaining walls may reduce the seismic safety reserves of retaining walls.

With the above analysis, it can be known that reinforced material type, seismic intensity, the internal friction angle as well as input wave frequency all exert large influence on the seismic stability of reinforced retaining walls, which should be taken into consideration of the anti-seismic design of reinforced retaining walls.

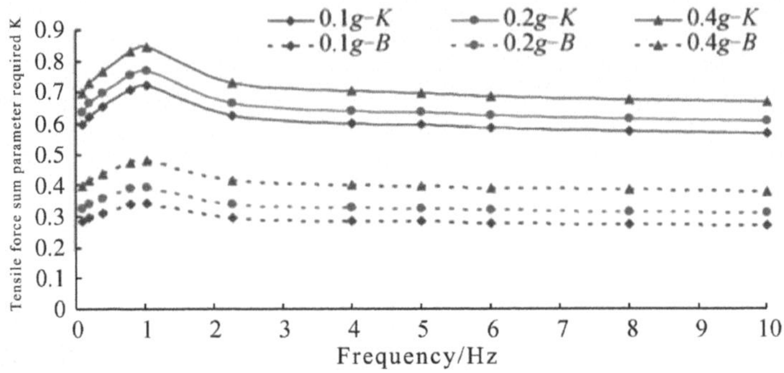

Fig. 5.88 Parameter K with $\varphi = 20°$ versus loading frequency

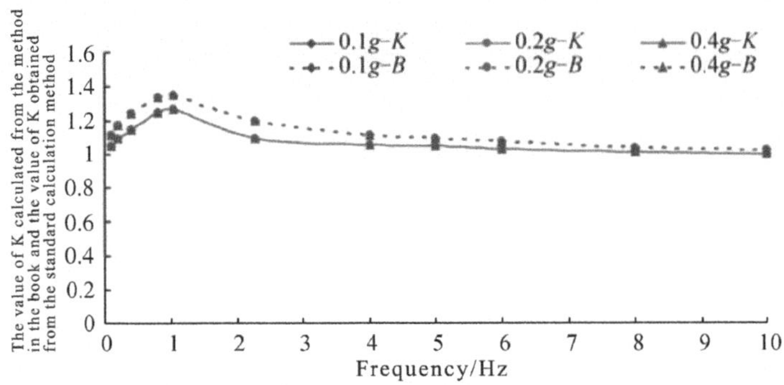

Fig. 5.89 Ratio of parameter K with $\varphi = 20°$ versus loading frequency

5.5.4 Correctness Verification of Time–Frequency Analysis Theory of Seismic Stability of Reinforced Retaining Structures

In order to verify the correctness of time–frequency analysis method of seismic stability of reinforced retaining walls, this section will only pay attention to PGA of seismic waves, but not the influence of seismic wave frequency on seismic stability of reinforced retaining walls. The concrete sample is as follows: The reinforced retaining wall height H is 5 m, the reinforced materials are ductile, unit weight γ of filling is 20 kN/m^3, cohesive force c is 0, and 18 layers of lacing wires are set up with the same spacing, with anti-pulling friction coefficient of 0.75, which studies sine waves with internal friction angle of filling $\varphi = 15°$, 20°, 25°, and 30°; the input wave frequency ω is 15 Hz, and combining with Sect. 5.5.3, it can be known that this frequency exceeds the influencing range of seismic wave frequency, and the detailed calculation results are shown in Table 5.34.

Through the comprehensive analysis of Table 5.34, it can be known that when paying no consideration to the influence of seismic wave frequency, the results of

Table 5.34 Comparison of K value calculated by time–frequency analysis method with that calculated by other methods

PGA	$\varphi = 20°$			$\varphi = 25°$			$\varphi = 30°$		
	Time–frequency analysis method	Reference	Standardized method	Time–frequency analysis method	Reference	Standardized method	Time–frequency analysis method	Reference	Standardized method
0.1	151	159	156	137	147	144	123	118	117
0.2	161	171	169	146	157	154	145	138	136
0.4	177	189	184	161	173	169	157	150	145

Note The reference introduces a design method for seismic stability calculation of reinforced retaining walls, which simplifies sliding fracture surface into logarithmic spiral fracture surface and derives the results with horizontal slice method

time–frequency analysis method of seismic stability of reinforced retaining walls are close to that calculated by reference and standardized method, which fully explains that the calculation method proposed by this book is reliable and effective.

Bibliography

Building Seismic Safety Council (BSSC). NEHRP guidelines for the seismic rehabilitation of buildings (FEMA 273)[S]. Washington D.C.: BSSC, 1997.

Chen Hua. Experimental study on mechanical properties of reinforced earth retaining wall with geotechnical grille [D], China Academy of Railway Sciences, 2011.

Chen Huifa, Duan Lian. Seismic Design of Bridge Engineering [M]. Beijing: China Machine Press, 2007.

Chen Houqun. Seismic fortification level and the corresponding performance goals of dam [J]. Earthquake Resistant Engineering and Retrofitting, 2005(27):1–6.

Chen Guoxing, Wang Zhihua, Zuo Xi, Du Xiuli, Han Xiaojian. Development of laminated type of soil box for shaking table test [J]. Chinese Journal of Geotechnical Engineering, 2010(33):89–97.

Chen Guoxing, Wang Zhihua, Zai Jinmin. Experimental study on large scale shaking table model considering interaction between soil and structure [J]. Earthquake Engineering and Engineering Vibration, 2001, 21(4):117–127.

Chen Guoxing, Wang Zhihua, Zai Jinmin. Experimental study on TMD vibration table model considering SSI effect [J]. Journal of Nanjing University of Technology, 2002, 24(5):34–39.

Chen Qiang, Yang Changwei, Zhang Jianjing, Zhou Xinwen. Study on failure mechanism of reinforced earth retaining wall in Wenchuan earthquake [J]. Railway Engineering, 2010, 9:73–77.

Chen Yueqing, Lv Xilin, Li Zhenpei, et al. Experimental study on shaking table model of interaction system between layered soil, foundation and high-rise frame structure [J]. Earthquake Engineering and Engineering Vibration, 2001, 21(3):104–112.

CHOUDHURY D, TIMBALKAR SS. Pseudo-Dynamic Approach of Seismic Active Earth Pressure behind Retaining wall[J]. Geotechnical and Geological Engineering, 2006, 24(1):1103–1113.

Christopher B.B, Full-Scale shaking table tests and finite element analysis of reinforced soil retaining walls, dissertation, Columbia University, 2004.

Compilation Committee of Atlas of Highway Seismic Hazard in Wenchuan Earthquake. Atlas of Highway Seismic Hazard in Wenchuan Earthquake [M]. Beijing: China Communications Press, 2009, 5.

Du Xiuli. Theory and Method of Engineering Wave [M]. Science Press, 2009.

EUROPEAN STANDARD 1998 Eurocode 8: Design of structures for earthquake resistance Part 5: Foundations, retaining structures and geotechnical aspects [S].

Harris, H.G. Dynamic Modeling of Concrete Structure [R] Publication SP-73, ACI, Detroit, USA, 1982.

Hu Yuxian, He Xun. Fitting of artificial ground motion response spectrum considering phase spectrum [J]. Earthquake Engineering and Engineering Vibration, 1986, (2):37–51.

HUANG N E, SHEN Z, LONG S R, et al. The empirical mode decomposition and Hilbert spectrum for nonlinear and non-stationary time series analysis [J]. Proceeding of the Royal Society. 1998, 454:903–995.

ICHIBARA M, MATSUZAWA H. Earth pressure during earthquake [J]. Soils and Foundations, 1973, 13(4):75–86.

Industry standard compilation group of the People's Republic of China. TB 1002—2001. Code for design of railway subgrade retaining structure [S]. Beijing: China Railway Press, 2001.

J.H. Wood D.G. Elms SESMIC DESIGN OF BRIDGE ABUTMENTS AND RETAINING WALL [R] Road Research Unit Bulletin 1984, 2.

J. Koseki, F. Tatsuoka, Y. Munaf, M. Tateyama and K. Kojima A Modified Procedure to Evaluate Seismic Active Earth Pressure Considering Effects of Strain Localization in Backfill Soil [J], Soils Found, 1998, 209–216.

Jorge G. Zornberg, James K. Mitchell, Finite element analysis of geosynthetically reinforced soil walls with sloping backfills, dissertation, University of California Berkeley, 1993.

Japan Railway Technical Research Institute. Japan Railway Construction Design Standards [S]. 1998.

Jia Liang. Seismic action and dynamic stability analysis of reinforced earth retaining wall [D], Lanzhou University of Technology, 2011.

Jiang Jianqing, Zhou Yinsheng. Calculation of internal stability of reinforced earth retaining wall with the action of complex dynamic, Central South Highway Engineering [J], 32(1):51–54, 2007.

Jiang Qingjian, Yang Guolin. Horizontal slice method for seismic stability analysis of reinforced soil retaining wall, China Railway Science [J], 30(1):36–40, 2009.

Jin Yingchun. Calculation of internal stability of reinforced earth retaining wall under earthquake, Highway [J], (05):43–47, 1991.

Li Haiguang, Zhou Depei, et al. Design and engineering example of a new type of retaining structure. Beijing: China Communications Press, 2010.

Liao Yi. Investigation and Analysis on seismic damage of Highway Subgrade and the study of vulnerability in Wenchuan earthquake [M]. Southwest Jiaotong University, 2012.

Lin Yuliang, Yang Guolin, Zhao Lianheng, et al. Horizontal slices analysis method for seismic earth pressure calculation [J]. Chinese Journal of Rock Mechanics and Engineering, 2010, 29(12):2581–2391.

Liu Huabei. Analysis of internal stability of reinforced earth retaining wall of modular panel under earthquake. Chinese Journal of Geotechnical Engineering, 2008, 30(2):278–282.

Lu Tinghao. Active earth pressure formula considering cohesion and adhesion of the back of the wall [J]. Rock mechanics, 2002, 23(4):470–473.

Luo Qifeng, Wang Yumei. Structural performance design theory proposed in the summary of seismic damage in recent years [J]. Engineering seismology, 2001, (2):3–7.

Lv Xilin. Similar conditions for structural seismic model tests [A]. Structural seismic test (Zhu Bolong, editor in chief) [C]. Beijing: Seismological Press, 1989, 2:8–13.

Ma Hongwei, Wu Bin. Elastic dynamics and numerical methods [M]. Beijing: China Building Industry Press, 2000.

Magdi M, El Eman, Behaviour of Reinforced Soil Walls under Earthquake Loading, dissertation, Queen's University of Canada, 2003.

MEYMAND P. Shaking table scale model test of nonlinear soil-pile-superstructure interaction in soft clay [D]. Berkeley: University of California, 1998.

Ministry of Communications of the People's Republic of China. JTJ 004-89. Code for seismic design of highway engineering [S]. Beijing: China Communications Press.

Ministry of Communications of the People's Republic of China. JTG D30-2004. Specifications for Design of Highway. Beijing: China Communications Press.

Ministry of Communications of the People's Republic of China. JTJ015-91. Code for design of highway reinforced soil [S]. Beijing: China Communications Press, 1999.

Ministry of Railways of the People's Republic of China. TB 10025-2006. Code for design of railway subgrade retaining structure [S]. Beijing: China Railway Press.

Ministry of Railways of the People's Republic of China. GB 50111-2006. Code for seismic design of railway engineering [S]. Beijing: China Planning.

MYLONAKIS G, KLOUKINAS P, PAPANTONAPOULOS C. An alternative to the Mononobe-Okabe equations for seismic earth pressure[J]. Soil Dynamic and Earthquake Engineering, 2007, 27(10):957–969.

NOURI H, FAKHER A, JONES CJFP. Evaluating the effects of the magnitude and amplification of pseudo-static acceleration on reinforced soil slopes and walls using the limit equilibrium horizontal slices method[J]. Geotextiles and Geomembrances, 2008, 26(3):263–278.

Pang Zhenpeng. An introduction to engineering wave [M]. Beijing: Science Press, 1996.

SHA HGHOL I M, FA KHER A, JONES. Horizontal Slice Method of Analysis [J]. Geotechnique, 2001, 51(10): 8812885.

The Second Highway Survey and Design Research Institute Co., Ltd. of China Communications. 2008. Rules for design and construction of highway retaining wall [S]. Beijing: China Communications Press.

U. S. Department of Transportation and Federal Highway Administration. FHWA2 N HI2002043 Mechanically Stabilized Earth Walls and Reinforced Soil Slopes Design & Construction Guidelines [S]. Washington, D. C: National Highway Institute Office of Bridge Technology, 2001.

Wang Liqiang, Wang Yuanzhan, Chi Lihua. Seismic earth pressure of retaining wall and its distribution [J]. China Harbor Engineering, 2007(5):1–5.

Wang Yayong. Prospect of the standard of seismic design in China [J]. Building structure, 1999(6):32–36.

Xiao Shiguo, Zhang Jianjing, Ma Yaoxian. The investigation of gravity retaining wall damage in Wenchuan earthquake zone [J]. Journal of underground space and engineering, 7(1):174–178.

Xu Qiang, Zeng Yuping. Study on acceleration variation characteristics of landslide with creep characteristics and warning indicators before sliding [J]. Chinese Journal of Rock Mechanics and Engineering, 2009, (28): 1009–1106.

Yang Changwei, Zhang Jianjing. Study on time frequency analysis of seismic stability of rock slope with the action of SV wave, Chinese Journal of Rock Mechanics and Engineering, 2013, 32(3):483–491.

Yang Changwei, Zhang Jianjing. Analysis of the seismic collapse effect of the high and steep slope. Journal of Southwest Jiaotong University, 2013, 48(3):1–8.

Yang Youhai. Stability analysis of reinforced soil retaining wall under earthquake, Journal of Lanzhou Railway Institute (Natural Science) [J], 21(4):9–11, 2007.

You Shengming. Experimental study of reinforced earth retaining wall and analysis of deformation mechanism [M], Dalian University of Technology, 2009.

Zhang Jianjing, Feng Jun, Xiao Shiguo, Liu Changqing. Two key technical problems in the seismic design of retaining structures [J]. Journal of Southwest Jiaotong University, 2009, (44):321–326.

Zhou Depei, Zhang Jianjing, Tang Yong. Seismic damage analysis of road slope in Wenchuan earthquake [J]. Chinese Journal of Rock Mechanics and Engineering, 2010, 29(3):565–576.

Chapter 6
Seismic Dynamic Time–Frequency Theory of Layered Site

Seismic response analysis of soil layer is a key study of field seismic dynamic parameters determination and geotechnical seismic design. For the seismic response analysis of soil layer, the commonly used methods at present are time-domain elastic wave propagation theory analysis method, frequency-domain elastic wave propagation theory analysis method, and numerical simulation analysis method. For these three methods, the time-domain elastic wave propagation theory analysis method came out first, but actually, it could only give solutions to field with few cover layers (1–2), and it could not consider the influence of input wave frequency on field seismic response; by principle, frequency-domain elastic wave propagation theory analysis method fit for fields with any layers, but it could only consider the harmonic wave effects with given circular frequency, and it could not comprehensively consider the influence of frequency-duration coupling effects; although numerical simulation could conduct time–frequency analysis to seismic response under earthquake, it is complex to operate, especially for field with large difference in ground layer thickness. While seismic wave is a kind of non-stationary signal, and its frequency and amplitude gradually change over time. Therefore, for the above problems, this book, based on the shaking table test and elastic wave theory, uses HHT to generalize the analysis model for horizontal layered field, proposes time–frequency analysis method of seismic response of horizontal layered field under shear wave effects, and explains the correctness and the concrete application of this method through large-scale shaking table test and numerical analysis.

6.1 Shaking Table Test for Seismic Dynamic Characteristics of Layered Field

With large-scale shaking table test, this chapter makes three-layered tilting field models with different dip angles, and their dip angles are, respectively, 7.5°, 10°, and

C. Yang et al., *Three Dimensional Space-Time Analysis Theory of Geotechnical Seismic Engineering*, https://doi.org/10.1007/978-981-13-3356-9_6

12.5°. Through inputting seismic waves with different amplitudes in the three directions of shaking table surface, it studies the seismic dynamic response characteristics of layered tilting fields with small angles in dip direction, strike direction, vertical direction and direction vertical to slope surface, and its research results have guiding significance to recognize the seismic dynamic response rules of layered tilting fields with small angles.

6.1.1 Brief Introduction to Shaking Table Test

As the preparation work of this chapter is basically the same with **the similarity system, model container design, model container process, loading work conditions of the experiment, and experiment flow in** Chap. 3 **Shaking Table Test Study on Dynamic Characteristics of Rock Slope**, so it won't repeat here. It will only describe **the test model container design, configuration of similarity materials, and distribution of sensors**, which is as follows.

6.1.1.1 Design of Test Model Containers and Distribution of Sensors

The test model is composed of four models with different dip angles, and the size of each model is 2.2 m (L) × 2.2 m (W) × 1.6 m (H), and these four models are set inside a model container with size of 5.0 m × 5.0 m × 2.1 m. Styrofoam with thickness of 30 cm is lined to the four sides of model containers to avoid the influence of reflection effects of input waves from steel model containers on the test results. The model container is divided into four same sections, in which four test models with different occurrences are to be completed. The rock layer dip angle of the four test models are 0°, 7.5°, 10°, and 12.5°, and its thickness is 1.6 m. The rock layer of the model is composed of gravel, soft rock, and hard rock from top to bottom, with thickness of, respectively, 0.4, 0.6, and 0.6 m. Plan and profile map of the test models are shown as Figs. 6.1, 6.2, 6.3, 6.4, 6.5 and 6.6.

6.1.1.2 Configuration of Similarity Materials

Before configuring the similarity materials required by the test, it needs to preset target parameters of the three materials (gravel, soft rock, and hard rock) according to the test needs, so as to examine the test results. Especially for gravel, the major parameters are cohesive force C and internal friction angle φ, and its unit weight condition could be naturally satisfied. The configured soft rock and hard rock are actually made up by compressing the mixture of clay, quartz sand, plaster, barite powder, and water, and its features are more close to that of the compressed soil. Cohesive force C and internal friction angle φ are easy to be satisfied, but density is hard to reach, so controlling the unit weight of these two kinds of rocks through

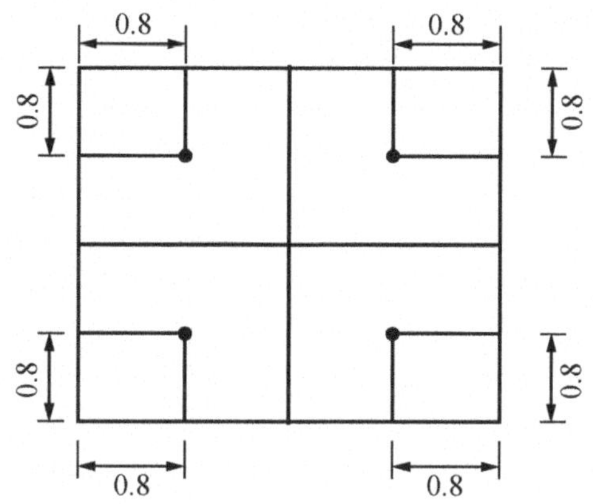

Fig. 6.1 Plan sketch of models

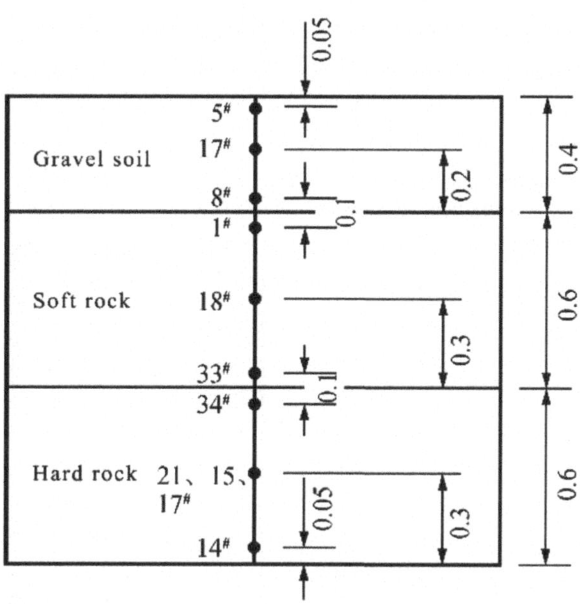

Fig. 6.2 Longitudinal profile map of 0° rock layer test model

compression is the main factor to consider in configuration. The finally decided physico-mechanical parameters of the test models are shown in Table 6.1.

This shaking table test configures three kinds of lithologic materials, namely gravel, soft rock, and hard rock, and their physico-mechanical parameters and configuration ratio are shown in Table 6.2.

When estimating the model volume, as there will be material loss in the model makeup process, so it adds 15% to the calculated volume of soft rock and 30% to the calculated volume of hard volume. The volume of gravel keeps the calculated results without addition. The test materials needed by makeup of test models are shown in Table 6.3.

Fig. 6.3 Longitudinal profile map of 7.5° rock layer test model

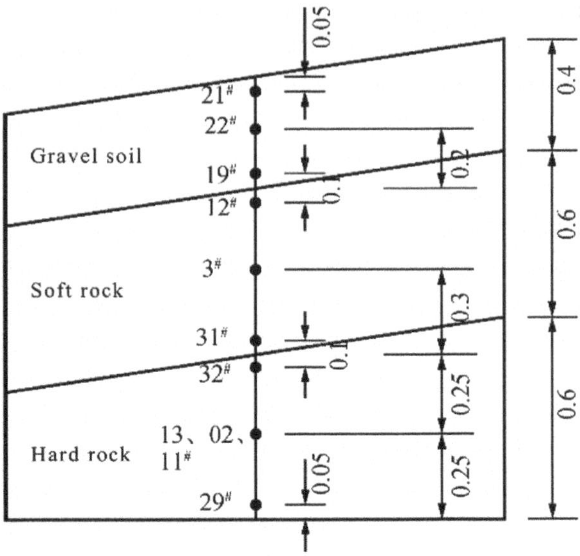

Fig. 6.4 Longitudinal profile map of 10° rock layer test model

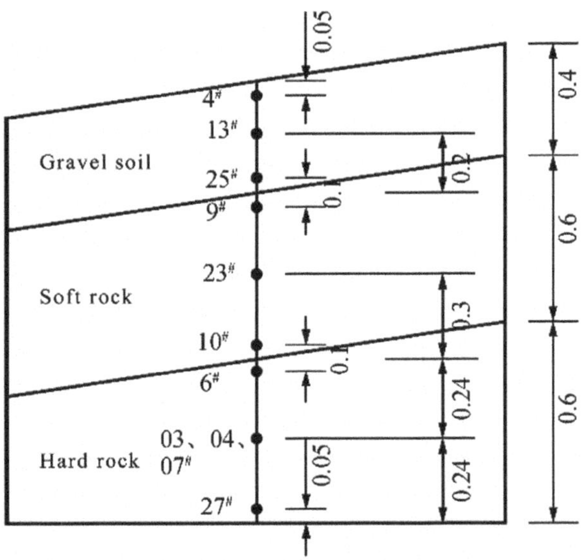

6.1.2 Analysis of Analysis of Shaking Table Test Results

In earthquakes, the major reason for damage of ground surface buildings is the amplification effects of stratum on seismic motion, and it is affected by many factors. This book uses acceleration amplification coefficients to represent the amplification effects of field on seismic motion. It firstly analyzes the measured frequency spectrum

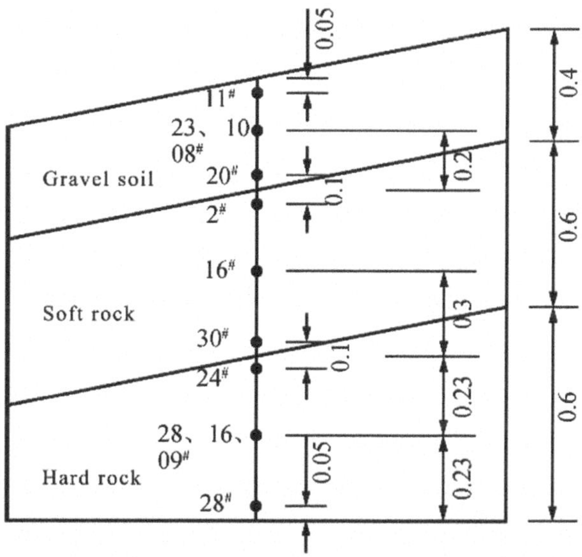

Fig. 6.5 Longitudinal profile map of 12.5° rock layer test model

Fig. 6.6 Longitudinal profile map of 12.5° rock layer test model

Table 6.1 Physico-mechanical parameters of the test model materials

	Density ρ (g/cm^3)	Moisture content ω (%)	Cohesive force c (kPa)	Internal friction angle Φ (°)
Gravel	1.9	13.95	8	25
Soft rock	2.2	7.826	104	41
Hard rock	2.3	9.655	200	45

characteristics and then explores the influencing rules of elevation, dip angle of lower bedrock and input seismic motion amplitudes on acceleration amplification coefficients. The calculation of acceleration amplification coefficients adopts PGA with period of 0.04 s in acceleration response spectrum. As there are many input

Table 6.2 Final configuration ratio of similarity materials

Gravel	Parameter	Density	Moisture content (%)		Cohesive force (kPa)	Internal friction angle
		1.9	13.95		8	25
	Configuration ratio (Mass ratio)	River sand	Clay		Quartz sand	Water
		33	35		18	12
Soft rock	Parameter	Density	Moisture content (%)		Cohesive force (kPa)	Internal friction angle
		2.2	7.826		104	41
	Configuration ratio (Mass ratio)	River sand	Plaster		Water	Clay
		7.5	2		0.9	2
Hard rock	Parameter	Density	Moisture content (%)		Cohesive force (kPa)	Internal friction angle
		2.3	9.655		200	45
	Configuration ratio (Mass ratio)	River sand	Plaster	Water	Clay	Barite powder
		5	2.5	1.4	3	4

work conditions of seismic motion in the test process, this book will select some representative work conditions to carry out data analysis.

It defines acceleration amplification coefficients as the ratio of measured peak value in acceleration time history of acceleration sensors on model tops with peak value of seismic wave input onto the shaking table surface. Given that large amount of results have been obtained in seismic dynamic response rules of horizontal layered fields, and they have been widely applied to the regional seismic safety evaluation and anti-seismic fortification of buildings, this book studies the seismic dynamic response characteristics of layered tilting fields with small angles with reference to the horizontal layered fields, and takes the ratio of surface peak ground motion amplification coefficient of tilting fields with that of horizontal layered fields as quantitative index of the study. If the ratio is less than 1, it means the amplification coefficient of corresponding field is weaker than that of horizontal layered fields, otherwise, it means the amplification coefficient of corresponding field is stronger than that of horizontal layered fields.

Table 6.3 Amount of model test materials

Upper gravel	Volume (m^3)	Density (kg/m^3)		Total mass (kg)		
	7.51	1900		14,269		
	River sand (kg)	Clay (kg)		Quartz sand (kg)		Water (kg)
	4804	5095		2620		1747
Middle soft rock	Volume (m^3)	Density (kg/m^3)		Total mass (kg)		
	8.64	2200		19,008		
	River sand (kg)	Plaster (kg)		Water (kg)		Clay (kg)
	11,490	3064		1379		3064
Lower hard rock	Volume (m^3)	Density (kg/m^3)		Total mass (kg)		
	9.762	2300		22,452		
	River sand (kg)	Plaster (kg)	Water (kg)		Clay (kg)	Barite powder (kg)
	7060	3530	1977		4236	5648
Gross weight (kg)	River sand (kg)	Plaster (kg)	Water (kg)	Clay (kg)	Barite powder (kg)	Quartz sand (kg)
55,729	23,354	7094	5103	12,395	5648	2620

6.1.2.1 Influence of Dip Angle on Peak Ground Motion Amplification Coefficient of Fields

It selects El Centro wave and Wenchuan wolong wave with seismic amplitudes of 0.08, 0.20, 0.32, and 0.50 g as the study work condition to explore the influencing rules of dip angle on seismic dynamic response of fields. As there is certain dip angle to the stratum of tilting fields, so the seismic dynamic response of fields in different directions may be different. In order to explore the directivity of seismic dynamic response of tilting fields, this section studies the seismic dynamic response rules of stratum dip direction, strike direction, and vertical direction.

The changing rules of field amplification coefficients with dip angles in *X* direction are shown in Fig. 6.7. The figure shows that with the increase of dip angles, the ratio of acceleration amplification coefficients in stratum dip direction also increases. When the dip angle is about 10°, the ratio of acceleration amplification coefficients is 1, which means that in the stratum dip direction, with the increase of dip angle, the amplification effects in dip direction of tilting direction enlarge, and when the dip angle is less than 10°,the amplification effects in dip direction are weaker than the horizontal layered fields, while when the dip angle is over 10°, the amplification effects in dip direction are stronger than the horizontal layered fields. The reason for this may be that the tilting field has certain degree of freedom of displacement in

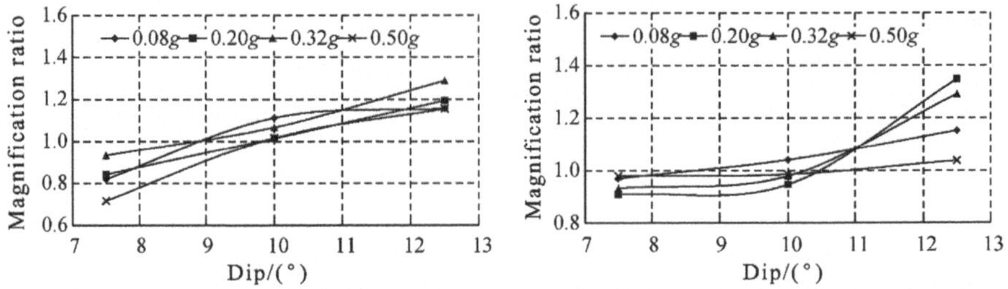

(a) Proportional magnification factor under El Centro wave action

(b) Proportional magnification factor under WenChuan wave action

Fig. 6.7 Influence of dip angle on peak ground motion amplification coefficient in stratum dip direction

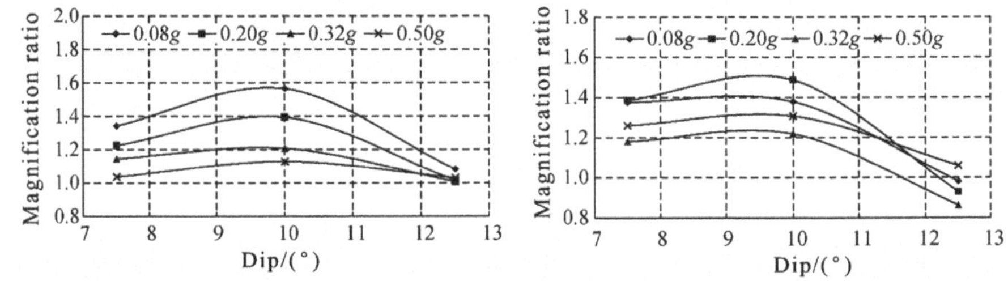

(a) Propagating Direction Amplification Coefficient under the Action of El Centro Wave

(b) Steering magnification factor under WenChuan wave action

Fig. 6.8 Influence of dip angle on peak ground motion amplification coefficient in stratum strike direction

dip direction, i.e. free face direction, and the lateral restraints in this direction are weaker than in other directions, and with the increase of dip angles, the degree of freedom of displacement in dip direction becomes larger, and turns out with more apparent amplification effects. Therefore, it can estimate that when the dip angle is over 12.5°, the amplification coefficient of tilting fields in dip direction will enlarge with the increase of dip angle.

The changing rules of field amplification coefficients with dip angles in *Y* direction are shown in Fig. 6.8. It could see from the figure that with the increase of dip angle, the ratio of amplification coefficient firstly increases and then decreases, and it reaches the maximum value when the dip angle is 10°. When the dip angle is over 10°, the amplification coefficient decreases with the dip angle increase, and when the dip angle reaches 12.5°, the ratio approximately equals to 1. The above rules show that when the dip angle is less than 12.5°, the amplification effects in strike direction of tilting field are stronger than that in horizontal layered fields, and when the dip angle is 10°, the amplification effects in strike direction of tilting field are the strongest.

The changing rules of field amplification coefficients with dip angles in *Z* direction (vertical direction) are shown in Fig. 6.9. It could see from the figure that with the increase of dip angle, the ratio of amplification coefficient in vertical direction firstly increases and then decreases, and it approximately equals to 1 when the dip angle

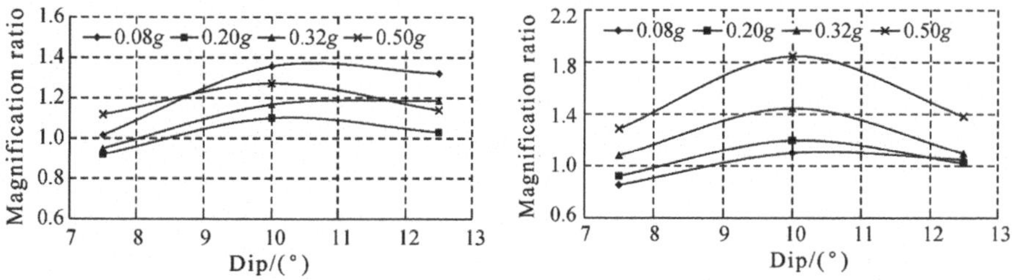

(a) Vertical magnification factor under El Centro wave action

(b) Vertical direction amplification factor by WenChuan wave

Fig. 6.9 Influence of dip angle on peak ground motion amplification coefficient in stratum vertical direction

is 7.5°. When the dip angle is 10°, the amplification coefficient reaches the maximum value, and when the dip angle is 12.5°, the amplification coefficient decreases compared with that when the dip angle is 10°, but it is still over 1. The above rules show that when the dip angle is less than 7.5°, the amplification effects in vertical direction of tilting field are stronger than that in horizontal layered fields, and when the dip angle continues increasing, the amplification effects in vertical direction of tilting field are stronger than that in horizontal layered fields. When the dip angle is 10°, the amplification effects in vertical direction of tilting field are the strongest.

6.1.2.2 Amplification Effects in Slope Surface Direction of Tilting Stratum

For slopes with possible sliding potential, studying the seismic response rules in possible sliding direction (slope surface direction) and in vertical direction to the sliding direction is of significant importance to know of the failure mechanism of slopes. The above analysis adopts three-dimensional Cartesian coordinate system, in which the vertical directions are vertical to the horizontal surface. The analysis in this section will adopt the above coordinate system by rotating θ degree, which is the dip angle, and that will make the Z'-axis of the coordinate system vertical to the stratum surface, X'-axis parallel with the slope surface and Y'-axis the same with Y-axis, which is as shown in Fig. 6.10.

According to the coordinate conversion shown in Fig. 6.10, it will get the coordinate conversion matrix:

$$\begin{bmatrix} X' \\ Y' \\ Z' \end{bmatrix} = \begin{bmatrix} \cos\theta & 0 & -\sin\theta \\ 0 & 1 & 0 \\ \sin\theta & 0 & \cos\theta \end{bmatrix} \begin{bmatrix} X \\ Y \\ Z \end{bmatrix} \tag{6.1}$$

Decompose the acceleration recorded in X, Y, Z direction in the experiment into X', Y', Z' direction, calculate the surface PGA amplification coefficients in X', Y', Z'

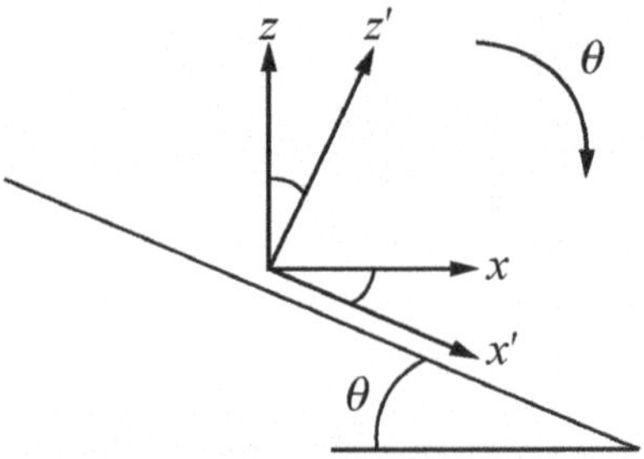

Fig. 6.10 Sketch map of coordinate conversion

direction when inputting Wenchuan wolong seismic wave with different amplitudes, and work out the changing rules of surface PGA amplification coefficients with dip angle increase, which is as shown in Fig. 6.11.

After the coordinate conversion, the changes of ratio of amplification coefficients in slope surface direction and in its vertical direction with amplification coefficients in horizontal layered fields with dip angle increase are shown as Fig. 6.11. It could see from the figure that, in slope surface direction, the ratios of amplification coefficients are all less than 1, and it reaches the minimum value when the dip angle is 10°; in direction vertical to the slope surface, the amplification coefficients are approximately all over 1, and it reaches the maximum value when the dip angle is 10°. The above shows that the amplification effects in slope surface direction of tilting fields are weaker than that of horizontal layered fields, and when the dip angle is 10°, the amplification effect in slope surface direction is the weakest and that in its vertical direction is the strongest.

It is noteworthy that the collapse damage of tilting slopes is caused by acceleration effects in both slope surface direction and in its vertical direction, and there exist slopes that come across sliding failure along slope surface with small angles, for example, the collapse in the northeast corner to Panzhihua Airport, whose leading edge dip angle is only 3°–10°. This research, focused mainly on layered tilting fields with small angles, studies the amplification effects in slope surface direction and in its vertical direction, and the relevant results will be reference to the recognition of slope failure mechanism under seismic effects.

6.1.2.3 Influencing Rules of Dip Angle on Response Spectrum

In order to explore the influencing rules of dip angles on response spectrum amplification effects, it selects Wenchuan wolong seismic wave and El Centro seismic wave with amplitude of 0.4 g as input waves and takes the response spectrum of gravel surface as study object. To highlight the influence of dip angles on response spectrum amplification effects, it defines the ratio of gravel surface response spectrum amplitude of tilting fields with that of horizontal layered fields as the response spectrum amplification coefficient, and it explores the response spectrum amplification effects in dip direction, strike direction, and vertical direction with different dip angles.

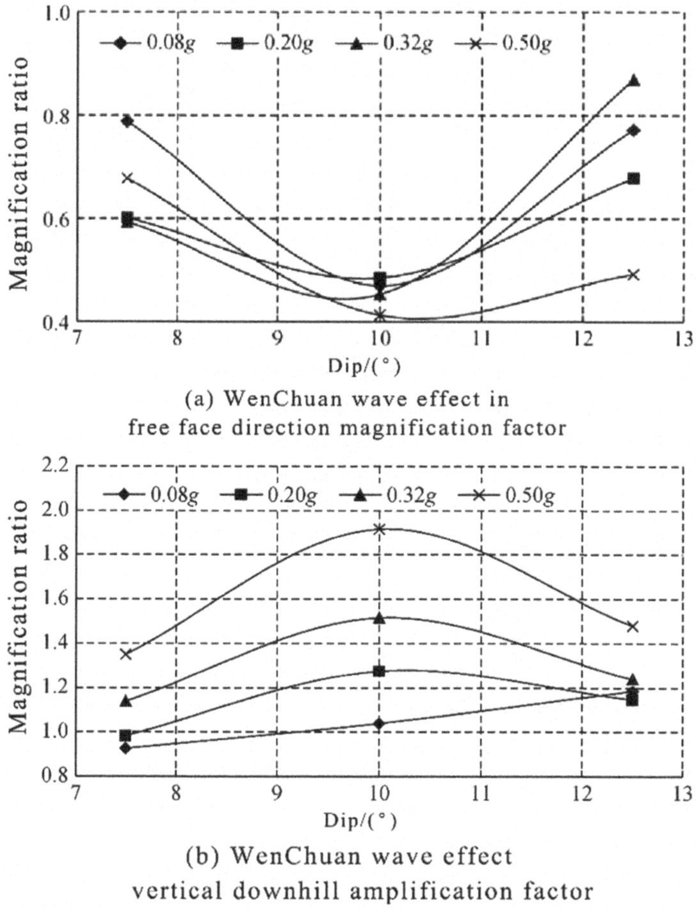

Fig. 6.11 Changes of amplification coefficients in slope surface direction and in its vertical direction with dip angle increase

Figures 6.12 and 6.13 show the changing rules of response spectrum amplification coefficients of Wenchuan wolong seismic waves and El Centro seismic waves with amplitude of 0.4 g in dip direction, strike direction, and vertical direction, which adopt semilog coordinate to represent.

It could be seen from the figure that in dip direction, strike direction, and vertical direction, when the dip angle is 7.5° or 10°, response spectrum amplification coefficient is approximately identical to 1, and it only amplifies slightly for the part of ≤0.1 s. When the dip angle is 12.5°, the response spectrum amplification effects are obvious in dip direction and in strike direction for the short period part ($T \leq 0.1$ s), while when $T > 0.1$ s, the response spectrum amplification coefficient suddenly decreases below 1 with sharp decrease of response spectrum amplification effects, and the response spectrum amplification coefficients stay at a comparatively

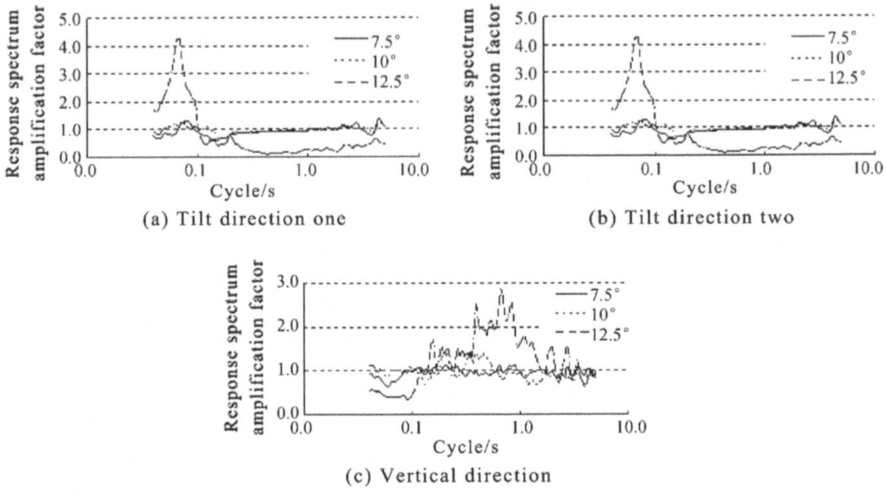

Fig. 6.12 Response spectrum amplification coefficients of Wenchuan wolong seismic waves with amplitude of 0.4 g

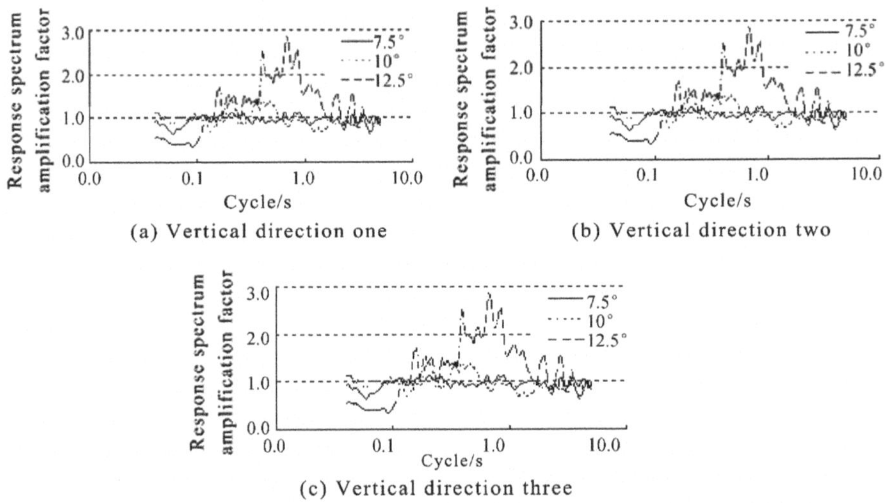

Fig. 6.13 Response spectrum amplification coefficients of El Centro seismic waves with amplitude of 0.4 g

lower level, which shows that when $T > 0.1$ s and the dip angle reaches 12.5°, the response spectrum amplitudes of tilting fields in dip direction and strike direction are lower than that of horizontal layered fields. For the vertical direction, it could be seen from Figs. 6.10 and 6.11 that, for fields with dip angle of 7.5° or 10°, the response spectrum amplification coefficients fluctuate around 1, which shows that the response spectrum amplification effects are not obvious, i.e. the ratio of response spectrum of tilting fields with that of horizontal layered field hardly enlarges. But for fields with dip angle of 12.5°, the fields show clear amplification effects in horizontal directions (dip direction and strike direction) at around $T = 0.8$ s, and the amplification coefficient of Wenchuan wolong seismic waves reaches 2.8, and that of El Centro seismic waves reaches 2. That may be because dominant period of horizontally input seismic waves and test models with dip angle of 12.5° is around 0.8 s, which leads to obvious amplification effects of test models with dip angle of 12.5° in horizontal directions (dip direction and strike direction).

6.2 Time–Frequency Analysis Theory Seismic Dynamic Characteristics of Horizontal Layered Fields

6.2.1 The General Thinking and Basic Assumption

The time–frequency analysis method derivation of seismic response of horizontal layered fields is mainly as follows: firstly, generalize the analysis model of horizontal layered fields, select any two adjacent soil layers to study, and use elastic wave theory to solve the instant stress components generated by reflection and transmission when seismic waves propagate through the soil layers, and adopt stress between the soil layers and continuity conditions of displacement to build the recurrence equation of soil layer shear stress; then use HHT combining with the boundary conditions of ground surface and bedrock to solve the seismic response like shear stress and acceleration of any soil layer. The derivation of this method will base on the following hypothesis: ① the closely connected soil layers are homogeneous and isotropic; ② due to that the seismic waves continuously coming across refraction and transmission when transmitting from seismic source upwards, which causes they close to the vertical direction when reaching the earth surface, so this chapter selects seismic waves transmitting vertically upward from bedrock to soil layers as input wave; ③ because wave-type conversion will occur when seismic wave arrive at the ground surface with its major energy transferring into surface wave moving along the ground surface, but the surface wave attenuated along the depth direction, so this section will pay no attention to the influence of surface waves on soil layer seismic response.

6.2.2 Generalized Models

Suppose that the thickness of each soil layer of horizontal heterogeneous fields is h_1, h_2, h_3 … hn, in which the $N-1$th soil layer covers the homogeneous semi-infinite space on bedrock, the dynamic shear modulus of soil layers is G, viscous damping coefficient is C, the correspondent damping ratio is λ, the soil mass density is ρ, and the propagating speed of shear waves are V_{s1}, V_{s2} … V_{Sn}. The soil mass adopts

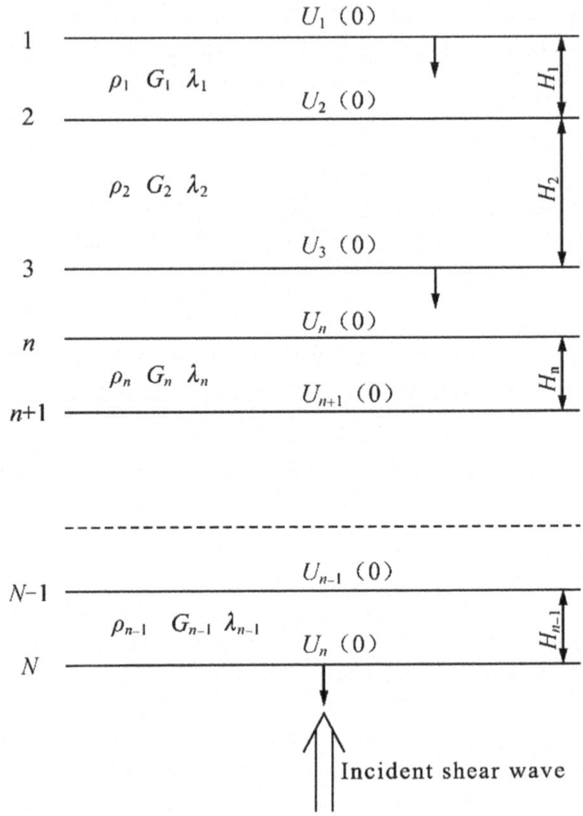

Fig. 6.14 Ideal one-dimensional wave model

linear cohesive elastic model. At the same time, the entire soil layer system adopts local coordinate system and set the origin of Z-axis on the upper surface of each soil layer with positive direction downward, and the input seismic waves are shear waves propagating vertically, which is shown as Fig. 6.14.

6.2.3 *Formula Derivation*

For linear cohesive elastic soil layers, the one-dimensional wave equation is:

$$\rho \frac{\partial^2 u}{\partial t^2} = G \frac{\partial^2 u}{\partial z^2} + C \frac{\partial^3 u}{\partial t \partial z^2} \tag{6.1}$$

The potential function of input waves, reflected waves, and transmitted waves shown in Fig. 6.14 could be uniformly represented by elastic displacement. The details are as follows:

$$u(z,t) = U(z)\mathrm{e}^{i\omega t} \tag{6.2}$$

In the formula, $u(z,t)$ is the potential function of input waves, reflected waves, and transmitted waves; $U(z)$ represents the displacement amplitudes of input waves, reflected waves, and transmitted waves along Z direction; ω represent the instant frequency of input waves, reflected waves, and transmitted waves.

It can be known through comprehensive analysis of Formula (6.1) and (6.2) that:

$$(G + iwC)\frac{\mathrm{d}^2U}{\mathrm{d}x^2} = -\rho w^2 U \tag{6.3}$$

Make

$$G^* = G + iwC \tag{6.4}$$

Then G^* could be called complex dynamic shear modulus, as $wC = 2\lambda G$, then

$$G^* = G(1 + i2\lambda) \tag{6.5}$$

At the same time, make $k^2G^* = \rho w^2$ and substitute it into Formula (6.3), and it can get:

$$U(z) = E\mathrm{e}^{ikz} + F\mathrm{e}^{-ikz} \tag{6.6}$$

In Formula (6.6), the first item is the seismic waves propagating inversely to Z direction, i.e. input seismic waves propagating downward along the soil layer; the second item is the seismic waves propagating along Z direction, i.e. reflected seismic waves. Formula (6.6) is suitable for any soil layer.

As the shear stress of the mth layer of any soil layer $\tau_m(x,t) = G_m\frac{\partial u}{\partial x} + C_m\frac{\partial^2 u}{\partial t\partial z} = G_m^*\frac{\partial u}{\partial z}$, for the mth and $m+1$th soil layers, it should satisfy displacement continuity conditions and stress continuity conditions on the interface, i.e. $\tau_m|_{z=h_m} = \tau_{m+1}|_{z=0}$, $u_m|_{z=h_m} = u_{m+1}|_{z=0}$, from which it could get the recurrence formula of input and reflected wave amplitudes:

$$E_{m+1} = \frac{1}{2}(1 + a_m)E_m\mathrm{e}^{ik_mh_m} + \frac{1}{2}(1 - a_m)F_m\mathrm{e}^{-ik_mh_m} \tag{6.7}$$

$$F_{m+1} = \frac{1}{2}(1 - a_m)E_m\mathrm{e}^{ik_mh_m} + \frac{1}{2}(1 + a_m)F_m\mathrm{e}^{-ik_mh_m} \tag{6.8}$$

In the formula, h_m and h_{m+1} are the thickness of the mth and $m+1$th soil layers, and a is the complex impedance ratio of waves, whose definition is as follows:

$$a_m = \sqrt{\frac{\rho_m G_m^*}{\rho_{m+!}G_{m+1}^*}} = \frac{k_m G_m^*}{k_{m+!}G_{m+1}^*} \tag{6.9}$$

It defines the amplitude vector $\{H_m\}$ of the mth soil layer is

$$H_m = \begin{bmatrix} E_m \\ F_m \end{bmatrix} \tag{6.10}$$

The conversion relation of the amplitude coefficients of the mth and $m + 1$th soil layer could be obtained by rewriting Formula (6.7) and (6.8) into:

$$\{H_{m+1}\} = [T_m]\{H_m\} \tag{6.11}$$

In which, the conversion matrix $[T_m]$ of adjacent layers is:

$$[T_m] = \begin{bmatrix} \frac{1+a_m}{2}\mathrm{e}^{ik_m h_m} & \frac{1-a_m}{2}\mathrm{e}^{-ik_m h_m} \\ \frac{1-a_m}{2}\mathrm{e}^{ik_m h_m} & \frac{1+a_m}{2}\mathrm{e}^{-ik_m h_m} \end{bmatrix} \tag{6.12}$$

By Formula (6.11), it could get the conversion relation of the amplitude coefficients of top layer with any layers:

$$\{H_{m+1}\} = [\overline{T_m}]\{H_1\} \tag{6.13}$$

In which, the 2 × 2 order transferring matrix is:

$$[\overline{T_m}] = [T_{m-1}]\ldots[T_2][T_1] = \begin{bmatrix} t_{11} & t_{12} \\ t_{21} & t_{22} \end{bmatrix} \tag{6.14}$$

Therefore, it could be known through comprehensive analysis of Formula (6.9) and (6.12) that, $[\overline{T_m}]$ depends on the features and thickness of each soil layers as well as the duration, frequency, and amplitudes of input seismic waves. On the mth layer surface with coordinate of z, the displacement amplitude is

$$U_m(z) = E_m\mathrm{e}^{ik_m z} + F_m\mathrm{e}^{-ik_m z} \tag{6.15}$$

While the shear stress in the soil layer is:

$$\tau_m(z, t) = G_m^* \frac{\partial u}{\partial z} = ik_m G_m^*(E_m\mathrm{e}^{ik_m z} - F_m\mathrm{e}^{-ik_m z})\mathrm{e}^{iwt} \tag{6.16}$$

The boundary condition on free surface is as follows:

$$\tau_1(0, t) = 0 \tag{6.17}$$

It can get $E_1 = F_1$, and at the same time, if the amplitude vector of the first layer wave is known, $\{H_1\} = \{E_1, F_1\}^T$, and substituting Formula (6.16) and (6.17) into Formula (6.11), it will get:

$$E_m = e_m E_1 = (t_{11} + t_{12}) E_1; \tag{6.18}$$

$$F_m = f_m F_1 = (t_{21} + t_{22}) F_1 \tag{6.19}$$

Therefore, E_1 could be determined by the input conditions of seismic waves on the semi-space above bedrocks. Supposing that the semi-space above bedrocks is homogeneous, then the wave amplitude of input waves EN is the vibration amplitude of all the uplink waves in this homogeneous semi-space, so it can be known through Formula (6.18) and (6.19):

$$E_1 = E_N / e_N; \tag{6.20}$$

$$F_N = f_N \cdot E_N / e_N \tag{6.21}$$

Thus, the seismic displacement amplitude, shear stress, shear strain, and horizontal acceleration amplitude of the *m*th soil layer are, respectively:

$$U_m(z) = (e_m \mathrm{e}^{ik_m z} + f_m \mathrm{e}^{-ik_m z}) E_N / e_N \tag{6.22}$$

$$T_m(z) = ik_m G_m^* (e_m \mathrm{e}^{ik_m z} - f_m \mathrm{e}^{-ik_m z}) E_N / e_N \tag{6.23}$$

$$\Gamma_m(z) = ik_m (e_m \mathrm{e}^{ik_m z} - f_m \mathrm{e}^{-ik_m z}) E_N / e_N \tag{6.24}$$

$$A_m(z) = -w^2 \times (e_m \mathrm{e}^{ik_m z} + f_m \mathrm{e}^{-ik_m z}) E_N / e_N \tag{6.25}$$

6.2.4 Analysis of Seismic Wave Time–Frequency Effects

Based on the above derivation, it can be known that the time–frequency effects of seismic waves mainly embody in the elastic displacement amplitude $U(z)$ and frequency ω of input seismic waves. Therefore, this dissertation the following analysis process of seismic wave time–frequency effects: firstly, break up seismic wave time–frequency into several of IMF (time history curve) through Empirical Mode Decomposition (EMD for short); secondly, carry out Channel Switch, i.e. switch multi-channel signal composed of multiple IMF into single-channel signal made up of single IMF; thirdly, solve the instant frequency of each channel, and draw the time–frequency curve of each IMF. Lastly, bring each IMF and their instant frequency into Formula (6.22)–(6.25) to work out the results and superimpose the results, so we can get the seismic response of this horizontal layered field.

Based on the above discussion, the time–frequency analysis method for seismic response effects of horizontal layered fields proposed by this book could well complement the shortcomings of time-domain and frequency-domain elastic wave analysis methods, which could be explained as follows: (1) it could reasonably consider multiple soil layers and their thickness and their physico-mechanical parameters, which makes up the few soil layers (1–2 layers) of analysis methods given by time-domain elastic wave propagating theory; (2) it could reasonably consider the influencing factors like amplitudes, frequency and duration of seismic waves, which makes up the shortcoming of analysis methods given by frequency-domain elastic wave propagating theory which only harmonic effects at a given circumferential frequency can be considered.

6.3 Solving Process of Time–Frequency Analysis Method of Seismic Response for Horizontal Layered Fields

Based on the above discussion, this book summarizes the solving process of time–frequency analysis method of seismic response for horizontal layered fields, which is as follows:

(1) According to the study object, sum up appropriate analysis models of horizontal layered field and decide the physical parameters such as soil layer distribution, thickness, and mechanical parameters.
(2) Decide the design response spectrum curve of this region according to *Code for Seismic Design of Buildings*, and synthesize artificial ground motion time history with seismic ground motion time history fitting method so as to decide the seismic ground motion acceleration time history of this region.
(3) Decide time-history curve of displacement based on seismic ground motion acceleration time-history curve, work out each IMF of seismic waves and the correspondent frequency time-history curve with HHT.
(4) Substitute the obtained results into Formula (6.22)–(6.25) to calculate and superimpose the results, so we can get the seismic response of this horizontal layered field.

6.4 Shaking Table Test and Numerical Simulation Verification of Time–Frequency Analysis Method of Seismic Response for Horizontal Layered Fields

6.4.1 General Situation of Shaking Table Test and Numerical Simulation

In order to verify the correctness of solving method for time–frequency-domain of seismic response of horizontal layered fields and to concretely explain its application, this book conducts large-scale shaking table test and numerical simulation, in which the situation of shaking table test has been described in Sect. 6.1. While as the shaking table test model is small, so if too much observing points are set up on it will lead to crowd of measured points, which may influence the test effects. Therefore, in order to more accurately and deeply verify the correctness of time–frequency analysis method for seismic response of horizontal layered fields, this book will build

Table 6.4 Physico-mechanical parameters of hard rock, soft rock, and gravel

Model name	Unit weight (kN/m^3)	Shear wave speed (m/s)	Cohesive force (kPa)	Internal friction angle (°)	Shear modulus (MPa)
Hard rock	23	943.45	632	45	2047.24
Soft rock	22	741.7	328.64	41	1210.3
Gravel	19	404.6	25.28	25	311.1

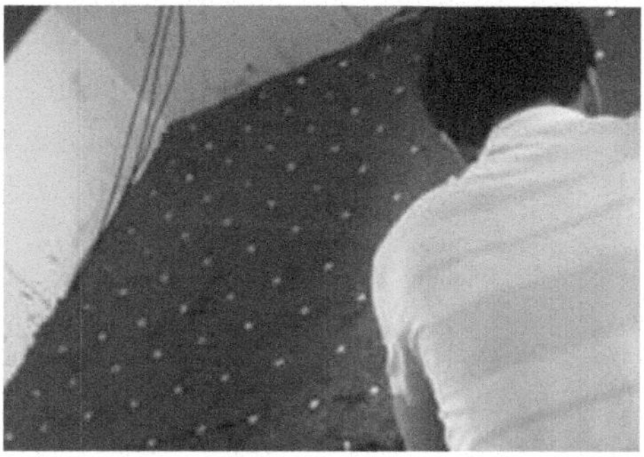

Fig. 6.15 Shaking table test model

numerical analysis model based on the shaking table test and verify its correctness with the shaking table test results. Then, restore the shaking table test model into the prototype field according to the similarity system and further build the numerical analysis model of horizontal layered fields, exert Wenchuan wolong seismic waves with PGA = 0.1, 0.2 and 0.4 g and set up horizontal acceleration observing points in the middle and on the top if each layer. The specific mechanical parameters of the rock are shown in Table 6.4, the shaking table test model is shown in Fig. 6.15, the numerical analysis model is shown in Fig. 6.16 and the time-history curve and Hilbert spectrum of input seismic acceleration, respectively, are shown in Figs. 6.17 and 6.18.

6.4.2 Time–Frequency Analysis of Horizontal Layered Fields

Based on the analysis of time–frequency effects of seismic waves, this book conducts time–frequency analysis to the input waves. The details are as follows: firstly, input Wenchuan wolong seismic waves, see Fig. 5; secondly, carry out EEMD decomposition and get every IMF, see Fig. 6.19; lastly, solve the instant frequency of every IMF, see Fig. 6.20.

Horizontal layered field solving method based on time–frequency analysis method solving approach of horizontal layered fields, substitute the relevant parameters of each soil layers and the above calculation results into Formula (6.25) to work out the acceleration time history of each measured point so as to obtain the PGA. See

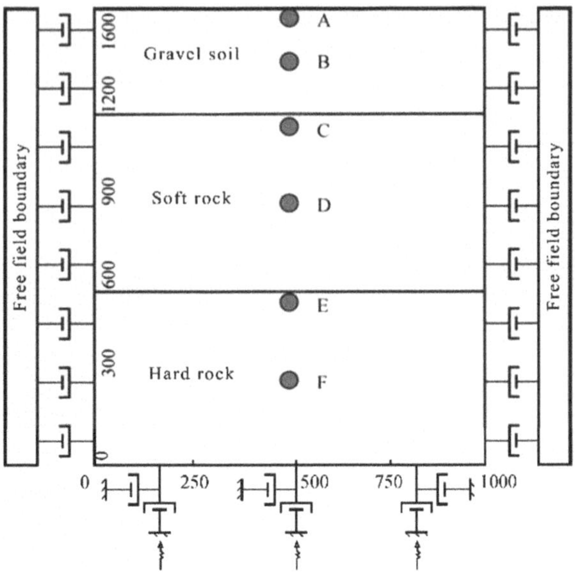

Fig. 6.16 Numerical analysis model

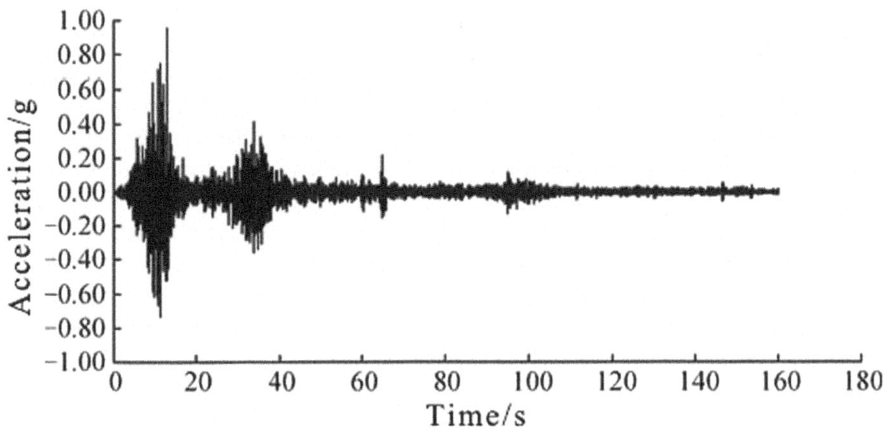

Fig. 6.17 Time-history curve of input waves of shaking table

Table 6.2 for detailed calculation results. At the same time, in order to more clearly compare the difference of calculation results of acceleration time-history curve and that of numerical calculation, this book selects the acceleration time-history curve containing PGA at 12–17 s. See Fig. 6.21 for detailed calculation results.

Through comprehensive analysis of Table 6.5 and Fig. 6.21, it can be known that under the effects of seismic wave with PGA = 0.1, 0.2, and 0.4 g, the errors of calculation results of peak seismic motion of horizontal layered fields with the numerical analysis results are all less than 10%, with minimum error of 0.43%. At the same time, the acceleration time history curve and response spectrum curve show that

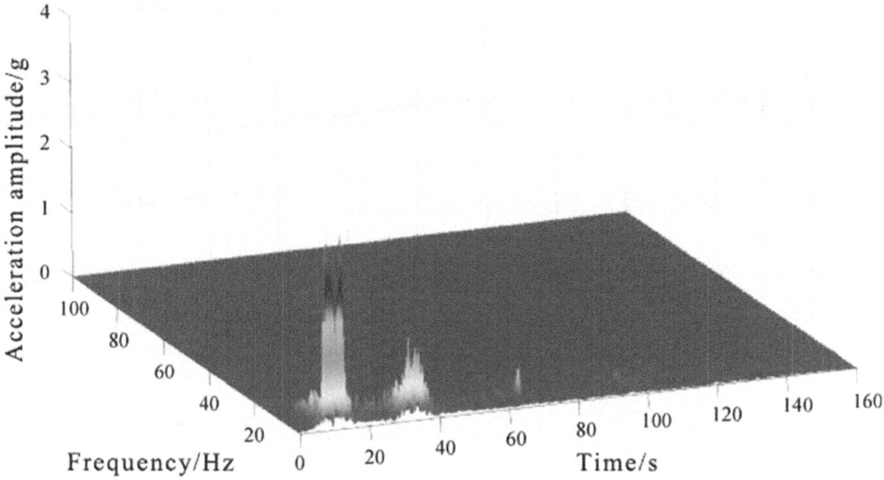

Fig. 6.18 Hilbert response spectrum of input waves (original wave)

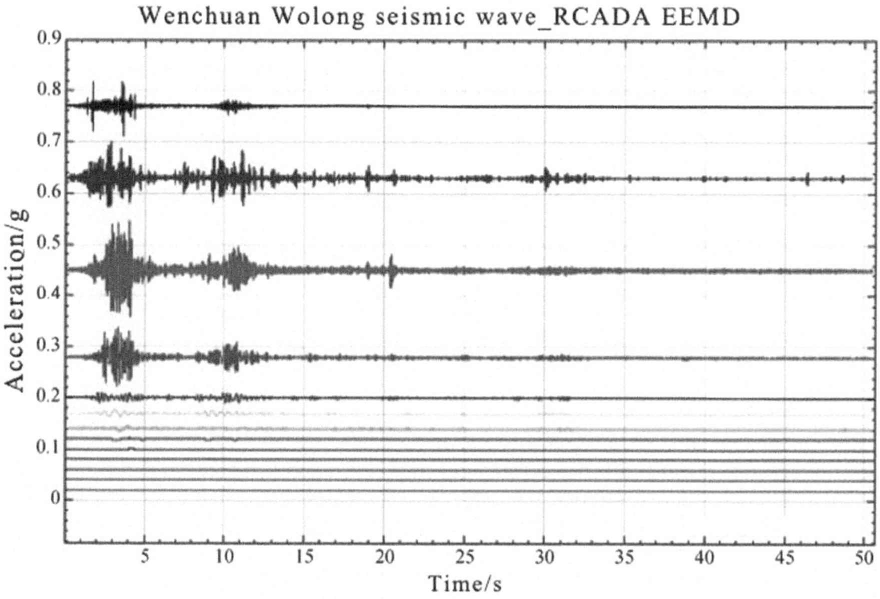

Fig. 6.19 Acceleration time history curve of IMF

there is slight difference to the calculation results and the numerical analysis results, which mainly embodies in slight lag of numerical analysis results with comparison to the calculation results, but the wave shape of the two curves are basically the same.

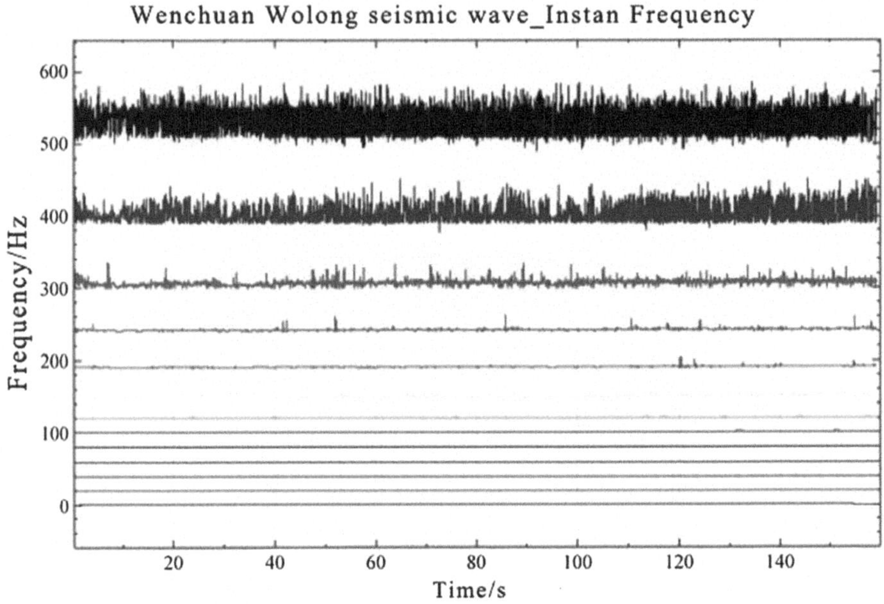

Fig. 6.20 Instant frequency map of IMF. *Note* The *Y* coordinate in Figs. 6.19 and 6.20 only represents physical characters, but not size

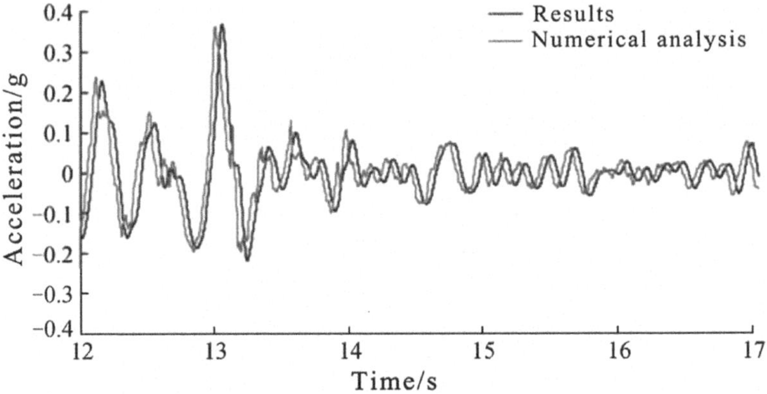

Fig. 6.21 Acceleration time-history curve at ground surface of input wave with PGA = 0.2 g (Point A)

This may be due to the following reasons: first, the time–frequency analysis method of seismic response of horizontal layered fields in this book is derived based on the elastic wave theory, which didn't fully consider the nonlinear features of rock–soil mass. Second, the time–frequency analysis method of seismic response of horizontal layered fields pays no attention to the time and space effects in wave prop-

Table 6.5 Comparing results of PGA

	Numerical analysis (calculation results) (g)			Error (%)		
Elevation	Input wave with PGA= 0.1 g	Input wave with PGA= 0.2 g	Input wave with PGA= 0.4 g	Input wave with PGA= 0.1 g	Input wave with PGA= 0.2 g	Input wave with PGA= 0.4 g
300	0.121(0.115)	0.211(0.202)	0.399(0.412)	4.958678	4.265403	−3.25815
600	0.132(0.124)	0.232(0.231)	0.416(0.434)	6.060606	0.431034	−4.32692
900	0.154(0.159)	0.268(0.273)	0.442(0.455)	−3.24675	−1.86567	−2.94118
1200	0.176(0.173)	0.302(0.316)	0.461(0.486)	1.704545	−4.63576	−5.42299
1400	0.228(0.210)	0.363(0.373)	0.518(0.557)	7.894737	−2.75482	−7.52896

Note In Table 6.5, the error = (numerical analysis results − calculation results)/numerical analysis results

agation process, such as partial coherence effects, traveling wave effects and local field effects, etc. But in general, the calculation results of the time–frequency analysis method solving approach of horizontal layered fields are basically identical with the numerical analysis results, which fully verify the correctness of the calculation results and feasibility of calculation method in this book.

Bibliography

Barker. T.B, Chen, B.N, Hague, P.F, Majain, Wong K.L, Understanding the time-lapse seismic response of a compacting carbonate field, Offshore Sarawak, Malaysia [J], Society of Petroleum Engineers, 2008, (3):1853–1866.

Chen Guoxing. Geotechnical Earthquake Engineering [M]. Beijing: Science Press, 2007.

Chen Guoxing, Chen Zhonghan, Ma Kejian. Principle of seismic design of engineering structures [M]. Beijing: China Water & Power Press, 2002.

Chen Lei, Chen Guoxing, Mao Kunming. 3D refined nonlinear analysis of seismic response characteristics of frame metro station under near-field strong ground motion of large earthquake, Chinese Society of Civil Engineering [J], 2012, 34(3):490–496.

Chen Zuyu. Stability analysis of soil slope-the principle, method and process [M]. China Water & Power Press, 2005.

Dong Jianping. Discussion on several problems in system time domain analysis [J], Journal of Shanghai Normal University (Natural Science), 1999, 3:100–103.

Du Xiuli. Theory and Method of Engineering Wave [M], Science Press, 2009.

Edward L W. Static and Dynamic Analysis of Structures: A Physical Approach with Emphasis on Earthquake. Fourth Edition [M]. USA: Computers and structures Inc, 2004.

Hu Yuxian, He Xun. Fitting of artificial ground motion response spectrum considering phase spectrum [J]. Earthquake Engineering and Engineering Vibration, 1986, (2):37–51.

Huang N.E Shen Z. Long S R.et al. The empirical mode decomposition and Hilbert spectrum for nonlinear and non-stationary time series analysis [J]. Proc R. Soc Lond. 1998 A 454:903–995.

Liu Baojian, Xie Yongli. Frequency domain analysis of soil dynamic characteristics under stochastic dynamic load [J], China Journal of Highway and Transport, 1999, 12(4):7–12.

Ohtsu, Uesugi, Shinpei. ANALYSIS OF SH WAVE SCATTERING IN A HALF SPACE AND ITS APPLICATIONS TO SEISMIC RESPONSES OF GEOLOGIC STRUCTURES [J]. Engineering Analysis, 1985, (2):198–204.

Wu Zaiguang, Han Guocheng, Lin Gao. One dimensional stochastic seismic response analysis of nonlinear ground [A]. Proceedings of the National Symposium on the seismic resistance of soil structures and foundations [C]. Xi'an: Xi'an Science and Technology Press, 1986.

Zhang Kexu. Soil Dynamics [M]. Beijing: Seismological Press, 1989.

Zhou Xiyuan, Xu Wei, Fu Shengcong. Overview of earthquake engineering [M]. Beijing: Science press, 1977, 4–5.

The manufacturer's authorised representative in the EU is Springer Nature Customer Service Centre GmbH, Europaplatz 3, 69115 Heidelberg, Germany. If you have any concerns regarding our products, please contact ProductSafety@springernature.com

Printed and bound by CPI Group (UK) Ltd, Croydon, CR0 4YY
28/07/2026
02179215-0003